MINNESOTA STUDIES IN THE PHILOSOPHY OF SCIENCE

T0256643

Minnesota Studies in the
PHILOSOPHY OF SCIENCE

HERBERT FEIGL AND GROVER MAXWELL, GENERAL EDITORS

VOLUME V

*Historical and Philosophical Perspectives
of Science*

EDITED BY

ROGER H. STUEWER

FOR THE MINNESOTA CENTER FOR PHILOSOPHY OF SCIENCE

UNIVERSITY OF MINNESOTA PRESS, MINNEAPOLIS

© *Copyright 1970 by the*
UNIVERSITY OF MINNESOTA

All rights reserved

PRINTED IN THE UNITED STATES OF AMERICA

Library of Congress Catalog Card Number: 57-12861
ISBN 0-8166-0592-0

PUBLISHED IN GREAT BRITAIN, INDIA, AND PAKISTAN BY THE
OXFORD UNIVERSITY PRESS, LONDON, BOMBAY, AND KARACHI
AND IN CANADA BY THE COPP CLARK
PUBLISHING CO. LIMITED, TORONTO

Foreword

The historian and the philosopher of science would appear to have this at least in common, that they wish to understand science and to know what can truly be said about it—about its origins and development, about its structure and explanatory power. But disagreement is possible from the beginning. The conceptual boundaries of science as defined by development do not coincide neatly with its conceptual boundaries as defined by structure; in the former case boundary problems arise between science and hermeticism, for example, in the latter between science and other forms of rational activity, to take only one area of potential misunderstanding. The historian might seem to have an advantage of sorts, as being nearer to the empirical source: if he does his work well he will not apparently be open to correction by the philosopher on the facts of the case, whereas a historical correction might invalidate a philosophical analysis by showing that what it claimed to analyze corresponded to no historically recognizable facts. And yet the matter is not so simple. If the facts had been as represented, the analysis might have been brilliant; and one might say that it would have been worth inventing the facts in order to make the analysis possible. And again, since the facts are in general facts about theories, rather than facts about the world, a philosophical analysis might even show that the historian had got them wrong, without in the least challenging the historical evidence.

Faced with the complex character of their interaction, specialists in the history and the philosophy of science might opt for either of two diametrically opposed strategies, the choice between which need have nothing to do with professional affiliation. On the one hand they might urge an ever closer relationship between the two disciplines, arguing that their common devotion to science requires them to help each other as much as possible, not only to throw light on one another's problems but also to lend mutual assistance in avoiding elementary mistakes—factual mistakes on the part of philosophers, conceptual mistakes on the part of historians. This position is likely to be taken by two classes of worker, at opposite ends of the scale of generality: by philosophers who work on individual figures in the history of science (Galileo and Newton, for example), and by historians who work on philosophically self-conscious scientists (Mach and Einstein, for example); or by philosophers and historians who work at a global level, concerned with patterns of discovery and of rationality, with the principles rather than the details of historical change, with historiography or the history of the philosophy of science. On the other hand they might urge a total separation of the two disciplines, arguing that these can only get in one another's way, that the kinds of refinement that historians might bring to a philosopher's understanding are philosophically irrelevant, and vice versa. This position is likely to be taken by workers at the middle level of generality: by philosophers interested in the structure of explanation, the logical problems of induction and probability, simplicity and the choice between competing theories, etc., and by historians interested in the development of particular theories, particular techniques, or particular instruments, over a more or less short time scale, on the part of scientists with experimental or mathematical gifts but with little or no philosophical sophistication (which, it can plausibly be argued, would only have hindered them).

Current trends seem to favor the second of these approaches. History of science and philosophy of science, although both can trace their intellectual ancestry from much earlier stages, have come into their own professionally and on a wide scale only in the last few decades. It is true that they have often done so—at least in books and institutions—hand in hand; the expression "history and philosophy of science" has become a familiar part of the vocabulary of administrators and textbook writers. But this easy association of names by now conceals, if it has not done so from the beginning, a considerable divergence of professional interest, and this divergence has resulted in a lowered level of mutual tolerance. In the early stages of the development of a field approximate categories may suffice, but as time goes on and more and more workers enter it a kind of technical sophistication takes place; problems are subdivided, conceptual distinctions insisted upon. When historians of science employ philosophical categories, or philosophers of science historical ones, they naturally tend to do so at a level of refinement already surpassed by their contemporaries in the other discipline, and since the expansion of the fields seems at this point to be following an exponential curve the rate of divergence accelerates and the overlap of competence and interest is in danger of vanishing.

This expansion also introduces a new consideration. I began by saying that historians and philosophers of science shared a common allegiance to science, and it has often been remarked how many of the pioneers in these fields came to them from active careers in science. The present organization of academic life, however, makes it less and less likely that specialists in the history and philosophy of science will ever have been scientists; they will have had some scientific training, no doubt, but few will have had firsthand experience of research under laboratory conditions or of the creative aspects of theory construction. A historian who goes into the history of science primarily because of his interest in history, and a philosopher who goes into the philosophy of science primarily because of his interest in philosophy, may have little in common even with their own colleagues who entered these fields primarily because of their interest in science. Not that experience of science is enough by itself to confer historical or philosophical competence, as the later careers of a number of distinguished scientists have testified; nor, for that matter, that it is necessary or even desirable—only that experience and motivation must be taken into account, and the more various they are the less antecedent reason for a community of interest.

And if, after all, the philosopher uses historical episodes only as convenient paradigms for his own reconstructions, so that he can easily change the illustration or even drop it altogether if it turns out to be historically inaccurate, the history of science may be genuinely irrelevant to his work; and similarly, *mutatis mutandis*, for the historian who uses philosophical material only to round out a story whose main lines would remain unchanged even under a different interpretation. The question of the interaction or collaboration between the disciplines is therefore a genuinely open one, to be answered only in terms of actual attempts to carry such collaboration through in specific cases. It is not so much a question of the interaction between history and philosophy in the work of a single scholar, of which there are some familiar examples, but of their interaction in the persons of scholars from either side.

One of the organizations operating under the joint title U.S. National Committee for the International Union of History and Philosophy of Science, having had this problem drawn to its attention, decided a year or

Foreword

so ago to organize a meeting which would put the matter to a test. The meeting was held in Minneapolis from September 11 to September 14, 1969, with the collaboration of the Minnesota Center for Philosophy of Science and with the generous support of the Center and of the Council for Philosophical Studies. One of its results was this book.

The original intention of the planning and program committees was to begin with limited and well-defined topics in science (field theory, vitalism, Copernican astronomy) and invite for each a historian and a philosopher to demonstrate the way in which such a topic would be dealt with from the point of view and with the resources of his discipline. This might be thought of as a second-order activity (taking the world as of order 0 and science, which is about the world, as of order 1); only later, when some such actual cases were to hand, would the third-order activity of comment about the history and philosophy of science and their interaction be engaged in. In the event, however, this intention was thwarted by the usual intractabilities, and the discussion proceeded from the beginning on the second-order and third-order levels simultaneously. Opportunities for talking at cross-purposes thus grew geometrically. That, on the whole, they were more often avoided than not is a tribute to the spirit in which the participants approached the meeting. The thanks of the organizing committees are due to all of them. The digestion of what was actually said, and the judgment of the success of the experiment, I leave to the reader.

Peter Caws

HUNTER COLLEGE OF THE CITY
UNIVERSITY OF NEW YORK

December 1969

viii

Preface

All but two of the papers constituting this fifth volume of the *Minnesota Studies in the Philosophy of Science* were delivered or informally discussed at a conference organized to explore the relationships between the history and the philosophy of science which was held at the University of Minnesota, September 11–14, 1969. Professor W. C. Salmon (Indiana University) wrote his paper in response to issues raised at the conference, while Professor P. K. Feyerabend (University of California, Yale University, University of London), who was unable to attend the conference, nevertheless kindly submitted a paper for inclusion in this volume. The conference itself was marked by lively and prolonged exchanges of views; and in an effort to preserve at least some of the thoughts that were expressed, comments on the papers were solicited from the participants after they had left Minneapolis and had returned to their home institutions. The responses that were received also form a part of this volume.

As Professor Caws notes in his Foreword, the conference grew out of the deliberations of a subcommittee of the U.S. National Committee for the International Union of History and Philosophy of Science. Members of this subcommittee charged with the organization of the conference were Peter Achinstein (Johns Hopkins University), Peter Caws (City University of New York; chairman), Robert S. Cohen (Boston University), David B. Kitts (University of Oklahoma), and Melvin W. Kranzberg (Case Western Reserve University). The program committee consisted of Peter Caws (chairman), Edward Grant (Indiana University), Carl G. Hempel (Princeton University), John Murdoch (Harvard University), and Dudley Shapere (University of Chicago).

The conference participants were Peter Achinstein (Johns Hopkins University), Peter A. Bowman (Grinnell College), Gerd Buchdahl (University of Cambridge), Roger C. Buck (Indiana University), Peter Caws (City University of New York), Robert S. Cohen (Boston University),

Preface

Herbert Feigl (University of Minnesota), Paul Forman (University of Rochester), Adolf Grünbaum (University of Pittsburgh), Keith Gunderson (University of Minnesota), Mary Hesse (University of Cambridge), Erwin N. Hiebert (University of Wisconsin), David B. Kitts (University of Oklahoma), Arnold Koslow (City University of New York), Imre Lakatos (London School of Economics and Political Science), Laurens Laudan (University of Pittsburgh), Grover Maxwell (University of Minnesota), Ernan McMullin (University of Notre Dame), John Murdoch (Harvard University), Edward Rosen (City University of New York), Wesley C. Salmon (Indiana University), Kenneth F. Schaffner (University of Chicago), Henry Small (American Institute of Physics), Howard Stein (Case Western Reserve University), Roger H. Stuewer (University of Minnesota), Arnold Thackray (University of Pennsylvania), Richard S. Westfall (Indiana University), and Harry Woolf (Johns Hopkins University).

I speak for all of the participants when I express my sincere gratitude to the scholars who gave their time and energy in organizing and planning the conference and to Professor Herbert Feigl, director of the Minnesota Center for Philosophy of Science and Regents' Professor of Philosophy, who gave valuable help in planning the conference and then guided it to a successful conclusion; he also gave generously of his advice in the preparation of this volume. The financial support received from the Council for Philosophical Studies and from the Carnegie Corporation of New York, through part of its grant to the Minnesota Center for Philosophy of Science, is acknowledged with pleasure and gratitude. Finally, thanks are due to the staff of the University of Minnesota Press, to Mr. Robert Anderson, who taped the conference proceedings, and to Mr. Richard K. Gehrenbeck, who prepared the index for this volume.

Roger H. Stuewer, *Associate Professor*
MINNESOTA CENTER FOR PHILOSOPHY OF SCIENCE
SCHOOL OF PHYSICS AND ASTRONOMY

January 1970

Synopsis

The following brief summaries provide an introduction to each of the papers in this volume.

1. *Beyond Peaceful Coexistence.* HERBERT FEIGL. This paper presents the substance of the author's introductory remarks at the Minnesota conference. In it he attempts (1) to draw some clear lines of demarcation between historical and philosophical studies of science; (2) to show the indispensability of logicomethodological analyses for—at least—the "internal" history of science; and (3) to point out the urgent need for closer collaboration of both types of specialists. An episode from the recent (early twentieth-century) history of chemistry is suggested as one interesting example in which historians and logicians of science might fruitfully join forces in research that might lead to a better understanding of the positivistic (anti-atomistic) and the realistic (atomistic) tendencies in scientific theorizing.

2. *The History and Philosophy of Science: A Taxonomy.* ERNAN MC-MULLIN. This paper explores the various ways in which HS and PS may be related, and argues that for certain sorts of problem in PS only an approach which relies heavily on HS for its evidence is likely to succeed. It is pointed out that the temporal dimension of science has been largely ignored by twentieth-century philosophers of science largely because of the dominance of formal-logical techniques in the analysis of scientific method. In the last decade, however, this has begun to change; philosophers of science are now beginning to take HS very seriously.

The author shows that it is necessary to distinguish between two different senses of science and two correspondingly different ways of writing history. Three different approaches to PS are noted, the basis of the division being the type of warrant upon which each primarily depends. There are also different domains of PS, epistemological and ontological. There is a particular tension between the logical and the historical approaches to the understanding of science as a human activity; what they seek, and the modes of analysis they employ, are altogether different. It is sometimes not easy to separate logical-philosophical from social-psychological analyses of science; and the relations between them are worth exploring in more detail than is customarily given to this topic.

The paper ends by discussing the question of the ontological significance of the models possessed by the scientist. McMullin illustrates with a case history the fact that an adequate answer to this classic question can be given only by careful analysis of certain developmental aspects of science.

3. *Bayes's Theorem and the History of Science.* WESLEY C. SALMON. Assuming that the historian is legitimately concerned to distinguish rationally well-grounded from irrational developments in the history of science, this paper defends the distinction between the *context of discovery* and the *context of justification*, and goes on to argue that an adequate understanding of the distinction demands a satisfactory account of scientific confirmation. Although there is no universally accepted analysis of confirmation, many of the leading contemporary candidates agree in rejecting the simpleminded hypothetico-deductive schema, and come much closer to conforming to a "Bayesian" approach. Since the "Bayesian" account differs from the H-D schema by admitting "prior probabilities," the historian's assessment of the acceptance or rejection of a given hypothesis may depend crucially upon the way in which he views confirmation—i.e., the status he accords to plausibility considerations. On the H-D theory, plausibility arguments can have at most a heuristic value, while on the "Bayesian" approach, plausibility arguments constitute an integral part of the context of justification. Moreover, if prior probabilities are empirical, the history of science provides the data upon which their evaluations depend.

4. *Inference to Scientific Laws.* PETER ACHINSTEIN. This paper examines philosophically what is meant by inference *to* scientific laws, as contrasted, for example, with the hypothetico-deductive scheme in which consequences are drawn, or inferences are made, *from* scientific laws in order to test those laws. Hanson's (and Peirce's) retroductive mode of inference is also criticized. Finally, the meaning and force of the distinction between the context of discovery and the context of justification is discussed; Achinstein rejects the idea that there are two separate logics operative, proposing instead that one ask in which, if either, context a scientist's reasoning takes place. To illustrate his arguments, and to display their relevance to both

philosophers and historians of science, Achinstein examines Gay-Lussac's reasoning in connection with his discovery of his law of combining volumes of gases, as well as the related work of Avogadro.

5. *Science: Has Its Present Past a Future?* ARNOLD THACKRAY. In this paper a first analysis is made of the historical, sociological, and intellectual roots of those historiographic assumptions that have characterized Western history of science for a generation. Thackray argues that this analysis enables us to understand why idealist history now commands an unhealthy hegemony over the field. The latter part of the paper explores some possible alternative modes of the history of science. These modes would allow both a greater interaction with other areas of history and a greater sensitivity to the complex social issues that now invite critical thought from the historian of science.

6. *Hermeticism and Historiography: An Apology for the Internal History of Science.* MARY HESSE. The implications of the concepts of "internal" and "external" history of science are examined in this paper, and a distinction is made between the effect of internal and external factors on the occasion of scientific development and the effect on its character. Further, influences on character are distinguished into "rational," "irrational," and "nonrational" (or causal). Miss Yates's work on the hermetic tradition in the scientific revolution suggests the questions "Can rational and irrational factors be distinguished conceptually and historically, and the internal history of science be defined in terms of rational factors?" and "Is internal history relatively autonomous?" An examination of various philosophical analyses of science reveals that no conceptual constraints can be put upon the type of influence admitted into "rational science," so the answer to the first question seems to be negative. It is then argued that the answer to the second question depends entirely on each historical case. In particular, recent claims for the importance of external influences on seventeenth-century science are examined, and it is concluded that as far as evidence presented to date is concerned, the received internal tradition appears to be relatively autonomous.

7. *Was Copernicus a Hermetist?* EDWARD ROSEN. The general thesis of this paper is that a satisfactory history of science must be "internal," but also must take account of "external" factors; it must combine comprehension of the scientific subject matter with a broad understanding of the historical period. This thesis is illustrated by discussing one of Archimedes' discoveries, and, especially, by analyzing the validity of Miss Yates's con-

tention that Copernicus in his scientific work was strongly influenced by magical beliefs associated with the hermetic-cabalist tradition.

8. *Philosophy of Science: A Subject with a Great Past.* PAUL K. FEYERA-BEND. This paper, without going into too much detail, tries to indicate that contemporary philosophy of science deals in a highly sophisticated and precise manner with problems which have nothing to do with science at all, while the philosophy of Galileo, or Descartes, or Newton advanced the sciences and led to many useful discoveries. A brief account of Mach's philosophy is also given, and it is argued that most of his twentieth-century followers misrepresent him.

9. *Mach's Philosophical Use of the History of Science.* ERWIN N. HIE-BERT. This paper attempts to show how Mach used the history of science to illuminate (clarify, expose, and analyze) scientific puzzles that he encountered in his work as a professional physicist. Mach's epistemological studies of the physiology and psychology of sensations, his principle of economy of thought, his monism of method, and his radical and empirical skepticism all contributed to the formulation of a theory of cognition. Antimetaphysical in tone, Mach claimed that his scientific epistemology was borne out empirically by the study of the history of science, when, in fact, it rested upon certain presuppositions devoid of empirical content. Nonetheless, Mach exerted a great influence upon several generations of natural scientists and philosophers of science—and primarily on the basis of his provocative, critical, and perceptive historical analyses.

10. *History of Science and Criteria of Choice.* GERD BUCHDAHL. In this paper a contradiction is explored between Newton's use of gravity as an explanatory concept and his view that action at a distance (in a material sense) is not ontologically basic. Various responses to the ensuing dilemma, on the part of both Newton and later writers, are evaluated by means of a triadic schema of components of choice of hypotheses, labeled "conceptual," "constitutive," and "architectonic." The responses are shown to involve a variety of escape routes from Newton's dilemma, formalist, phenomenalist, mechanistic, as well as teleological, corresponding to reconstructions of the constitutive and architectonic components. Finally, a reconstruction of the conceptual component is shown to lead to a reinterpretation of the notion of "inconceivability" as "contingent," and (alternatively) of the concept of matter as reducible to a field of force. A general discussion of inductive criteria is appended.

Synopsis

11. *Non-Einsteinian Interpretations of the Photoelectric Effect.* ROGER H. STUEWER. As an illustration of the interplay between theory and experiment that takes place in a period of transition between creation and acceptance of a scientific hypothesis, this paper examines in some detail the origin of Einstein's light quantum hypothesis, its application to the photoelectric effect, and the origin, nature, and validity of a number of alternate classical interpretations of the photoelectric effect proposed between 1910 and 1913 by H. A. Lorentz, J. J. Thomson, Arnold Sommerfeld, and O. W. Richardson. All these physicists believed that finding a classical interpretation of the photoelectric effect was tantamount to casting very serious doubt on the validity of Einstein's light quantum hypothesis. Actually, however, this is not the case, because, as Einstein proved, his light quantum hypothesis is a necessary consequence of the validity of Planck's law. Present knowledge indicates that a non-photon interpretation of the photoelectric effect is indeed possible, but, in common with the earlier alternate interpretations, this does not affect the validity of Einstein's light quantum hypothesis.

12. *On the Notion of Field in Newton, Maxwell, and Beyond.* HOWARD STEIN. The notion of field plays a significant role—in some ways similar, and in some important ways different—in a series of scientific investigations, theories, and speculations. In this paper, this role is examined first in the work of Newton, where it is argued that the notion of field (although not, of course, under that name) functioned in crucial ways in Newton's own research, in the program he developed for all of physics, and (in a way that contrasts strongly with the traditional view of Newton's philosophy) in his most fundamental views of the constitution of the world. A briefer examination is made of the concept of the field in Maxwell's electrodynamics and its subsequent development; it is argued that this development led to the breaking up of Newton's special program for physics, but not of his fundamental view of the constitution of the world. (That fundamental view has, however, been overturned by quantum mechanics and quantum field theory.) In the concluding section of this paper, some points of philosophical analysis developed in the body of the paper are made the basis for a general comment upon certain methodological issues in the philosophy of science.

13. *Outlines of a Logic of Comparative Theory Evaluation with Special Attention to Pre- and Post-Relativistic Electrodynamics.* KENNETH F. SCHAFFNER. This paper attempts to develop a tricategorical logic for asses-

Synopsis

sing the merits of competing scientific theories. The problems which give rise to such an attempt come from both philosophy and history of science, and are variously described as difficulties with scientific objectivity, the relations between theory and experiment, and certain problems involving priority of discovery in science. The logic is based upon some suggestions made by the physicist Heinrich Hertz in the context of assessing competing "images" (*Bilder*) of mechanics. Hertz's categories are generalized and reinterpreted to accord more closely with twentieth-century scientific, historical, and philosophical discoveries. A modified Hertzian schema employing the categories of "theoretical context sufficiency," "experimental adequacy," and "relative simplicity" is applied to a specific case of theory competition in the history of science, viz., the dispute in the sector of the electrodynamics of moving bodies between Lorentz's "absolute" theory and Einstein's special theory of relativity. It is concluded that in 1905 Lorentz's theory ranked higher in theoretical context sufficiency, Einstein's higher in relative simplicity, and that each theory accounted satisfactorily for the relevant experiments. In several years' time, however, changes in the experimental situation and in the theoretical context began to favor Einstein's theory. The paper closes with some brief considerations on the relations of the history of science and the philosophy of science in the light of what was discussed in the paper, and concludes that the two disciplines can and should be made more relevant to each other.

Contents

Contents

Contents

MINNESOTA STUDIES IN THE PHILOSOPHY OF SCIENCE

Beyond Peaceful Coexistence

As Professor Peter Caws has so perceptively and succinctly pointed out in the Foreword to this volume, there are various sources and types of misunderstanding between the historians of science and the philosophers of science. I believe, as he does, that these misunderstandings can be overcome. Properly conceived, the two approaches toward a better comprehension of the scientific enterprise are not merely complementary. They interpenetrate, they illuminate each other. I am inclined to plead for a "merger" of at least some aspects of the historical-sociological-psychological studies of science with the logical-methodological-philosophical aspects. True enough: orientations, trainings, predilections of the two "parties" have been rather divergent, particularly in recent times of increasing specialization. And since confession is (said to be) "good for the soul," I will admit that for a long time, I (along with quite a few other philosophers of science of recent times) have been a "sinner." Some of us have been satisfied with a "smattering of ignorance" in regard to the historical development of the sciences, their socioeconomic settings, the psychology of discovery and of theory invention, etc. A few of us, though proud of our *empiricism*, for some time rather unashamedly "made up" some phases of the history of science in a quite "a priori" manner—at least in public lectures and classroom presentations, if not, even, in some of our publications. "This is the way Galileo, or Newton, or Darwin, or Einstein (for example) must have arrived at their ideas" was not an uncommon way of talking—through our respective hats. Even if the sources were not always complete, and not always accurate, they were available, but we rarely consulted them. Most of us have come to repent this inexcusable conduct.

The purposes of the following brief observations are (1) to discuss as clearly as I can the lines of demarcation between the historicosociological and the logicophilosophical approaches; and (2) to clarify the proper aims of a philosophy of science for our age of rapid change (or "revolution") in science, and in this connection to criticize some—to my mind—regrettable tendencies of philosophical interpretation, some of which have originated from genuine, and some from questionable, historical insights.

1. It should not be necessary to elaborate once again the distinction and the differences between the *historicosociological* narratives, analyses, causal accounts of the origins, developments, conflicts, and Zeitgeists of scientific ideas on the one hand, and the *logicomethodological* reconstructions of scientific knowledge claims, on the other hand. Hans Reichenbach's by now classical distinction between "studies in the context of discovery" and "studies in the context of justification," though in need of some qualifications, remains (and, as I see it, *should* remain) at least a most important first approximation if we are to retain even a minimum of clear thinking in these badly confused matters. From the historical-sociological-psychological point of view we search and re-search the sources in order to discern the ("external") circumstances that have been influential in the evolution of scientific thought. Nor is this all. Good history of science must be at least equally concerned with the ("internal") conditions of the discovery, the reasoning, the testing, the acceptance or rejection of scientific knowledge claims. It is in this latter area that historians and philosophers of science largely coincide in their interests. This means that the "good" historian of science must devote a great deal of attention to the meaning and the justification of scientific knowledge claims—in the very period or phase of their development which is under historical scrutiny in a given case. Paraphrasing a well-known saying of Kant's, my point is simply that history of science without philosophy of science would be "blind," while philosophy of science without regard to history (i.e., analysis of specific cases in their cultural setting) would be "empty."

Let me illustrate very briefly: Historians of recent physics are legitimately concerned with the question whether or not Einstein in 1904–5 knew about (or at least implicitly was aware of) the results of the Michelson-Morley (or related) experiments. But it should be equally clear that in a logical reconstruction of the special theory of relativity, those results play the role of confirming evidence, or (if, with Karl Popper, we wish to put it the other way around) of disconfirming the ether hypothesis. Or take another, even more recent example. The temptations of Lamarckism in our century may be historically understandable in view of the socialist ideologies held at the time by the men succumbing to them (e.g., Kammerer in Austria; Lysenko in Russia); but the scientific evidence spoke

overwhelmingly against the inheritance of acquired phenotypical characteristics. Neo-Darwinism, buttressed by modern genetics, in the opinion
of most biologists, won out over the neo-Lamarckian doctrines. Moreover,
actual experimental mistakes (or even deception, e.g., in the tragic case of
Kammerer who had trusted his assistant who "cheated"; Kammerer committed suicide!) were detected, though for a few years the Lamarckian
ideas were quite fashionable. The initial rejection of the Einsteinian ideas
on relativity in Soviet Russia (the different case of Nazi Germany is hardly
worth mentioning) can only be understood on the basis of prevailing
philosophical ideologies, especially represented by Lenin's opposition to
the "empiriocriticism" of Mach et al. Examples like these—and there are
dozens in the history of recent science—clearly indicate that a full understanding of the nature, validity, and acceptance of scientific theories requires close cooperation and mutual supplementation of the historical and
the philosophical approaches.

2. In view of the undeniable role of sociocultural factors in the growth,
life, and death of scientific theories, historically minded scholars have occasionally felt that there is no need for any sort of philosophy (or logic or
methodology) of science. Furthermore the exponential growth of the sciences, the frequent "revolutions" in their very framework (i.e., the conceptual presuppositions of the sciences), have led—I think misled—some
prominent scholars to a rather "anarchistic" position in the philosophy of
science. Worse yet, the entire enterprise of the philosophers of science has
been called into serious doubt. Although I might be suspected of defending "vested interests" (or, in the opinion of the critics, undeserved privi-
leges) I shall outline briefly why I consider the philosophical approach
absolutely indispensable.

First, by way of what almost amounts to an *argumentum ad hominem*
addressed to the skeptics, I do want to mention that great creative scientists of all ages, but particularly of recent times, have been their own philosophers of science. Consider, for example, Riemann, Helmholtz, Maxwell,
Mach, Ostwald, Hilbert, Poincaré, Einstein, Bohr, Schrödinger, Born,
Bohm, et al. Actually this modern tradition begins at least with Galileo,
Descartes, and Newton. (We could go back to Aristotle, of course, who
for his time—if not for all time—was clearly one of the greatest scientist-
philosophers.) Also, I shall merely list a few of the philosophically articulate recent biologists, psychologists, and social scientists, though clearly
G. G. Simpson, J. B. S. Haldane, E. Mayr, C. L. Hull, E. C. Tolman, W.

Köhler, M. Wertheimer, Max Weber, *et al.*, deserve much more than "honorable mention." Philosophy of science, in our century, has scored its most impressive achievements in the works of G. Frege, B. Russell, A. N. Whitehead, R. Carnap, H. Reichenbach, C. D. Broad, and a host of younger scholars partly under the influence of these prominent leaders or of the towering scientist-philosophers listed before.

Now as to the *argumentum ad hominem.* Consider, just as prime examples, the writings of Galileo and Newton. Their formulations usually came "after the fact" of their discoveries—experimental or theoretical. This is prototypical of all *good* philosophy of science. In Newton's formulation of his assumptions (postulates), definitions (explications), and *regulae philosophandi* (precepts), we find one of the truly great masters giving us a rational reconstruction of his theoretical achievements. Newton's *Principia* does contain an informal axiomatization of his basic ideas in ("classical") mechanics and of his law of gravitation. Beginning with the critical analysis given by Ernst Mach, more incisively searching explications and axiomatizations have been undertaken. Depending on what sort of questions are asked and to what degree of exactitude and completeness one aspires, we have relatively informal systematizations (as in Newton) through the "halfway houses" (such as Mach's) to the strictly formalized axiomatizations of more recent vintage (McKinsey, P. Suppes, H. A. Simon, *et al.*). Depending on whether one seeks mathematical elegance and compression, as in Caratheodory's axiomatization, or whether one seeks a representation that makes as explicit as possible the logical relations of the basic assumptions to their respective supporting observational-experimental evidence, as pursued in Reichenbach's axiomatization of the theory of relativity, we have a whole spectrum of forms of logical reconstruction. Each fulfills one or another kind of desideratum.

It would be absurd to criticize such axiomatizations from the point of view of fidelity to the historical development. These logical reconstructions are admittedly artificial in that they attempt to provide a "snapshot" of a set of scientific assumptions, assertions, conjectures, derivations—in *idealized* form. The purpose of such exercises is to make as clear as the given type and level of aspiration demand precisely what it is that a scientific theory (or law, or hypothesis, etc.) maintains and to what extent it is substantiated (confirmed or corroborated) by a specified body of evidence. Surely this is operating by "hindsight." But if there is to be drawn a line of demarcation between positive science and the philosophy of science, it is precisely the "meta-level" on which the philosophy of science is pursued; the "object level" is, of course, science (at a given stage and in a given context) itself.

I am not trying to suggest that philosophical reflection cannot *assist in* blazing new trails for science. It has always played that role and presumably will continue to do so. It is only in order to make the division of labor more explicitly clear that I suggested the distinction made above. In the actual development of the sciences conceptual analyses, even conceptual reforms and revolutions, usually go hand in hand with the expansion and growth of scientific knowledge. Nevertheless, the *least* a philosopher of science can (and should) try to achieve is clarity in regard to the meaning and the justification of scientific knowledge claims. He need not be confined to the kind of primarily passive role that critics of art (or literature or music) are usually bound to play. The philosopher of science (if he is thoroughly competent in the science under consideration) may well function in an advisory capacity—provided he is sensible and tactful enough neither to "dictate" prescriptively nor to "nip in the bud" scientific creativity.

Despite the profound differences among philosophers of science (consider for example Reichenbach, Carnap, and Popper—and their respective "disciples"), they do have in common the conviction that there is always a nucleus of objective factuality in well-confirmed scientific knowledge claims. Dazzled by the (indeed) radical changes ("revolutions") in the theoretical "superstructure" of the sciences, various scholars (especially Michael Polanyi, P. K. Feyerabend, N. R. Hanson, Thomas Kuhn) have, I believe (each in his different way), misled their readers by stressing the dependence of science upon "passionate commitment" to theoretical presuppositions, the relativity of conceptual frameworks or even of their perceptual bases; they have contributed to the current vogue of skepticism and subjectivism.

Although I personally abandoned, long ago, whatever adherence I had to positivistic, instrumentalistic, or operationalistic philosophies of science, I still see in the empirical laws (deterministic or stochastic, as the case may be) the major objectively (or, if you prefer, intersubjectively) certifiable content of the sciences (natural and social!). Theories are "superstructures" in the sense that we usually have more possibilities of choice in regard to them than we have on the level of empirical or experimental laws. Just consider the thousands of "empirical" constants in phys-

7

ics and chemistry. While, of course, each of them may be subject to correction (in some cases—as, for example, in regard to the isotopes—very radical revisions), on the whole, and by and large, these constants—and the empirical laws in which they figure so pivotally—represent the most precise and the most reliable knowledge mankind has yet attained. Philosophers of science (and I am not excluding myself) have very understandably been fascinated by, and have focused their prime attention upon, the *theories*. But theories come and go, whereas most empirical laws seem here to stay! Of course, there are many theories that have been very well established (until further notice!) by impressive lines of converging evidence, often from qualitatively quite heterogeneous areas of empirical evidence. Naturally, such theories are of the greatest interest also to the philosopher of science. All I am urging is that we should not be blinded by the shift in theories to the hard core of empirical regularities in the light of which the validity of theories must be appraised.

It is simply not true that all empirical knowledge is "contaminated" by theories. Among the countless examples that are ready at hand, let me mention just a very few: The phenomenon of the Brownian motion can be *described* independently of the explanations given by Einstein and Smoluchowski in 1905. The chemical laws of constant and multiple proportions (i.e., of compounding weights) can be formulated without the atomic theory for which, of course, they formed the primary evidence nearly 200 years ago (Dalton!). The laws of the propagation, reflection, and refraction of light (as in geometrical optics) can be formulated quite independently of any theory regarding the nature of light (particles, waves, or "wavicles"!). The speeds of chemical reactions as dependent on concentrations, temperature, etc., can be asserted without reference to any micro-theory; etc., etc.

It is not pertinent to say that the very observations that are made in physical or chemical laboratories (or observatories) presuppose theories in regard to the instruments of observation, the processes of observation, etc. Of course, the assumptions regarding the continued existence of some of our apparatus, and of some regularities involved in the procedures of measurement and experimentations, are indispensably presupposed. But it is not *these* assumptions (or hardly ever!) which are under scrutiny when we are testing a given theory. To refer to an example from the psychological domain on psychoanalytic research, it is *of course* assumed that it is the genidentically very *same* person who returns to the "couch" again and again. While in science fiction or mystery stories the contrary might be entertaining or interesting, the assumed genidentity is not under scrutiny when psychoanalytic hypotheses are being tested. As Reichenbach pointed out a long time ago, science progresses by successively "securing" its various knowledge claims. For example, the optics of telescopes, microscopes, spectroscopes, interferometers, etc., is indeed presupposed in the testing of astrophysical, biological, etc., hypotheses. But these presuppositions—while, of course, always in principle open to (in rare cases actually in need of) revision—are comparatively so much better established than the "farther out" hypotheses that are under scrutiny. Epistemologically, I suggest, it is the domain of the elementary, rather directly testable empirical laws (instead of the "given," be it conceived as sense data or as perceptual Gestalten) that is the testing ground to which we should refer in the rational reconstruction of the confirming or disconfirming evidence for scientific theories. (This much for philosophy of science; the knowledge claims of common sense that are here presupposed do require resorting to some data of immediate experience together with some principles of interpretation or construal for their justification.)

Finally, a few words on some misinterpretations stemming from predominant concern with the history and especially the psychology of scientific knowledge. In the commendable (but possibly utopian) endeavor to bring the "two cultures" closer together (or to bridge the "cleavage in our culture") the more tender-minded thinkers have stressed how much the sciences and the arts have in common. The "bridges" that M. Polanyi and others have been trying to build are passable only in regard to the psychological aspects of scientific and artistic creation (and appreciation). Certainly, there are esthetic aspects of science (and there can hardly be any essential impossibility of a scientific approach in regard to the sociopsychological features of scientific construction and invention). But Einstein, quoting Boltzmann, said, "Let elegance be a matter for cobblers and tailors." If the same Einstein relied on "beauty," "harmony," "symmetry," and "elegance" in constructing (with the help of tensor algebra) his general theory of relativity, it must nevertheless be remembered that he also said (in a lecture in Prague, 1920—I was present then as a very young student): "If the observations of the red shifts in the spectra of massive stars don't come out quantitatively in accordance with the principles of general relativity, then my theory will be dust and ashes."

Art does not stand or fall by tests of this kind—for the simple reason that

9

Herbert Feigl

the aims of the arts (whatever those aims are taken to be) do not (at least not *primarily!*) consist in the providing of information. But it is the primary aim of the empirical sciences precisely to find out what happens under what conditions (or what correlations there might be among various relevant variables) and how to understand (i.e., explain) these regularities. For a perhaps somewhat oversimplified formula I recommend: What is primary in the appraisal of scientific knowledge claims is (at best) secondary in the evaluation of works of art—and vice versa.

Such considerations may help in reconciling ourselves to the unavoidable divergence of the arts and the sciences. Talk of "two cultures" is in any case a gross oversimplification. As has been remarked by others, scientists even within the same discipline (e.g., modern mathematics) have extreme difficulties of communication with one another—the degree of specialization has become so tremendous!

In conclusion, permit me to mention just one episode in the history of recent science that seems to me worthy of the (collaborative!) attention and investigation by scientifically well-informed historians and philosophers of science: The peculiarly late (and hardly justifiable) opposition to the atomic theory by Ernst Mach and especially Wilhelm Ostwald, and the critical counterattacks by Ludwig Boltzmann and Max Planck, furnish —to my mind at least—a veritable treasure trove for such an endeavor. As far as I know the curious attempts by Franz Wald and Wilhelm Ostwald to derive the laws of constant and multiple proportions from classical ("phenomenological") thermodynamics have not been scrutinized by the logicians of science. In this connection and with this historical background it should be worthwhile to review the entire issue of instrumentalistic versus realistic philosophies of scientific theories. There are many other examples that could be mentioned as excellent prospects for fruitful collaborative work between the two "parties," and all demand that they achieve something much better than mere "peaceful coexistence."

NOTE ON REFERENCES

A fairly comprehensive and almost up-to-date bibliography of books on the philosophy and history of science is contained in H. Feigl and C. Morris, *Bibliography and Index*, vol. II, no. 10, *International Encyclopedia of Unified Science* (Chicago: University of Chicago Press, 1969). The works of most of the authors mentioned in the foregoing essay are listed there.

The peculiarly anachronistic attempts at deriving without atomic theory the basic chemical laws of combining are to be found in the following: F. Wald, "Bausteine zu einer neuen chemischen Theorie," *Annalen der Naturphilosophie*, 5 (1906), 271–291; F. Wald, "Mathematische Beschreibung chemischer Vorgänge," *Annalen der Natur-*

philosophie, 8 (1909), 214–265; W. Ostwald, *The Fundamental Principles of Chemistry*, trans. H. W. Morse (London: Longmans, Green, 1917). Hans Reichenbach's important axiomatization of the theory of relativity has at last appeared in English translation: *Axiomatization of the Theory of Relativity*, trans. Maria Reichenbach (Berkeley: University of California Press, 1969).

The History and Philosophy of Science:
A Taxonomy

The temptation to ignore the temporal dimension of science, to treat it as irrelevant to the proper understanding of what science is, has always been strong among philosophers. One can detect its influence just as surely in the accounts of the nature of science given by the logical empiricists of our own century as in those put forward by the Greek thinkers of the fourth century B.C. There was more excuse for the Greeks, of course, because the empirical science of their day was still a rudimentary affair, and its developmental aspects had not had time to manifest themselves. Besides in the metaphysical climate of their time, it seemed quite obvious that the most authoritative methodological ideal for a physical science was given by the newly discovered axiomatic science of geometry. The deductive elaboration of principles themselves "self-evident" (that is, bearing their own sufficient intrinsic warrant) gave just the sort of certain knowledge of unchanging structures that the prevailing notions of Form led philosophers to seek as the goal of epistēmē, of "knowing" in its fullest sense. Such epistēmē would obviously of itself be timeless. The concepts in which it was expressed were identical in their intelligible content with the forms of natural things (Aristotle), or were imperfect images of a realm of unchanging Form, the imperfection deriving from the instability of the sensible world itself (Plato). Science in either of these views does not have a history, strictly speaking; the tentative groping that precedes the formulation of concept or axiom is in no way reflected in the final product, and the only specifiable methodology for science is clearly that of logical demonstration (syllogistic demonstration, if one makes some simple assumptions about "scientific"—i.e., universal—propositions).

The logical empiricists who have dominated the philosophy of science throughout much of the present century do not, of course, share the Greek ideal of *epistēmē*, or the metaphysics of Form underlying it. Yet there is one important sense in which the Carnap of *The Philosophical Foundations of Physics* is a kindred spirit with the Aristotle of the *Posterior Analytics*. Each reduces the philosophy of science to a *logic* of science, to a study of science considered as a formal system. The only fruitful methodological issues, therefore, concern the way in which the different propositions in the system are related to one another, the types of inference used to validate one proposition on the basis of others. What is sought is a logical theory of confirmation which will allow one to justify a "scientific" proposition by applying a set of logical rules (whether deductive as with Aristotle, or inductive as with Carnap) to the propositions constituting the "evidence."

It is assumed by Aristotle and Carnap alike that one can make a sharp cut between that which is to be proved or justified (what Carnap calls the "hypothesis") and the evidence for it. The latter is supposed to be somehow "given"; the concepts in which it is expressed are taken to be unproblematic. Furthermore, no question is asked about how the hypothesis itself is derived in the first place, about the modifications of concept or the postulates of structure that may have been needed in order to arrive at it.[1] In defense of so dubious a set of assumptions and so drastic a limitation of goal, it is argued that only thus can purely logical modes of analysis be used, and some over-all methodological pattern established. The danger, of course, of such a procedure is that the impressive formal structures at which one arrives may be of very little service in aiding one to understand how scientists actually operate; they may turn out to be nothing more than exercises in logic, ingenious and interesting in their own right, and occa-

---

[1] The notion of "retroduction" which Hanson and others have taken over from Peirce in their discussion of whether or not there is a "logic of discovery" is a disagreeably ambiguous one. If "inference" is taken to be a logical method of *derivation*, there is no such thing as retroductive (or abductive) inference. Deduction and induction can be used to derive propositions that differ from the starting premises, but retroduction cannot. Insofar as it is a *method*, it is a method of confirmation only, and is in fact equivalent to what is more often called nowadays the "H-D method." If H (hypothesis) implies P (an empirically testable proposition), and P is known to be true, under certain circumstances this can serve as a warrant for taking H to be plausible (or "confirmed"). The basic pattern of retroduction is thus $H \rightarrow P, P, \therefore$ plausibly H. But this is not a mode of inferring H, in the sense of allowing us to discover it in the first place. The real crux of retroduction is to know how one is to arrive at the hypothesis contained in the major premise $H \rightarrow P$. If 'retroduction' is taken to mean the mode by which one hits upon H, given P, it is not logically specifiable and cannot be treated as a formal rule or set of quasi-formal rules after the analogy of deduction and induction. See E. McMullin, "Is There a Scientific Method?" *Proceedings of the Minnesota Academy of Sciences.* 34 (1967), 22–27.

sioned to be sure by the formal properties of empirical science, but too remote from the thought sequences that constitute "science" as the practitioners know it to warrant their being called "philosophy of science" in anything other than an honorific sense.[2]

It would be sufficient as a check on the dangers of logical "escapism" to pay close attention to science in the concrete as it is practiced here and now, in all of its diversity and variety. But in this paper, a somewhat more specific thesis will also be argued. To understand science in its modes of concept formation, in its methods of confirmation, and above all in its ontological implications, it is essential to pay attention to what may be called its "temporal dimension." It is not enough for the philosopher to consider a "slice" of scientific work at a moment in time. Rather, he has to trace the sequences by which concepts are gradually modified in the course of time, the way in which the fertility of a hypothesis over the course of time serves to confirm its validity, and the manner in which a model can continue to guide research over a long period so that one can legitimately suppose it to provide an approximate insight into the real structure of the object studied. In order to do this, the philosopher has got to draw upon the developmental aspects of science for his evidence. It is not merely a matter, then, of understanding the history of science in terms of philosophic categories, of making use of the philosophy of science to illuminate the history of science. Rather, what we are saying is that the philosopher must pay attention to the actual course that science has followed if he is to do his own job of understanding what science is. The connectives that he will be studying here will be historical, not logical, ones. They cannot be discovered by following some pattern of logical inference. They can only be known by determining as best one can what has *actually* happened in the course of scientific investigation in the past. If I am right in this, the relationship between the philosophy of science and the history of science is much closer than has usually been supposed.

But before we come to this specific issue, it will be necessary to give a

---

[2] The saying among critics of the formalist trend in recent American philosophy that "logic is the opium of the philosopher" does not, of course, imply that the philosopher can dispense with logic in his quest. Rather it is presumably meant to suggest that logical system building can all too easily become an end in itself, the philosopher's way of escape into a pleasant world of his own construction where he can gradually forget the messy and logically irreducible realities that gave rise to his system building in the first place. It is worth adding that an equally undemanding world can be reached by taking the opposite route. An unconcern for logic and for the taxing demands of correct category use is a different sort of "opiate" but one with even more serious consequences for responsible philosophy.

fairly general taxonomy of the ways in which the two enterprises of history of science and philosophy are related. These are so various that they will have to be carefully catalogued, if our main point is to emerge with any clarity.

1. Two Senses of 'Science'

When one speaks of the philosophy or the history of "science," what is meant by the term 'science'? There are two principal senses, very different in their implications for philosopher and historian alike. Science may be regarded as a collection of propositions,[3] ranging from reports of observations to the most abstract theories accounting for these observations. Let us call this S_1. S_1 is the end product of research, the careful statement in approved technical terms of something that has been empirically determined to be so, and perhaps also of a tentative explanation of why it is so. S_1 ordinarily contains only those definitions, theories of the measuring instruments involved, and the like, that are needed to allow another scientist, within the bounds of a research paper or book, to grasp the "data," to test their reliability if need be, and to evaluate claims made to generalize or explain them.[4] The *Principia* of Newton would be an example of S_1, as would the average paper or letter in the *Physical Review* today. It will be noted that S_1 does not contain an account of how discoveries were made, of the various false starts, of the ways in which concepts were gradually

[3] Carnap's example of taking it to consist of *sentences* rather than propositions has been followed by many. His main reason for doing this is that sentences can be exhibited; sentence tokens are perceptible singulars, comfortably unmysterious by comparison with propositions, which to him have a dangerously platonic and otherworldly status. But this leads to an altogether counterintuitive view of what science is (or indeed what logic is); it becomes dependent on the language in which the sentences are expressed (thus 'The velocity is . . .' will differ from 'La vitesse est . . .'), and even upon the word order chosen (since any change in word order or in the choice of words will make the sentence different, even though the meaning may remain unchanged). Thus H. Kyburg in his recent *Philosophy of Science: A Formal Approach* (New York: Macmillan, 1968), p. 17, takes the elements of his "science" to be English sentences. Though this is a consistent usage, the nominalistic reasons usually given in its favor seem greatly outweighed by the inconveniences of supposing a French text of physics to give us a different "science" from that given by an accurate English translation of it. A change in sentence form is not a change in the "science" conveyed; only a change in the *meaning* of what is said affects it as science.

[4] The formalist account of science supposes that S_1 should also contain all the so-called "rules of interpretation" that are needed to establish a semantics for the science, considered as an empty calculus. But this is never carried through in practice, and would impose an altogether formidable and perhaps impossible requirement. It is not necessary for the scientist to provide these "rules" explicitly because he assumes that most of his terms and procedures are at least partially understood in advance.

15

modified to fit the new problem, of the various extrascientific factors that influenced the author to adopt the theory he is proposing.

S_2 includes all of these. It is "science" considered as the ensemble of activities of the scientist in the pursuit of his goal of scientific observation and understanding. It includes the various influences that affect him significantly, perhaps unknown to himself, in this pursuit. It contains all the propositional formulations, both provisional and "finished," with the reasonings *actually* followed (not just those ultimately reported). In short, S_2 is everything the scientist actually *does* that affects the scientific outcome in any way. S_2 contains S_1; it is, however, far broader and vaguer than S_1. It is not just propositional, for it includes the building of apparatus, the making of measurements, the half-conscious speculation, the rough sketch —all brought into some sort of unity by the aim of accurately describing or explaining some feature of our experience.

It would be impossible ever to convey S_2 fully, even in the case of a relatively simple piece of scientific research. And no one tries to do so because it is S_1 in which everyone (including the scientist himself) is interested. S_1 is the measure of his achievement; it is that part of S_2 which is intersubjective, communicable, in some hopeful sense permanent. Because of its vagueness and singularity, S_2 will be difficult to comprehend; the effort to grasp it may well seem unrewarding or even futile. In the permanent record of the textbook, it is S_1 that figures, and usually in an artificial form that gives practically no clue to the real sequence of events and considerations. S_2 is, for the most part, soon forgotten; indeed, even to begin with, much of it may never have been made explicit. The interest of S_2 is only this, that in a very definite sense it serves to explain how S_1 came to be formulated in the first place.

2. History of Science

And this, of course, is of special concern to the historian. Thus, he will have to take at least some account of S_2. But there are very different ways of going about writing history of science. As historiography,[5] its first responsibility is to establish what the facts were: who said what, and what he

---

[5] There is an unfortunate ambiguity in the English word 'history.' It signifies both the sequence of events and what is written about the events. Thus, "history" of science (in one sense of the term) is about the "history" of science (in the other). The technical term 'historiography' is sometimes used for the former, but is rather cumbrous. We shall rely on context for clarification. HS below will, however, always mean the written account, the product of the historian.

16

meant by it, and what reasons he adduced. But after that, a considerable difference of emphasis is possible. At one extreme is chronicle, an establishing of the "facts" with a minimum of interpretive addition; at the other is "overview" or "applied history" where history is used to make a philosophical, theological, or political point, and the goal is discovery of an overall pattern rather than determination of contingent singulars. These divergent aims manifest themselves among historians of science as among other historians. But because what they are giving is a record not of battles or of treaties but of *ideas*, intelligibly linked with one another, they are forced to some extent, at least, to be interpretive. The historian of science is by definition a historian of ideas.

This suggests yet another sort of emphasis, the "history of ideas" approach now canonized by the establishment of departments and doctorates under that title in many universities in the United States. The historian of ideas has a methodologically very complex task. He has to trace a concept like *matter* or *force* or *democracy* through the writings of one or more people, subordinating the contingent historical particularities to the main aim of grasping what the concept meant and how this meaning was progressively modified. The danger of this approach (as "professional" historians are quick to emphasize) is that it may entirely subordinate history to a quite different sort of enterprise in which the connectives between, or developments of, ideas are created by the writer himself, rather than laboriously recovered from the intractable past. Ideas have a permanence and a transparency that persons and historical events lack. Thus it is tempting when tracing, let us say, the development of Newton's concept of force to pay more attention to the logical implications or plausible modifications of the concept as we see them than to the actual sequence as it occurred in Newton's own thought. The history of ideas can easily become a logical and analogical development whose dynamism lies in the ideas themselves and in the creativity of the person constructing it, rather than in the partial records of the free decisions and semi-opaque mental constructions of men long dead. The connectives of history are not always those of logic or analogy. We may wish to use a historical instance as the starting point for the exploration on our part of an idea; this is perfectly legitimate and often very revealing. But of course it is not history. The fact that it is on the whole easier to write than is "real" history need not prejudice one against its validity as a genre, but ought to serve as a warning that the two genres ought to be carefully kept apart. Their characteristic modes

17

of evidence differ: to support a claim made in the history of ideas proper, one will have final recourse to the documents of the past; to support an assertion in the "logic of ideas," one will call upon considerations of logical inference, philosophical principle, analogical similarity, and so forth.[6]

It has seemed worthwhile to dwell on this distinction in some detail because when philosophers of science turn their attention to the history of science, it is very frequently to construct a "logic of ideas" in this sense. What bothers historians of science about this is that it often seems to them to be *masquerading* as history; it makes use of the great scientists of the past as lay figures in what *seems* to be a historical analysis but really is not. They are manipulated to make a philosophical point which, however valid it may be in itself, was really not theirs, or at least is not really shown (using the proper methods of the historian) to have been theirs. Though names of scientists and general references to their works may dot the narrative, what is *really* going on (in the sense of where the basic evidence for the assertions ultimately lies) is not history. The historian has to guard

---

[6] A similar difference in emphasis may be noted in the historiography of philosophy. There is an unmistakable difference of approach between, say, the average article on Kant or Aristotle written in the United States or England and one written in Germany or Italy. The "analytic" philosopher (to use a dangerously loose label) will draw upon the writings of a philosopher of the past in order to explore various philosophic options in a systematic way. The emphasis is likely to be upon conceptual clarifications and interrelations, on the *doing* of philosophy. History thus functions here as an occasion for further philosophizing of a conceptual and analytic sort. The principal criterion for a good piece of writing in this genre is the illumination it brings to some characteristic philosophical thesis or set of theses. Whether or not the historical figure whose work is being discussed was aware of the implications attributed to his position, whether or not his views shifted in the course of his lifetime, whether he was as careful as he should have been in his choice of words to convey the points he wished to make, all of these questions, while not unimportant to the "analytic" historian of philosophy, will be incidental by comparison with the central concern, which is at bottom not so much historical as conceptual-philosophical.

At the other extreme is the historian who painstakingly tries to reconstruct the thought of some philosopher of the past in all its historical singularity; he will insist on working with the texts in their original language (this need not be of nearly so much concern to the "analytic" historian); he will attempt to reconstruct the cultural and intellectual milieu out of which the philosopher wrote, and trace the various stages of his thought. The criteria of a good piece of work here are the degree of assurance given that this *is* indeed what the philosopher said and meant, and the quality of the understanding afforded of how he came to say it. This sort of historiography focuses on the historical singular in all its contingency, whereas the other abstracts from this contingency in order to construct analyses of some permanent philosophic interest. The important thing to note here, as in the case of history of science, is that *both* genres of research are perfectly legitimate; they fulfill different functions, each of them an indispensable one. It is only when debates arise (as they so frequently do) about how "history of philosophy" *ought* to be undertaken that one comes to realize how easily the use of a single label can lead one to assume that there is (or ought to be) a single methodologically well-defined enterprise corresponding to it.

18

against rejecting this genre entirely, simply on the grounds of its not being history. What he ought rather demand is that it clarify its credentials, and avoid the suggestion that its warrant is a historical one, i.e., certain things that happened in the past.

Newton scholars, for instance, sometimes get annoyed when they read a piece by a philosopher who singles out a passage in the General Scholium or in the Queries, say, and analyzes its implications in detail, without adverting to the fact that the passage went through many drafts before attaining final form, or perhaps was later modified or even repudiated by Newton. The crucial question here is whether the philosopher is using Newton as *evidence* or as *illustration*. Is he claiming the fact that Newton argued in a particular way as a support for a more general claim about the nature of science, thus explicitly relying on the authority of what Newton said or did? Or is he merely taking a passage from Newton's work to illustrate a philosophic claim which stands in its own right (so that if a historian can show that Newton did not really write the passage or later repudiated it, the philosopher's point is left basically unaffected)?

Another sort of problem arises when a historian of ideas uses a methodological notion (the hypothetico-deductive method, for instance) or first-order concept (like *mass*) to discuss the work of some scientist of the past, even though the concept or distinction was only elaborated in explicit form in later times. The critic may urge the danger of anachronism here. Thus, even the title of Clagett's classic work, *The Science of Mechanics in the Middle Ages*, has been criticized on this score. Likewise, one will find reviewers objecting on principle to discussions of Plato's concept of matter, of Galileo's notions of virtual work or momentum, of Descartes's use of hypothetico-deductive modes of confirmation. They argue that these writers were not explicitly aware of such concepts, in fact had no terms corresponding to them; thus, it must inevitably mislead readers to find scientific work described and analyzed in terms that were only developed later on. Even the translation into a modern language of earlier texts (a translation into English, say, of a medieval Latin commentary on Aristotle's *Physics*) would seem to face the same objection; terms like 'force' and 'velocity' have been sharpened so much even in ordinary usage that there are simply no one-word equivalents in English today for many of the basic terms of medieval natural philosophy.

The solution, of course, is that the reader ought to be warned, indeed ought to realize without need for warning in the case of a translation, that the precisions and implications of the modern terminology cannot be carried backwards in time without a specific warrant for doing so in each case. The use of contemporary *methodological* distinctions is clearly valid when analyzing the actual methods followed by early scientists; to say that Galileo made use of a hypothetico-deductive method of confirming that the lunar surface is like that of the earth is an illuminating and accurate way of rendering in modern terms what Galileo *did*. It does *not* necessarily imply that he had a hypothetico-deductive theory of confirmation explicitly worked out.[7] The tracing of a concept is a more difficult and riskier affair. One cannot tie oneself to the use of a particular term as the necessary and sufficient condition for the effective presence in a writer's thought of a particular concept. Thus, it would be overly narrow to hold that there could be no question of a concept of matter in the pre-Socratics just because they had no specific agreed-upon term for matter. The criteria here are too complex to enter into in detail,[8] but in general one has to see a scientific or philosophical concept as a response to a certain problematic or set of problematics, and it is often possible to trace the prehistory of such a problematic and analyze responses to it analogous with the later more specific ones associated with a particular term like 'space' or 'mass' or 'inertia.'

[7] Take a further illustration. In my essay "Empiricism and the Scientific Revolution" (in *Art, Science and History in the Renaissance*, ed. C. Singleton, Baltimore: Johns Hopkins Press, 1968, pp. 331-369), I attempted to contrast two modes of evidence, one "intrinsic" (where a scientific statement is supposed to carry conviction in its own right, by virtue of its conceptual interconnections, once it is fully understood) and the other "extrinsic" (where one calls on something extrinsic to the statement itself—an experiment or series of observations, for example—in support of the statement). These are sometimes labeled the "rationalist" and "empiricist" notions of evidence, but this is somewhat misleading since the so-called "rationalist" theory (as one finds it in Aristotle, for example) is by no means independent of specific experience by means of which the concepts originally came to be understood. One can trace the tension between these two ideals of evidence through Greek and early medieval times (when the "intrinsic" mode dominated) and the late medieval period (when the nominalists argued for an "extrinsic" mode) to the seventeenth century (when there was a gradual shift to the "extrinsic" mode, although the major scientific figures of the century still on the whole took the "intrinsic" mode to be the ideal, even though rarely attained in practice). This is a valid framework for an elucidation of the science of that period; one simply scrutinizes the actual methods used, and the occasional remarks on method made (though these latter were often belied by the former). It would be invalid only if one were to imply that the writers themselves were fully aware of what the distinction connoted, of what, for example, the empiricist theory of proof later came to be.

[8] I have discussed them in the Introduction to *The Concept of Matter in Greek and Medieval Philosophy* (Notre Dame, Ind.: University of Notre Dame Press, 1965), and again in more detail in the Introduction to *The Concept of Matter in Modern Thought* (Notre Dame, Ind.: University of Notre Dame Press, in press).

3. Two Approaches to the History of Science

Perhaps the most important difference of emphasis among contemporary historians of science corresponds roughly to the distinction drawn in the previous section between two senses of the term 'science.'[9] HS_1 is the history of science considered as a set of general propositions, together with the properly "scientific" evidence thought to count in their favor. It focuses on published work, on final versions, on pieces of research that ultimately proved successful. And it considers them not in the order in which they were carried out, but in the order in which they were ultimately presented to the world, as specific already-formulated hypotheses with their supporting evidence clearly delineated. HS_1 will clearly appeal to the scientist, because it tells how the body of propositions today called "science" gradually came to be, at what point particular concepts or theories originated, who was responsible for them, how one work influenced the ideas of another (as shown, for instance, by references to the later work). If a historian has a considerable expertise in science (so that he can venture confidently from the safe shallows of Newtonian mechanics into the deeps of electromagnetic theory or general relativity), he is quite likely to prefer HS_1, because he has the ability to elucidate the conceptual implications and connections in the original work, evaluate the sort of evidence that was used, and discuss the gradual modification of the theories and concepts involved. Such a historian may even be a trifle disdainful of the "biographical" approach of HS_2, considering it a resort for people who really don't understand the taxing mathematical, logical, and experimental complexities of the work under study.

HS_2 takes a broader aim. It attempts to understand how the specific pieces of S_1 came to be; it makes much of the unpublished draft, the penciled correction; it seeks out the social, psychological, religious influences affecting the author, as well as implicit affiliations of a scientific or philosophical kind. The documents it uses will not be just the published S_1, but also letters, unpublished manuscripts, and surveys of the philosophy, the theology, even the politics of the day. The practitioner of HS_2 will be conversant with caches of correspondence in various corners of the world; he

---

[9] The distinction drawn here between two approaches to the history of science is similar in many ways to that between "external" and "internal" history drawn elsewhere in this book, in Mary Hesse's paper, for example. These terms are so vague in this context, however, that a variety of opinions tend to develop about what is properly "external" to science. And of course what is adjudged to be "external" in one epoch (as Dr. Hesse notes) may very well not be in another.

will be well read in the general historical background of the period he studies. He will be a man of travel grants, with his own laboriously assembled microfilm archive. As the National Science Foundation has long since discovered, HS_2 is one of the most expensive types of scholarship! The man with the history degree, with general training in historical research, is likely to lean toward HS_2, just as the man with a primary degree in science is somewhat more likely to opt for HS_1.[10] The man of HS_2 may be tempted to consider HS_1 extremely narrow and not very helpful. He will incorporate in his own work all the HS_1 he considers relevant, but the emphasis given it and the interpretations given to its methodological and logical structures are likely to be very different from those given by a "straight" HS_1 man.

Between HS_1 and HS_2, there is a continuum of possible intermediates. Making the distinction as sharp as I have done above may well seem rather artificial. Yet it does serve to suggest how far the "extremes" in historiography of science may be from one another. One has only to think of standard works of history of science as remote from one another in genre as are E. T. Whittaker's *History of Theories of Aether and Electricity* and Pearce Williams's *Michael Faraday*. Even the most cursory glance at these two books shows that their aims are markedly different. The first is concerned with science as a body of theories supported by experiment. To understand how it has been gradually built up over the years is to understand better the vast, and of itself nontemporal, conceptual network, "science," in terms of which the universe is partially categorized and explained. The second book, on the other hand, is concerned with science as a human activity; it attempts to understand Faraday's science as something to which all aspects of the great scientist's background and personality contributed. Its aim is thus to explain how this *particular* piece of science occurred, what the relevant causal lines leading to it were. Whether these lines are capable of generalization, whether they occur in all or most cases of scientific research, is of lesser concern to the author, though he does have some suggestions to make on that score. And where Whittaker describes Faraday's scientific results and analyzes their significance, Williams asks how he arrived at them, what the false leads were, and how his religious beliefs affected his estimate of the relative plausibilities of the various explanatory hypotheses that occurred to him.

---

[10] These are, of course, outrageous sociological generalizations which would need much more evidence in their support than the random sample of historians known to one person. But I am fairly sure that they are not *too* wide of the mark.

It is obvious that the latter genre is much closer to that of the "ordinary" historian who reconstructs the contingent singular past event and tries as best he can to say why it happened the way it did. HS_2 is, in fact, like "ordinary" historiography in its general method of approach; it differs mainly in the internal conceptual complexity of part of its material. HS_1 is very different from "ordinary" history writing because it leaves aside most of the clues the historian would think relevant, and concentrates only on those features which can be logically or methodologically reduced to some pattern. It is still history, because it attempts to isolate a fragment of S_1 as it was first stated, precisely as it was first stated. But it leaves aside as irrelevant all the contingent historical circumstances surrounding this statement; only those circumstances are chronicled which are logically related to it, necessary therefore if one is to understand what S_1 meant, or what its precise warrant was. Thus the aim of HS_1 is not exclusively historical; in part at least, it is to understand S_1 as a piece of science, whereas HS_2 seeks to understand why the historical event, which is the formulating of S_1, happened as it did. In a certain sense, then, HS_1 terminates in a universal, where HS_2 terminates in the historical singular. The criteria for a good piece of HS_1 involve our understanding of the piece of science as science, whereas for HS_2 it would be our understanding of how that "piece of science" happened when and how it did.

4. Three Approaches to Philosophy of Science

In attempting to define what is meant by a "philosophy" of science, the first problem one encounters is the notorious vagueness of the term 'philosophy.'[11] Unlike historiography, which is relatively well defined in its method and in the types of evidence on which it draws, "philosophy" can in practice be anything from a cloudy speculative fancy to a piece of formal logic. The term has become almost hopelessly equivocal in modern usage; even in academic contexts, despite the unity implied by a label like 'Department of Philosophy,' there can be the widest divergence concerning what the aims and methods of the "philosopher" should be. Five strands might be roughly separated. Something may be called "philosophy" because of (1) its concern with the "ultimate causes" of things; or (2) the immediate availability of the prescientific or "ordinary-language"

[11] I have argued elsewhere that a failure on the part of those writing what they call "philosophy of science" to say what the term 'philosophy' means for them leads to this label (at present an honorific one) being used to cover ever broader areas of thought. See "Philosophies of Nature," New Scholasticism, 43 (1968), 29–74.

23

Ernan McMullin

or "core-of-experience" evidence on which it rests; or (3) the generality of the claims it makes; or (4) its speculative character, allied with difficulties in confirmation, particularly empirical confirmation of any kind; or (5) its "second-level" character, the fact that it is concerned with other first-level disciplines rather than with the world directly. In practice, some ill-defined combination of these criteria will usually be operative. It is the last (and most recent) of these senses that seems most relevant to the notion of a "philosophy" of science. It is "philosophy" just because it *is* a second-order critical and reflective enterprise. The label 'philosophy of science' is of course of very recent origin, even more recent than the separation of the domains of "science" and "philosophy" from which it takes its origin.

There are, it would seem, two quite different ways in which one could set about constructing a reflective philosophy of science (*PS*). One could look outside science itself to some broader context, and in this way derive a theory of what scientific inquiry should look like and how it should proceed. We shall call this an "external" philosophy of science (*PSE*), because its warrant is not drawn from an inspection of the procedures actually followed by scientists. *PSE* will often appear as normative, because it can serve to pass judgment on the adequacy of the methods followed in a particular piece of scientific work, or even in scientific work generally. Since it does not rest upon any analysis of the strategies actually followed by those who would regard themselves as "scientists," it need not be governed by current orthodoxies in this regard. Thus, *PSE* need not take account of the history of science, except as it furnishes illustrations. *PSE* in no way rests upon *HS*, though it must obviously give *some* sort of plausible reconstruction of *HS* if it is to be taken seriously. If a *PSE* diverges radically in its implied norms from what scientists actually appear to be doing, it is likely to be challenged, and its starting point may be called into question. Yet a surprisingly large divergence can be tolerated; it will be said simply that the "science" under discussion falls short of what "science" ought to be. One thinks, for example, of the account of the nature of science given by Aristotle in his *Posterior Analytics*, so obviously and widely at variance with what might have been inferred from his own extensive contributions to the science of biology.

There are two main types of "external" warrant for an account of the nature of science:

1. *PSM*: If one views science as the ideal of human knowing, or as one specific type of human knowing, it is plausible to suppose that its nature

24

can best be understood by beginning from a general theory of knowing and being. This was essentially the starting point from which both Plato and Aristotle commenced in their discussion of the nature of science; to a large extent it was still the framework within which Descartes constructed his *Discourse on Method.* Such a *PS* can begin from an epistemological or from a phenomenological starting point; it will derive from a more general "metaphysical" theory, therefore; hence the label '*PSM.*' Since it is a *PSE,* the "metaphysics" here should not be a science-based one (otherwise the warrant would not be extrinsic). When we speak of a *PSM,* therefore, it will be assumed not only that its warrant is basically a "metaphysical" one (another admittedly vague label), but also that it is prior to any analysis of the actual procedures followed in science.

2. *PSL:* To the extent that science is thought of as a logical structure of demonstration or of validation, *PS* becomes akin to a formal logic, whether a deductive logic of demonstration (like the Aristotelian theory of syllogism) or an inductive logic of confirmation (such as that constructed by Carnap). Such a *PS* can be judged as one would any other purely formal system, in terms of consistency, simplicity, and so forth. Only the most general specifications of what would constitute "demonstration" or "inductive evidence" may be needed to get the system construction under way. There may be very little reference to present or past scientific practice; it is not suggested that this logic is the one actually followed by scientists in their work of discovery or of validation. Rather, it is a reconstruction, an idealized formal version of what, for example, proof *really* amounts to in science, whether the scientist knows it or not. It may be interpreted normatively as suggesting how, for instance, scientists *should* proceed when faced with two competing theories. Or it may be intended *only* as rational reconstruction of a general logic that is intrinsic to scientific inquiry, though not capable of being made operationally specific enough to serve as a methodological manual for the scientist wondering how best to do his work.

The best known recent instance of a *PSL* of this latter type is Carnap's inductive logic. This is a formidably complex and logically fascinating formal system relating various types of confirmation in a mathematically expressible way. But no one has been able to suggest how the basic "measure" utilized by Carnap (that of degree of support of a hypothesis, H, on the basis of evidence, e) could be related to any actual hypothesis/ evidence situation in empirical science. Thus, though Carnap's logic has been (and continues to be) of great interest to logicians, it is not clear that it has led to an understanding of what goes on in scientific inquiry. Yet it qualifies as a *PSL* in intention, at least; the reason for undertaking it, and the general conceptual framework of *hypothesis, evidence, plausibility,* in terms of which it was developed, derived from empirical science. But the justification for it as an intellectual construction lies in its logical interest, rather than in any insight it provides into the actual procedures of the scientist.

Discussions of the nature of science up to the seventeenth century were nearly always "external" in character, though one occasionally finds in the later medieval and Renaissance periods some analysis, for instance, of the actual methods of "composition" and "resolution" followed by scientists. The theory of science was based on a prior metaphysics or on an autonomous logic.[12] And even though the pioneers of the scientific "revolution" purported to be drawing upon new sources for their methodology, they were still much closer to the *PSM* and *PSL* of the Greek tradition than they were willing to admit. Though Bacon, Boyle, Huygens, and many others depended on their knowledge of the practice of science in their analyses of methodological and epistemological issues, it was only in the nineteenth century that writers like Whewell and Mill took this new source of *PS* with complete seriousness.[13] It is easy to see why the astounding successes of the new mechanics, and the beginnings of a new era in biology, geology, and chemistry, should make it for the first time plausible that if one wished to understand the nature of science, one should look at what scientists actually *do.* No longer did "science," a stable knowledge of the world, seem a remote ideal; in terms of practical success, it had clearly been achieved already.

3. *PSI:* In contrast, therefore, with *PSE* is a philosophy of science which relies for its warrant upon a careful "internal" description of how scientists actually proceed, or have in the past proceeded. The function of different methodological elements (law, hypothesis, predictive validation, etc.) is studied not in the abstract, but in the practice of the scientists them-

---

[12] These were combined in the dominant Aristotelian account of science of this period. See, for example, A. C. Crombie, *Robert Grosseteste and the Origins of Experimental Science, 1100–1700* (Oxford: Clarendon Press, 1953), chapters 4, 11; and E. McMullin, "The Nature of Scientific Knowledge: What Makes It Science?" *Philosophy in a Technological Culture,* ed. G. McLean (Washington, D. C.: Catholic University of America Press, 1964), pp. 28–54. See also the first half of my "Philosophies of Nature."

[13] See my "Empiricism and the Scientific Revolution."

selves.[14] This approach presupposes that one can already identify competent scientists and successful pieces of research. *PSI* is based on what scientists do rather than upon what they say they are doing; when contemporary scientists set out to give an account of the nature of scientific method, they can sometimes be as remote from scientific practice as were Aristotle or Descartes. They may have some sort of idealized *PS* in mind, an oversimplified isolation of one procedure, perhaps, or even a *PSM* in disguise.[15] A *PS* constructed by a scientist is not necessarily *PSI*, and if it is *PSI*, it is not necessarily accurate *PSI*. The evidence on which *PSI* is based is a descriptive account of the procedures by which empirical science is built; though the testimony of scientists is of primary importance in achieving such a description, such testimony cannot be taken without question, especially if there is reason to suppose that the scientist allows a *PSE* or an overly simplified *PSI* to color what he has to say of his own procedures.

By comparison with *PSE*, *PSI* is a relatively empirical undertaking, not very different in this respect from an empirical science itself. If one wishes to give a *PSI* analysis of the role of models in science, one begins from a carefully documented review of how scientists have made use of models. *PSI* thus differs in several important respects from *PSE* (whether of the traditional *PSM* or *PSL* varieties). It is expressly second-level, in that it takes another intellectual discipline as its object of study. It presupposes an already-functioning methodology, whose pragmatic success is a sufficient warrant of its adequacy as a heuristic. There is no need to ask what science *ought* to look like, in some abstract sense. The very success of modern natural science in prediction and control gives a sufficient reason for taking it as an object of analytic epistemological study in its own right. Furthermore, the claims made in *PSI* are relatively easily confirmed, as a rule; they can usually be settled by an analysis of the interrelations of some elements of descriptive methodology. There is not much affinity, in conse-

---

[14] A good example would be Leonard Nash's recent work, *The Nature of the Natural Sciences* (Boston: Little, Brown, 1963).

[15] Examples are not hard to find. One recalls the "pointer-reading" account of scientific method on which Eddington built his elaborate "Fundamental Theory"; Bridgman's operationalism also comes readily to mind as an illustration. A recent delightful example is an article by the biochemist J. R. Platt: "The New Baconians," *Science*, 146 (1964), 347–353. He reduces scientific inquiry to what he calls a "Baconian" method of "strong inference" which he compares to climbing a tree, each fork corresponding to a choice between alternative hypotheses; the decision on which way to go at each fork is made on the basis of crucial experiments ("clean results"). He attributes the recent rapid advance of biochemistry to its fidelity to this simple method, and suggests that other parts of science could enjoy equal success if only they could see the methodological light.

27

quence, between the practitioner of *PSI* and the metaphysician or moralist. (There is just as little affinity, but for different reasons, between the exponents of *PSM* and *PSL*.) This may help to explain the not infrequent tensions between philosophers of science and other philosophers; the closer to *PSI* the former are, the more likely they are, for example, to plan their conventions in conjunction with those of scientists or historians of science rather than those of philosophers.

Why are *PSI* and *PSL* with their heavily empirical or formal emphases called "philosophy" at all, then? It might seem that *PS* in either of these two genres could just as readily be called "science of science" or "logic of science," or be given an entirely new label. The main reason for retaining the name of "philosophy" is that the logical analysis of method and the drawing out of conceptual implications characteristic of both *PSI* and *PSL* present obvious analogies with the techniques traditional to the philosopher. Granted that the type of evidence called on and the mode of confirmation employed are rather different, there is still a sufficient family resemblance based on the procedures followed. And there is also a sufficient cross-relevance between *PS* of the *PSI* or *PSL* variety and other parts of "philosophy" to make it desirable that they should be studied in conjunction with one another. Besides which, we have already noted the modern tendency to describe all second-order critical discussions, whether they are of art, of history, of literature, of law, as "philosophy of . . ."

In any discussion of the relevance of history of science to philosophy of science, it makes a very great difference which type of *PS* one has in mind. Clearly, history of science may be of little concern to a practitioner of *PSE* (whether *PSM* or *PSL*), though he cannot be wholly unconcerned about serious divergences between his own account of the nature of science and the course science has actually followed. And he may want to draw upon *HS* for illustrations and indirect support. But the philosopher whose interest is *PSI* has to take history of science very seriously. It furnishes not merely examples but the basic evidence from which his inquiry has to begin. More exactly, *PSI* can begin either from a historical review or from an account of contemporary practice (or both). But even if a *PSI* practitioner prefers to focus on the details of contemporary practice, leaving the historical dimension of this practice out of account, he cannot draw any sharp distinction between past and present, and thus will have to admit the potential relevance of *HS* to what he is doing, whether he chooses to make use of it or not.

It might be argued that all there is of methodological import in the history of scientific development is likely to find a place somewhere in contemporary scientific practice, so that explicit recourse to the past history of science is unnecessary to the philosopher of science. If he bases his analysis on what scientists are currently doing, he is taking advantage of the learning process that has gone on in science itself over the centuries, as scientists have gradually become more expert in how to go about their experimental and theoretical researches. A pragmatic type of validation procedure has, after all, been at work in science itself; the methodology of today's physicist is by no means the same as that of Galileo.

While this is true up to a point, it will be argued below that *PSI* has to take into account the *developmental* aspect of science, the characteristic ways in which a theory, for instance, is modified in the face of successive anomalies. To do this properly, it will not be enough to examine the science of a particular moment; one will have to follow it over a period, even a considerable period. Besides, it may be important to note the ways in which the procedures of the scientist have changed since Galileo's time and to ask why these changes have occurred. Furthermore, historical distance allows one to isolate and understand much better the influences at work in a piece of scientific research (as in any other human activity). The philosopher may learn more about the nature of explanation in dynamics from a careful analysis of, say, the writings of Newton and his contemporaries than from a review of contemporary relativistic dynamics, not only because the simpler seventeenth-century context may reveal features of method that are more difficult to uncover today, but also because the variety of influences at work on Newton, as well as the different nuances his thought took on in successive drafts of his work, permit a more detailed analysis than would ordinarily be possible in the case of some contemporary piece of research. In summary, then, the history of science is relevant to *PSI* for two different sorts of reason: (1) because it provides complete case studies, of a kind one could not recover from contemporary science; (2) more fundamentally, because it allows one to study science in its all-important temporal dimension.

5. *HS* and Some Philosophers

The distinction drawn above between *PSM*, *PSL*, and *PSI* ought not be taken to imply that any given piece of *PS* conforms to one and only one of these patterns. In practice, one finds philosophers of science calling upon all three sorts of criteria, sometimes even in the same piece of writing, and intermingling them in very complex (and not necessarily consistent) ways. Nevertheless, it is often possible (and when possible, helpful) to characterize a piece of writing in PS under one or other of the categories above, depending on which of the three types of warrant seems to dominate in it. There is no reason why an author could not combine logical, metaphysical, and descriptive-empirical elements in constructing a philosophy of science. But it is of paramount importance that he not be misled (or that he not mislead the reader) about what the balance between them in his argument really is.

In particular, it is easy for an author to suppose that what he is presenting is *PSI* when it is in fact *PSE*. This is all the more likely to happen today; because of the sheer weight of evidence available on what the procedures of the scientist are, it is hazardous to put forward any philosophy of science nowadays without some attempt, at least, to make it look like *PSI*, that is, to make it appear to derive from a familiarity with current scientific practice or from an intimate knowledge of the history of science. Yet if, in fact, the genre of writing is really that of *PSM* or *PSL*, there is an obvious danger that the wrong criteria of evaluation will be applied.

Philosophers of science of even the most "external" sort have always made *some* reference, at least, to what they believe the scientific practice of their day to be. But they have not usually turned their attention to *HS*; in the logical-empiricist tradition which has dominated much of the work in *PS* of our century, virtually no attention has been paid to *HS* until recently, on the grounds presumably that the logical structures which were the philosopher's concern exhibited themselves readily in any random slice of contemporary scientific inquiry. It did not seem necessary or even desirable, therefore, to undertake first the difficult work of the historian of science as a means of carrying out the task of the philosopher of science.

This has changed in the last decades, and now one is beginning to find case histories dotted here and there throughout the journals of *PS*. But the change has brought with it some methodological headaches. How exactly *should HS* be incorporated in the philosopher's work? What weight should be given it? To what extent ought it be regarded as normative? In order to illustrate some of the difficulties that can arise, it may be useful to glance at two lengthy recent monographs. Both were written by pupils of Karl Popper, and this prompts one to ask an initial question: how should

Popper's own influential work be characterized, as *PSE* or as *PSI*?[16] No one has concerned himself more with the demarcation between "science" and "metaphysics" than has Popper, so that the distinction between *PSM* and *PSI* is a valid and important one to make in situating his own work. Does his theory of falsification start from an analysis of the ways in which scientists actually evaluate alternative theories? When he says that of two competing theories the one to be preferred is the one with more "empirical content," does this reflect a discovery he made by observing what sort of consideration influences scientists who have to choose between theories? The answer would on the whole seem to be that it does not. Popper's *PS* derives mainly from a general theory of rationality, with a considerable amount of logical analysis thrown in. It is, in fact, a good instance of a *PSE*, more specifically a *PSM*. The frequent references to scientific practice and even to instances from the history of science serve to define the conceptual-logical problem: By what sort of criteria must one suppose scientific growth to be guided if it is to be regarded as a "rational" process? The answer to this problem cannot, in his view, be found in an inspection of *HS*, both because it is very difficult to reconstruct what criteria actually *did* weigh in any historical instance of the replacement of one theory by another, and also because scientists are fallible in their methods. They can, for instance, make mistakes in holding on to established theories longer than they should, thus evincing the very reverse of the "critical attitude" that should characterize the good scientist. This "critical attitude," the attitude of someone who has comprehended the subtle methodology Popper proposes for the theoretical scientist, is an ideal to be sought after, therefore, not one which can necessarily be discovered by watching scientists go about their work. It is ultimately grounded upon considerations of the phenomenological "don't you see that it must . . ." variety, rather than upon a chronicle of the strategies that "successful" scientists have followed.

Some recent essays by Imre Lakatos, developing a variant of the Popper *PSM*, illustrate in a quite explicit way one possible (but ultimately rather equivocal) role of *HS* in a philosophy of science. He discusses two well-

[16] It may be worth reminding the reader once again that this distinction is not a sharp one, both because considerations of "external" and "internal" sorts are likely to be interwoven and also because a "metaphysics" today (in this context, an account of rationality) cannot help but be influenced, no matter how a priori or external its methodology may be, by some implicit notions on how scientists go about their work. The point of the question above is, then: what type of warrant appears to be the primary one in Popper's own practice as a philosopher of science?

known historical instances of "research programs" (i.e., general hypotheses which oriented research over a considerable period): Prout's hypothesis that the atomic weight of every pure element is a whole number and Bohr's light-emission hypothesis. Lest the reader be misled by the pages of detail in which he appears to be following the historical course of these hypotheses, he prefaces his remarks with an interesting methodological note: "A historical case-study to my mind must follow the following procedure: (1) one gives a rational reconstruction; (2) one compares this rational reconstruction with actual history, and tries to appraise actual history critically, and rational reconstruction self-critically. Thus any historical study must be preceded by a heuristic study: history of science without philosophy of science is blind. In this paper it is not my purpose to go on seriously to the second stage."[17] What Lakatos says is that in these papers it is not his purpose to write "actual history," but rather to give a "rational reconstruction" of history in the light of his general theory of science. His purpose is to illustrate this theory: "The dialectic of positive and negative heuristic in a research programme can best be illuminated by examples. Therefore I am now going to sketch a few aspects of two spectacularly successful research programmes."[18]

His aim in these examples is clearly not to provide *evidence* for his theory. Nor (despite appearances) is it to "illustrate" the theory in the ordinary sense, i.e., by pointing to actual sequences in the history of science which can be illuminated by applying the theory to them. The notion of a "rational reconstruction" (which is to precede the attempt to find out what exactly *did* happen) precludes the idea that these examples are to serve as historical illustration in the ordinary sense. Rather, they are imaginary or quasi-imaginary examples, recounting what *ought* to have happened in the course of development of physical hypotheses such as Prout's or Bohr's, if the author's theory of heuristic had been followed in all cases by the protagonists. Lakatos is not claiming that this is what *did* happen, only that ideally it is what *ought* to have happened. What is being "illuminated" here, therefore, is not the "historical example" but the theory itself. The "illustration" is a constructed one, after the fashion of a

---

[17] "Falsification and the Methodology of Scientific Research Programmes," in *Criticism and the Growth of Knowledge*, ed. I. Lakatos and A. Musgrave (Cambridge: Cambridge University Press, 1970), section 3c. This is a longer version of "Criticism and the Methodology of Scientific Research Programmes," *Proceedings of the Aristotelian Society*, 69 (1968), 149–186.

[18] This and the next two quotations are from the beginning of section 3c of the paper.

textbook problem meant to illustrate some point made in the text. To the extent that the anecdote conforms to "actual history" (and this for Lakatos is a separate problem), the discussion could, of course, serve as an "illumination" of history. But he is not concerned with this issue.

That this is really what Lakatos means becomes even clearer in the course of his discussions of the Prout and Bohr hypotheses. He begins his discussion of the former by saying that Prout "knew very well that anomalies abounded but said that these arose because chemical substances as they ordinarily occurred were impure, that is, the relevant experimental techniques of the time were unreliable." This sounds definite enough, but then one's attention is caught by a dismaying footnote: "Alas, all this is rational reconstruction rather than actual history. Prout denied the existence of any anomalies." The same happens with the Bohr story; a footnote tells us that the account of the discovery of electron spin given in the text is a "rational reconstruction," which does not correspond to the actual sequence of events.[19] Or again, to illustrate a "progressive theoretical problem-shift," he gives a detailed "imaginary case" involving a deviation in a calculated planetary path and the series of attempts made to explain it.[20]

From all this, one might think it could confidently be concluded that HS, qua history, plays no role either as warrant or as illustration in Lakatos's PSM. He does leave open a comparison of his "rationally reconstructed history" with "actual history" (in the first quotation above), allowing a modification in the former as one possible outcome. Yet it is assumed that this sort of comparison is a separate and optional enterprise; his PSI can apparently be constructed without it. This is PS at its most "external"; the role assigned here to "history" is likely to fill even the broadest minded historian with foreboding! But even more troublesome is the quite different role assigned to HS in his critique of methodological ("naive") falsificationism.[21] He notes the difficulty of rejecting any particular theory of rationality in science before a general theory of rationality is constructed.

---

[19] Ibid., section 3c, toward the end.
[20] Ibid., section 2a (3).
[21] Lakatos distinguishes between three types of "falsificationism," all of them associated with Popper's name. "Dogmatic falsificationism" is the simplified version which supposes that science grows when theories are refuted by facts. Popper never defended this in his published work; hence it is attributed to "P₀," ("pseudo-Popper"). "Methodological (naive) falsificationism" is a modified form of conventionalism, in which "refutation" still plays the central role but in a very weakened form. It is found especially in Popper's Logik der Forschung (Vienna: Springer, 1934); hence "P₁," ("proto-Popper"). "Sophisticated falsificationism" is the third option, the one Lakatos himself defends; it involves estimates of the "empirical content" of a theory and emphasizes

Instead of going on to specify such a theory, in some preliminary way at least, he unexpectedly makes an appeal to HS as the means of excluding various incorrect theories of rationality: "If we look at history of science, if we try to see how some of the most celebrated falsifications happened, we have to come to the conclusion that either some of them are plainly irrational, or that they rest on rationality principles radically different from [those of naive falsificationism] . . . Indeed, it is not difficult to see at least two crucial characteristics common to both dogmatic and methodological falsificationism which are clearly dissonant with the actual history of science . . ."[22]

Note that HS now appears as warrant for what the philosopher of science may assert. In this event, a "rational reconstruction" will not do, and may easily lead to circularity. If a PS were to be constructed along the lines suggested by the last quotation, it would clearly be a PSI, not a PSM, as the rest of the Lakatos monograph would lead one to expect. The uneasiness the reader feels with the over-all methodology of the monograph is due mainly to the equivocal role assigned to HS, at once emphasized and called upon as evidence, yet systematically "reconstructed" in the service of a prior theory of rationality.

One final illustration of the difficulties that philosophers can get into when they try to use HS in the service of their PS can be found in a recent monograph by Paul Feyerabend,[23] written partly by way of reaction against the Lakatos PS above.[24] Unlike Lakatos, Feyerabend claims to base his account of scientific method on the actual history of science as well as on "abstract" considerations. His monograph, he says, is "mainly historical. Abstract considerations are used only sparingly, and in the form of com-

corroboration rather than refutation (it is, indeed, not quite clear why it is called "falsificationism"). This is the view of "P₂," "proper-Popper." (These labels are mine; they are easier to remember than 'P₀,' 'P₁,' and 'P₂'!) Lakatos personifies his three Popper figures in a vivid way ("Popper₂ can easily get rid of Popper₁'s untenable falsificationist elimination rule"), and concludes that "the real Popper is a strange mixture of Popper₁ and Popper₂, and the only way to understand him is by cutting him into two" ("Falsification and the Methodology of Scientific Research Programmes," Appendix). Lakatos's method of "rational reconstruction" is obviously not confined to HS; it extends to the work of philosophers also. This use of lay figures with the names of real people seems a distracting and unnecessarily vulnerable way of writing philosophical analysis.

[22] "Falsification and the Methodology of Scientific Research Programmes," section 2b.

[23] "Problems of Empiricism, II," in The Nature and Function of Scientific Theory, ed. R. G. Colodny (Pittsburgh: University of Pittsburgh Press, 1970).

[24] Which he believes to be defective precisely because of the "tremendous abyss" (section 1) between it and "certain important episodes" in HS.

ments on the historical material. This material shows, I think, that there is something seriously amiss with the professional philosophy of science of today . . . [It] not only fails to adequately describe some of the most exciting episodes in the history of thought; it would also have given extremely bad advice to the participants."[25] This is explicit enough: he is about to elaborate a *PSI* based on the history of science, and is critical of other philosophers (notably the Popper group) for their failure to adopt the same approach.

For his historical material, he chooses the work of Galileo in cosmology. Upon this he bases a general thesis about the nature of science: "Progress [in science] is made by the gradual accumulation of views that are absurd or refuted [relative to the status quo], views which though undermined by reason and fact still support each other to such an extent that they finally supersede everything else."[26] Any attempt to see in this "progress" something "rational" (describable in some sort of methodological pattern) is thus doomed to failure: "My discussion of Galileo has, therefore, not the aim to arrive at the "correct method" [of science]. It has rather the aim to show that such a "correct method" does not and cannot exist. More especially, it has the limited aim to show that counterinduction [ignoring facts that do not fit] is very often a reasonable move."[27] Feyerabend does not explore the significant term 'reasonable' here; he does not ask what norms of "reasonableness" might have guided Galileo (assuming that he *did* ignore facts). Instead he is mainly concerned to reject the conventional empiricist and falsificationist accounts of science. It is simply *not* the case (he argues) that the transition from the Aristotelian-Ptolemaic cosmology to the Copernican-Galilean one "consisted in the replacement of a refuted theory by a more general conjecture which explained the refuting instances, made new predictions; and was corroborated by the observations carried out to test these new predictions."[28]

Instead of this account of Galileo's achievement, Feyerabend suggests a diametrically (and provocatively) different view: "While the pre-Copernican astronomy was in trouble (was confronted by a series of refuting instances) the Copernican theory (as found in Galileo's work) was in even greater trouble (was confronted by even more drastic refuting instances); but being in harmony with still further inadequate theories it gained

[25] Section 1.
[26] Section 1.
[27] Section 13.
[28] Section 7, end.

35

strength and was retained, the refutations being made ineffective by ad hoc hypotheses and clever techniques of persuasion."[29]

He concentrates on two features of Galileo's Copernican arguments, his use of the new telescopic data (especially those concerning relative planetary brightnesses), and his indirect method of getting the reader to admit the claim that an observer will not perceive a uniform motion which he himself shares. In the case of the first, Feyerabend argues that "the telescope produced spurious and contradictory phenomena and some of its results could be refuted by a simple look with the unaided eye. Only a new theory of telescopic vision could possibly bring order into the chaos . . . Such a theory was developed by Kepler."[30]

Feyerabend assumes that "evidence obtained in accordance with the older Aristotelian views" would have been "bound to clash with the new astronomy." Yet the natural philosopher had an obligation (why?) to "preserve the new astronomy," and this meant "that he must develop methods which permit him to retain his theories in the face of plain and unambiguous refuting facts,"[31] until the development of the lower-level "auxiliary sciences" would allow the apparent "facts" to be reinterpreted.[32] This leads him to the broad generalization that "a new period in the his-

---

[29] Section 7, end.

[30] Section 5, end. It is not at all clear what service Kepler's theory would have rendered in support of the reliability of Galileo's data. In point of fact, it never does seem to have been called upon explicitly as an auxiliary warrant for the telescopic data the Copernican argument required.

[31] Section 8. He makes much of the famous passage where Galileo praises Copernicus for laying aside the apparently refuting "facts" concerning the variations in apparent planetary sizes, which were much smaller than they should have been according to his theory. But no instance is given of Galileo himself deliberately setting aside awkward astronomical "facts"; in his Copernican arguments (as he reminds the reader more than once), he seems to take inconsistency between fact and theory very seriously. To make the point that Galileo did not follow the classical inductive procedures of empiricism, it is not necessary to go to the other extreme and assert that he constantly ignored facts that counted against his views; it is quite enough to observe that the anamnesis he relies on so much was not a bringing of new experimental evidence but rather a reminder of familiar facts and a warning that it was more difficult than it seemed to describe them consistently and correctly.

[32] The claim that various "auxiliary sciences" of physiology and optics would have been needed (and were not available) in order to make Galileo's telescopic data sufficiently reliable to count as independent evidence is open to serious question. It is true that the first telescopes were quite imperfect, and that for a few years contradictory data were reported. It is also true that expectation plays a role in what is seen. But what is one to make of the fact that the astronomers of the Collegio Romano had verified all the observations Galileo reported in the Sidereus Nuncius (except his incorrect description of Saturn) as early as April 1611, i.e., within a year of their being made? And these men were as yet Aristotelian in their sympathies. The phases of Venus, the sunspots, the satellites of Jupiter, the variations in planetary diameter and brightness, all of these were

tory of science commences with a backward movement that returns us to an earlier stage where theories have smaller empirical content."[33]

Galileo's kinematic arguments are treated even more severely by Feyerabend; in his view, they are "propagandistic machinations," concealing a basic "change in experience, an invention of a new kind of experience."[34] Like Koestler, he sees in Galileo's whole approach to the physics of motion a set of deliberate "psychological tricks" which "obscure the fact that the experience on which Galileo wants to base the Copernican view is nothing but the result of his own fertile imagination, that it has been invented."[35] The "essence of Galileo's trickery" lies in his use of something that purports to be the Platonic method of anamnesis in order to introduce a "new kind of experience manufactured almost out of thin air . . . [which] is then solidified by insinuating that the reader has been familiar with it all the time."[36] In particular, by focusing on certain special physical contexts (like the deck of a moving ship), he can argue that we have in fact already implicitly accepted the nonperceptible character of shared motion. By generalizing this to include the earth, Galileo thus "conceals" the fact that he has really redefined experience itself. The "absolutist" interpretation of our motion claims (that we can perceive what is *really* at rest and *really* in motion) nevertheless is "empirically entirely adequate." No difficulties arise with it (as long as we keep away from contexts like ships and other moving systems). Thus: " 'Experience,' that is, the totality of all facts from all domains described with the concepts which are appropriate in these domains, *this* experience cannot force us to carry out the change which Galileo wants to introduce. The motive for a change must come from a different source."[37] And the source is twofold—a "metaphysical urge for unity of understanding" and a prior bias in favor of Copernicanism, which Galileo has accepted in advance of the evidence and is not pre-

observed by a variety of different astronomers within a few years of Galileo's announcement of them. The regular periods observed for the satellites and for the phases of Venus strongly indicated that they were no artifact of physiology or optics. There was surely no need of "auxiliary sciences," especially when all these telescopic data had been woven together by Galileo under a single consistent theory. These data were not, of course, "hard facts" independent of the support given them by incorporation in such wider theories. But to say this is not to say that there was no empirical or factual basis for their use by Galileo to decrease the plausibility of the Ptolemaic system almost to zero and to corroborate the Copernican view.

[33] Section 8.
[34] Section 14.
[35] Section 13.
[36] Section 16.
[37] Section 14.

pared, on any account, to give up.[38] Thus by letting "refuted theories support each other, he built a new world-view which was only loosely (if at all!) connected with the preceding cosmology (everyday experience included); he established fake connections with the perceptual elements of this cosmology which are only now being replaced by genuine theories (physiological optics; theory of continua), and whenever possible, he replaced old facts by a new type of experience which he simply *invented* for the purpose of supporting Copernicus."[39]

To evaluate this reconstruction of Galilean mechanics with the care it deserves would take us very far afield indeed.[40] And the intention of introducing it here was not to provide a critical analysis. Rather, it was to illustrate one quite characteristic manner of approach to the history of science on the part of philosophers. Still, something has to be said about Feyerabend's crucial "change-of-experience" mode of characterizing the Galilean move, in order to appreciate the relation between *HS* and *PS* in his world view. First, he has chosen his example well; it is quite difficult to find any other in the history of modern science where the transition from one theory to another could plausibly be described as a change in experience or as a shift in the meaning of the observation statements on which the two theories are based. It is worth noting this, because it suggests that one cannot validly derive a *general* view of the nature of scientific advance from this instance only or from it principally. The Ptolemaic-to-Copernican move was of a very special methodological kind (as the term 'revolution' is often loosely used to underline); before one could call upon it in support of a general theory of science, one would have to dissect with great care the epistemological and ontological tangles latent in it. The moral of this is, of course, that the philosopher who turns to *HS* for the main war-

---

[38] It is usually assumed that Galileo was a Copernican from his first days in Padua; in support of this his letter of 1597 to Kepler is quoted. Yet there seems to be good evidence for saying that he remained doubtful about the Copernican world view until the telescopic data of 1609–11 refuted the main alternative. Though authorities like Tannery and Hall have Galileo seeking evidence in physics and astronomy for a Copernicanism already intuitively seen to be correct, this is not convincingly borne out by the evidence one can piece together from his Paduan period (1592–1610). For a discussion of this issue, see the section "Copernicanism and the Origins of the *Discorsi*" in the Introduction to my *Galileo: Man of Science* (New York: Basic Books, 1967), pp. 20–24.

[39] Section 16.

[40] For the elements of a very different view, see my Introduction to *Galileo: Man of Science*, especially the section "Copernicanism and the Origins of the *Discorsi*"; "Empiricism and the Scientific Revolution"; and "A Turning-Point in Physics: Galileo's *Discorsi*," to appear in a forthcoming volume of the Pittsburgh series edited by R. G. Colodny.

rant for his views has to beware (as has anyone else who "uses" history) lest he be influenced in his choice of historical evidence by the very theory he is going to substantiate and which he has already implicitly adopted in advance. There is more than a danger of special pleading unless a real effort is made by the philosopher of science to review the wide diversity of types of theoretical change that the history of science has to offer.

More seriously, though, does even the Galileo case support Feyerabend's view about the nature of scientific change? It does not seem so. Apart altogether from the moral overtones of terms like 'trickery,' 'cheat,' 'conceal' (overtones for which no adequate logical and biographical justification is given), he has not made a convincing case for saying that Galileo's use of anamnesis constituted a disguised "change of experience." First, it was not the case that the Aristotelian assertion that the earth is immobile relied upon some special "experience" of the earth at rest; rather, it was a theoretical conclusion from an elaborate teleological argument involving the ideas of natural motion and natural place, and the entire Empedoclean physics of the four elements. Nor was it the case that Galileo was alleging in his support an "experience" of the earth's motion. Feyerabend equates "experience" with an "interpretation" of sense impressions, a "judgment" inferred from phenomena, and is willing to admit that there was no change in the "phenomena" when Galileo persuaded men to agree that the earth is moving. What has changed is the *interpretation* of the phenomena. But ought this be called a "revision of our experience"?

Feyerabend distinguishes between two "paradigms," involving two different "natural interpretations": the Aristotelian one, asserting that *all* motion is "operative" (produces perceptible effects), and the Galilean one, asserting that only *relative* motion is operative. And then he describes the transition to Galilean mechanics as a change in paradigm. But this will not do, for reasons that Galileo himself carefully brought out. A person in a uniformly moving system (like a boat) cannot say what the motions of the objects he perceives about him would seem like to an observer on an unseen distant shore. He can perceive only the *relative* motions of these objects; perceived motion is thus motion relative to the observer in such a case, and the further difficult question of the observer's own possible motion must then be asked. But how is it to be answered? Galileo argued that the Aristotelian had no consistent way of answering this in such a way that he could disprove the possibility that the earth itself might be a moving reference frame. If sense experience will not tell someone in the hold of a ship whether the ship is moving, how is it supposed to tell us whether the earth is moving?

The context to which attention is drawn is one already familiar to the Aristotelian. What Galileo argues is that since his opponent already interprets observations made in such a context in a "relativist" way, how can he consistently do otherwise in the case of observations made on the earth's surface? It is not the case that the Aristotelian maintains an "absolutist" paradigm for all contexts. If this had been the case, Galileo's anamnesis would not, in fact, have worked. The difficulty facing the Aristotelian is an inconsistency in his own "paradigm." He has to admit that he already makes use of different interpretations in different contexts, and he is unable to justify his taking the earth as a privileged system without introducing a speculative theory of motion, one which Galileo has little trouble in demolishing. There is nothing fictional or "manufactured" about the experience on which Galileo draws, then, in his analysis of how we already describe our experiences in moving carriages and the like.

It is true that Salviati is "leading" Simplicio, just as Socrates did the slave boy in the *Meno*, and that there is an element of rational reconstruction about the result, in the sense that Simplicio has been led to see that the assertion that the earth is really in rapid motion in an orbit around the sun, however counterintuitive it may seem at first sight, cannot be claimed to be false on the basis of immediate experience without getting into inconsistencies elsewhere in one's descriptions of perceived motion. Simplicio's ultimate agreement with this argument was not a matter of his being duped or of his unwittingly accepting an arbitrary or at best only partially justified reconstruction. Rather, it was a matter of his being brought to realize the implications and responsibilities of consistent usage inherent in even the simplest attempts to systematize everyday kinematic concepts. Though Simplicio may interpret his experience differently in consequence of the anamnesis arguments, the only alternative Salviati left open to him was to deny the possibility of any ultimate unity in our understanding of motion. But for Galileo (as for any other scientist then or now), this was not a real alternative.

Feyerabend's reconstruction of the Copernican-Galilean "revolution" does not, therefore, carry conviction. A fortiori, it cannot provide a basis for a general theory of scientific change. Though it serves a valuable purpose in directing our attention to the methodological complexity of Galileo's work and the impossibility of fitting it into conventional empiricist or falsificationist patterns, it fails for three reasons: (1) it exaggerates some details and ignores others; (2) it generalizes readily and without much analysis of possible contextual differences; (3) it is suffused with a moral passion that transforms history almost into melodrama. But this sounds very like what Feyerabend says of *Galileo's* method, and by implication of the method of science generally. This gives us the clue we need to understand what Feyerabend is doing. The method of inquiry he describes is not so much Galileo's as it is his own. It is demonstrably the method he follows in constructing the argument of the paper we have just analyzed. To accept the account of Galilean physics given there, and to see in it an acceptable paradigm of scientific inquiry, would be simultaneously to accept Feyerabend's own methods of analysis and reconstruction as "scientific" and satisfactory.

Despite appearances, therefore, Feyerabend's *PS* in this paper does not rest upon *HS*. He brings a prior notion of rational inquiry to bear upon the history of science with a view to finding there some support for his view. But the way in which he does this forces one to say that his *PS* is *not* grounded in history; whatever support it has, it must draw from elsewhere. It is, thus, a *PSE* not a *PSI*, and it is a *PSE* of a peculiarly risky sort in that by purporting to be a *PSI*, it is effectively exempted from exploring its real warrant.

A final example of a rather different genre is afforded by Bernard Lonergan's influential and difficult work *Insight*.[41] His aim is to provide a general theory of intelligence. In Part One of his book, he discusses some characteristic structures of scientific inquiry, like measurement and probability theory. It sounds as though he is building a general theory of insight on an analysis of insight in science, presumably because it is the area whose epistemology has been the most carefully explored. The author himself, indeed, often seems to assume that this is what he is doing. This would make of this part of his book a *PSI* on which a more general metaphysics could later be constructed. But on closer view, this is not in fact what is going on. The author is *not* attempting to describe actual scientific practice; he is deriving in a quite abstract way what the features of coordinate measurement or mechanical explanation *have* to be. At this point, it sounds like a *PSM* of a broadly Kantian sort. But this is not correct either. What Lonergan seems to be doing in his Part One is presenting a speculative analysis

[41] London: Longmans, Green, 1957.

of some common structures of scientific inquiry and then treating this analysis *itself* as an instance of *metaphysical* insight (into science, as it happens, but it could just as readily be art) in order to construct a general theory of insight and judgment. It would, therefore, be more correct not to use the label '*PS*' at all in this case.[42] This instance differs from the other two in that the "internal" element called upon by the philosopher as apparent warrant of his assertions is current scientific practice rather than *HS* itself. Yet it illustrates the same general point as did the others: that *PSE* is often presented as though it were a *PSI*. Both modes of doing *PS* are perfectly valid. The danger is, however, that such an approach may easily divert attention from the difficult and crucial issue of justifying what is being said.

Can the philosopher allow himself to be *entirely* governed by what happens (or has in the past happened) in scientific practice? Is there an analogy here between the philosopher formulating a theory to account for the procedures of science and a physicist formulating a theory to account for the behavior of gases? To press such an analogy, to suppose that everything a scientist does contributes positively to a theory of science is clearly wrong. Scientists (unlike gases) can make mistakes; there can be bad pieces of research. And scientists can gradually learn to do things better, so that later science could conceivably be more significant than earlier science. But is there any norm for what should count as a "good" or "bad" piece of research work? any norm, that is, prior to the construction of a *PSI*? If not, how is the practitioner of *PSI* to proceed? Can he leave aside those events in *HS* which don't fit in with his views, on the grounds that they were "bad" science, or at least untypical of the "best" science? Would there not be a danger of *petitio principii* in such a procedure? Would such a *PS* be genuinely internal?

This is a real difficulty for anyone who purports to be giving a *PSI*. Can a *PSI* be normative? Does not this implicitly convert it into a *PSE*? A *PSI* has no source of autonomous prescientific evidence which would allow it to judge the adequacy of a particular piece of scientific work. Nevertheless, a *PSI* can legitimately point out when such a piece of work departs from

---

[42] This reconstruction of the methodological status of part one of *Insight* is put forward tentatively; the work is a very complex one and has given rise to much controversy. For two different appraisals of its relationship to "orthodox" *PS*, see E. McKinnon, "Cognitional Analysis and the Philosophy of Science," in *Spirit as Inquiry*, ed. F. Crowe (New York: Herder and Herder, 1964), pp. 43–68, and E. McMullin, "*Insight* and the *Meno*," *ibid.*, pp. 69–73.

the "normal," from the strategies that have proved in the past most "successful." Since it purports to be giving an account of what actually goes on in science, this is as far as it can go. It could not, for example, mount a critique of "normal" procedure itself without becoming a methodologically different sort of undertaking, one intended to define the ideal rather than explore the actual. One last reminder is in order, that in most cases a *PS* will not fall neatly into either category: it will draw from above as well as below. It will be governed by unstated metaphysical presuppositions, logical considerations of consistency, esthetic values, as well as by some knowledge of what has been going on in science these three centuries past. Our purpose in separating these considerations, and in classifying the types of *PS* built on only one of them to the relative exclusion of the others, was to focus attention on an important but often-overlooked ambiguity: what counts as evidence in *PS*, and in particular what role *HS* plays in it.

6. Philosophy of Science: Three Areas of Inquiry

In the preceding sections, we have been speaking of *PS* as though it were a single well-defined enterprise. This is far from being the case. *PS* comprises all those philosophic inquiries that take science as their starting point or as their object of concern. When discussing the distinction between *PSE* and *PSI* above, we assumed that the problems of *PS* are *epistemological* in nature, so that one could turn either to a more general theory of knowledge or to an inspection of the procedures actually followed in science in order to solve them. But two other sorts of problem have also got to be taken into account; they belong to the domains traditionally called ontology and philosophy of nature respectively. The abbreviations '*ES*' (epistemology of science), '*OS*' (ontology of science), and '*PN*' (philosophy of nature) will be convenient here. *ES* would at one time have been regarded as part of logic. *OS* constitutes a relatively new problematic, although there are some hints of this problematic in Plato's thought and in later medieval discussions of astronomy and optics. *PN* would originally not have been distinguished from "physics" (natural philosophy) itself. The development of Newtonian science profoundly affected all three of these. *ES* was greatly enlarged and strengthened as science itself became more and more sophisticated and self-conscious in its methods. *OS* became a crucial issue only where there was a sufficient body of scientific theory to make a question about its ontological import unavoidable. *PN* became a separate domain only when "philosophy" and "science" themselves began to separate in the post-Newtonian period. Metaphysics and physics had always been distinguished. But a distinction between the "philosophic" and "scientific" approaches to an issue is of very recent origin. *ES*, *OS*, and *PN* have come to be grouped together in recent decades under the broad title of "philosophy of science," a title which would have made no sense in Newton's day.[43]

ES is concerned with science as a way of knowing (explaining, proving, discovering, measuring, conceptualizing, etc.). It is a general methodology of empirical science; it is not concerned with particular scientific theories or even with particular domains (biology, chemistry, etc.) except insofar as the difference of domain brings with it a difference of methodology.[44] Most of the published work in what is called "philosophy of science" today would fall into this category. Topics like the nature of explanation in science, the logic of confirmation or discovery, account for more than half of all the essays in current anthologies of *PS* in the United States (in the Pittsburgh, Minnesota, Delaware, and Boston series, for example).[45] Although in principle *ES* is a general theory of scientific method, it is ordinarily elaborated in the context of the most developed sciences, notably mechanics, from which in the past the ideal of scientific method has most often been elaborated. Of late, however, philosophers have begun to realize the negative effects of this concentration of *ES* upon what is in fact a quite untypical part of science. "Explanation" in mechanics means something quite different from explanation in a structural science like biology or chemistry or geology. With the change in *PS* already noted from external (*PSE*) to internal (*PSI*) modes of warrant, *ES* has broadened very

[43] In some countries (U.S.S.R., Germany), and in some philosophic traditions (especially those of Aristotle, Aquinas, Kant, and Hegel), this grouping is less common. A strong distinction would be drawn between "theory of science" ("critique of the sciences," etc.) on the one hand, comprising *ES* and *OS*, and *Naturphilosophie* (*PN*) on the other. In the International Congresses of Philosophy, these constitute two different sections, though the assignment of papers to one or the other becomes ever more arbitrary. In the Vienna Congress of 1968, whether one submitted a paper to the "Theory of Science" division or to the "Philosophy of Nature" division seemed to depend largely on one's country of origin or on one's own philosophical standpoint. See my Introduction to the *Naturphilosophie* section of the Congress *Proceedings* (vol. 4, pp. 295–305): "Is There a Philosophy of Nature?" The main reason this distinction is not emphasized by English and American philosophers is that they are skeptical of the possibility of an autonomous philosophy of nature.

[44] Quantum theory has, for instance, suggested to some philosophers that a special multivalued logic is required where noncommuting operators stand for physical parameters.

[45] See E. McMullin, "Recent Work in the Philosophy of Science," *New Scholasticism*, 40 (1966), 478–518.

much in scope and has grown in sophistication. Because science represents in some sense an ideal of human knowing, *ES* (whether of the *PSE* or *PSI* variety) is highly relevant to the more general issues of epistemology and metaphysics. In some recent instances, indeed, the position adopted in *ES* has determined the entire shape of philosophy, as with logical positivism.

A second area of *PS*, closely related to the first, is the ontology of science (*OS*), the exploration of the ontological relevance of the claims made by empirical science. *OS* reduces, in essence, to a single question: to what extent do the postulational structures of science reveal a "real" structure, whether of the world or of the human mind? Various philosophers have argued that although science makes our experience intelligible by formulating correlations that enable predictions to be made, we cannot infer from this that scientific theories have any ontological import. They may be no more than arbitrary fictions, convenient instruments of prediction. *OS* is concerned, therefore, not with the general structures of scientific knowing, nor with the specific physical structures that occur in nature, but with the question of how these are related to one another. What, in brief, does science tell us about the world? This question has been a crucial one for philosophers ever since the time of Hume, who was the first to defend a phenomenalist ontology which would deny an intelligible structure to nature, and therefore by implication refuse any sort of realist view of science. There has been a significant difference between the *OS* of scientists and that of philosophers: the former did not have to contend with Hume. For the most part, they have maintained a realist *OS* (with the exception of some physicists working in the area of mechanics, an atypical part of science, as we have already noted). The resolution of this ontological issue is quite crucial for contemporary metaphysics, especially for a metaphysics (like that of process philosophy) which derives part of its warrant from the results of scientific theory.

Empiricist philosophers have paid relatively little attention to *OS*, as a glance at standard United States handbooks of *PS* will show.[46] For an empiricist of Humean sympathies, *OS* is not even a meaningful issue. But apart from this, *ES* and *PN* are much more congenial from the point of

---

[46] One notable exception is E. Nagel's *The Structure of Science* (New York: Harcourt, Brace and World, 1961). For detailed work in *OS*, see, for example, J. J. C. Smart, *Philosophy and Scientific Realism* (London: Routledge and Kegan Paul, 1963); W. Sellars, "The Language of Theory," in his *Science, Perception and Reality* (New York: Humanities, 1963); G. Maxwell, "The Ontological Status of Theoretical Entities," *Minnesota Studies in the Philosophy of Science*, vol. III (Minneapolis: University of Minnesota Press, 1962), pp. 3–27.

view of "research," since there is an abundance of material to work on; new problems arise as new scientific theories are formulated. The philosopher of science who busies himself with *PSI* (whether *ES* or *PN*) can easily leave *OS* out of account altogether; by comparison with other parts of *PS*, the problem it poses tends to seem an intractable one. Yet it is *OS* that poses the most specifically "philosophical" issue of any part of *PS*; until one has faced this issue, all other findings in *PS* are suspended in the air.

The third area of *PS* depends quite sensitively for its characterization upon the position one adopts in *OS*. Many scientific theories appear to have far-reaching implications for traditional philosophic problems concerning the nature of mind, the relations of space and time, the nature of causality, and so forth. If one defends a realistic or quasi-realistic theory of science, then the implications of relativity theory, of the theory of evolution, of cybernetics, and the like, have to be taken seriously by any philosopher who wishes to understand the most general traits of the physical world. One can describe these implications as a "philosophy of nature," meaning that the scientific theories themselves allow us to formulate a properly "philosophic" cosmology. On the other hand, if a nonrealist *OS* be defended, what passes as *PN* in the other view is likely to be regarded as no more than a speculative extension of science, a series of conceptual clarifications, "philosophical" only in a very loose sense. Since a realistic view of science will be defended later in this paper, the title '*PN*' will be used; it will be assumed that the philosopher is not debarred from making statements on his own account about the physical world or about specific structures like time or mind.

PN is obviously very close in methodology to science itself. They seem, in fact, to form a continuum. The conventional modern distinction between philosophy and science, which has come to seem so basic, is not readily applicable here. How is one to specify a demarcation criterion that would mark "philosophy" off from "science" in such works of *PN* as Adolf Grünbaum's *Philosophic Problems of Space and Time?* Much will depend on whether one believes an autonomous *PN* to be possible. Is it possible to construct a *PN* prior to the deliverances of science, based on the "common core of experience" or on the structures of ordinary language, or on an analysis of the general structures of possibility of any knowledge of a physical world? A wide variety of philosophers (neo-Aristotelians, Marxist-

Leninists, phenomenologists, Kantians, Hegelians, etc.) maintain that such a *PN* can be developed.

They disagree, however, on how to interpret its relationship with science. Is it altogether autonomous, and thus unaffected by the growth of scientific knowledge? If so, can it perhaps even serve as a norm to judge the adequacy of the categories and methods of the scientist? This strong claim for *PN* can be found in some writers of the Hegelian, Marxist, and Husserlian traditions. A weaker claim would give the prescientific *PN* a limited autonomy only, allowing for the possibility that it might have to be modified in the light of advances in science. In other words, part of the warrant for an adequate *PN* would be the sciences of nature. A philosophy of nature would thus have to balance evidence of two rather different sorts, evidence from some prescientific source (e.g., common experience or the categorial "cuts" of ordinary language) as well as from science. Each of these would in effect be taken seriously as a partial warrant for philosophic assertions about nature; the testimony of one could, however, modify that of the other. The alternative to these two views of *PN* is one which would make it wholly derivative from science, i.e., would deny any source of evidence for a *PN* other than contemporary scientific theory and practice.

The status given a *PN* thus serves as an indicator of the distinction between "philosophy" and "science" a particular philosopher maintains, i.e., of the ways in which he chooses to define these two very vague and dangerously honorific terms. (1) He may deny the existence of a *PN* entirely, in which case all knowledge of nature, however speculative, is by definition "scientific," and "philosophy" is entirely confined to second-order questions about language or method. (2) He may allow a *PN*, but insist that it be entirely derivative from science. In this case, the distinction between "philosophy" and "science" is in terms of speculative character or generality or the like. (3) He may allow a partial warrant for *PN* prior to the constructive activity of science. This gives a very complex notion of "philosophy," since quite different types of evidence can be relevant to it, and it can make first-order assertions about the world, of higher generality than those of science but presumably in a continuum with these. (4) Finally, he may hold out for a completely autonomous *PN* prior to science, in which case he can draw a very sharp distinction between "philosophy" and "science" on the basis of the type of evidence on which each rests. Since he does not in this case admit science as a source of properly "philosophical" knowledge of nature, he will not have a *PS* concerned with nature; *PS* for him will cover at most only *ES* and *OS*. This rather summary and abstract taxonomy may suffice to bring out the wide variety of approaches that may be covered by our label "philosophy of nature."[47] It would take us much too far afield to evaluate these approaches, and in particular to investigate whether or not there *is* a genuinely autonomous "philosophical" mode of approach to nature different from that followed by the scientist. But perhaps enough has been said to suggest that this part of *PS* is more complex and controversial in character, methodologically speaking, than are *ES* and *OS*. The existence of a *PN* (if it be admitted) suggests that philosophy and science have somehow got to complement one another—or else compete with one another—in the construction of a total world picture.

The distinction just drawn between three different approaches to *PN* can also be expressed in terms of the "external-internal" division above, even though it was originally elaborated in the context of *ES* rather than *PN*. If the warrant for *PN* is independent of science, we have a *PNE*; if it rests upon the theories of science, it is a *PNI*. If there is some external source of evidence for a *PN* but the procedures and theories of science are also taken into account, we have a *PN* of mixed warrant (*PNM*).[48] *OS* can likewise be governed either by external or by internal considerations (or by a combination of the two). One could, for instance, develop a positivist *OSE* on the basis of a Humean phenomenalism. A "pure" *OSI* is less likely; one would not normally wish to base an ontology on an analysis of science exclusively. Metaphysical and epistemological considerations of a more general sort would presumably have to be taken into account in deciding what the ontological implications of scientific theory are. In the final sections of this essay, I will argue that an adequate *OS* has to take the developmental aspects of science very seriously. Some features of the history of theoretical models will be brought forward to support a realistic *OS*. Broader philosophical considerations for or against realism are not

[47] I have developed this schema in more concrete detail against the background of the major exponents of *PN*, past as well as present, in "Philosophies of Nature." I have argued there that this question of the type of warrant on which a *PN* is supposed to rest makes an illuminating basis of distinction between contemporary approaches to nature. The tension between the different possible approaches to knowledge of nature a philosopher may take up has been of very great importance in the Kantian and more recently the Marxist-Leninist schools. For the latter, see my review-article on David Joravsky's book *Soviet Marxism and Natural Science, 1917–1932* (London: Routledge and Kegan Paul, 1961), in *Natural Law Forum*, 8 (1963), 149–159.

[48] These are the *PN1*, *PN2*, and *PNM*, of my article "Philosophies of Nature." *PNM* seems to me the most defensible sort of *PN*; I have argued elsewhere that it plays an implicit but important role in the heuristics of science (section 9 of the Epilogue to *The Concept of Matter in Modern Thought*).

raised. It could be argued that the only way of accounting for the history of the Bohr model is by assigning a realistic import to the physical structures it postulates. One might in this way construct an independent OSI based on an analysis of HS, though in the long run one would probably want to broaden it to an OSM by introducing arguments of a more general sort in favor of realism in epistemology and ontology.

What is the relevance of history of science to the three domains of PS described above? We have just implied that an adequate OS has to be based at least partly on information drawn from HS: that the universe is such that the sort of hypothetical structural models used by scientists function in inquiry in the way they do is perhaps the most significant clue to ontology that we possess. On the other hand, HS has very little direct relevance to PN. One will not be concerned in the fashioning of a PN with the details of how particular theories come to be formulated and progressively modified; rather what counts is the best theory available now. HS is, as we have seen, highly relevant to the epistemology of science, if taken as an "internal" study (ESI).

Another way of conveying the difference between the ways in which the three main divisions of PS approach the study of science is to say, using the distinction between S_1 and S_2 elaborated in section 1 of this paper, that ES must be concerned with S_2 since the epistemologist has to take into account the widest possible range of evidence on how scientists proceed. PN on the other hand need take only S_1 into consideration; only the confirmed results of science have a bearing on the philosophy of nature. OS is concerned much more with S_2 than with S_1. The finished propositional product of the scientist is enigmatic in its ontological implications, as we shall see. Realist and positivist alike can interpret it to their taste. It is only when the temporal dimension of science, the developmental aspect of S_2, is taken into account that a decision can be reached on the central issue of OS.

7. Philosophy and Psychology

What is the relationship between the "philosophic" mode of investigating science and other systematic modes of understanding human activity such as psychology (including variants like psychoanalysis) and sociology?[49] The distinctions we have already drawn may help us to bring some

[49] Other social sciences, like economics and politics, are much less relevant because their "laws" and explanatory hypotheses concern facets of human behavior or of social structure that are remote from the doing of science. Though economic or political situa-

clarity to this question, one which is quite crucial to the currently disputed question of a "logic of discovery" in science.[50] The science we have in mind here is, of course, S_2. Psychology is not relevant to the understanding of S_1, but it may well tell us something of the conditions under which S_2 is furthered. Only ES is involved in this correlation with the work of the social sciences; psychology is clearly irrelevant to OS and PN. ES_2 is, it would appear, the possible point of contact between PS and other modes of "understanding science," understanding it, that is, specifically as a human activity. Understanding is usually thought to involve two "moments": the discovery of regular patterns and the explanation of why these regularities recur in the way they do. Our question, then, reduces to this: what are the principal ways of understanding the patterns that occur in the complex of activities we call scientific research?

The answer we shall hazard is that only two need be considered: the philosophical and the psychological. We can trace regular conceptual or propositional connections, whether these be strictly *logical* (governed by a specifiable formal rule) or analogical. We can examine the scientist's activity with a view to describing and interrelating propositions implicit in it, the beliefs that guide him, the data he has obtained, the hypotheses he advances, and so forth. We can trace the gradual modification of a concept (like the concept of ether in Newton's thought), where it is possible to give plausible conceptual grounds ("reasons") for the modification's having occurred the way it did. The "pattern" here is a relation between ideas or can somehow be associated with such a relation. The techniques are those of conceptual analysis. The *ideal* here would be that of a complete logical reduction, the discovery of a fully formal system which would simulate the theory or procedure under investigation. But this is rarely possible,

tions and motives can obviously influence scientific research, this influence would be best understood as a rule in psychological or sociological terms. If, for example, one were to ask why certain lines of research moved more rapidly than others in the United States over the last two decades, one would have to take into account the availability of federal grants for some types of research and not for others. Note that this is a *psychological* explanation (the efficacy of the economic motive) rather than an economic one. Scientific activity does not lend itself to what we would ordinarily regard as economic or political patterning. To understand the economic situation that made one type of research more desirable to the federal government than another may require quite a bit of economic or political analysis. But this analysis is likely to be of only marginal interest to someone who is trying to discover invariants in scientific activity *as such*.

[50] R. Blackwell in chapter 3 of his recent *Discovery in the Physical Sciences* (Notre Dame, Ind.: University of Notre Dame Press, 1969) distinguishes between four possible ways of patterning scientific discovery: logical, psychological, historical, and epistemological. In general, though, distinctions of this sort have not been analyzed by philosophers of science in any detail.

since scientists do not follow a strict (i.e., fully specifiable formal-deductive) logic of this sort in the more significant parts of their work. Indeed, to the extent that they do follow fully formal rules, their reasoning is unlikely to be significant, since it is only unfolding something already conceptually and propositionally given.[51]

A *psychological* pattern is, broadly speaking, some regularity in human behavior, thought to be attributable to the specific personality structure (intellectual abilities, emotional makeup, character, etc.) of the individuals exhibiting it.[52] Since this includes propositional "behavior" (thinking, writing, proving), logical patterns are not going to be easily distinguished from psychological ones.[53] Since people generally obey the *modus ponens* rule in their reasoning, is it not a psychological law as well as a logical one? This is a matter of definition, of course, but logical rules of inference are usually *not* regarded as properly "psychological" laws, even though they

[51] This important point I have tried to make in some detail in "Freedom and Creativity in Science," in *Freedom and Man*, ed. J. C. Murray (New York: Kenedy, 1965), pp. 105–130.

[52] It is not necessary to distinguish explicitly between sociology and psychology in this context. Sociology seeks information about the behavior of persons as members of specific groups, and about the interactions of groups considered as units. But insofar as one wishes to *explain* this behavior or these interactions, one must ultimately return to psychology. As a type of *information*, sociology is distinct from psychology. But as modes of *explanation*, they tend to become one. E.g., much has been written about the influence of religion on the growth of science in the seventeenth century. Correlations have been sought between creativity in science and Unitarianism (or Anglicanism or free thought or Christianity). But if such are found, one will still have to ask why a Unitarian *should* have been a better scientist, and the answer will have to be either conceptual (lying in Unitarian belief) or *psychological* (the type of personality structure commonly found among Unitarians). The former approach is the commoner among those philosophers and historians who have discussed this problem (see, for example, R. S. Westfall, *Science and Religion in Seventeenth Century England*, New Haven, Conn.: Yale University Press, 1958). In a rather disreputable recent book, *The Scientific Intellectual* (New York: Basic Books, 1963), Lewis Feuer argued that science is favored by a hedonistic and antiauthoritarian personality; he links this sort of personality with certain religious groups and with various social movements. Members of religious groups that are nonhedonistic and authoritarian are, he claims, unlikely to contribute to science. The difficulty about his argument is that when in his impressionistic tour of the history of science he meets with counterexamples (Newton being a rather obvious one!), he goes on to claim that a *second* source of scientific creativity lies in neurotic conflict of a Freudian sort. This renders his original thesis impregnable, since notable scientists who do not qualify as hedonists or libertarians are automatically labeled by him as neurotics, even in the absence of independent evidence to this effect! This is the sort of writing that brings anguish to philosophers and historians of science alike. But a start has been made with sociology-psychology of science of a more responsible sort, such as one finds in the work of Derek Price or Bernard Barber.

[53] The whole question of how logic and psychology are to be situated relative to one another is so dismayingly vast, so amorphous in its ramifications, that these remarks have to be considered as no more than loose generalities.

are clearly patterns of a quite basic sort in the operation of the human psyche. The reason for not including them in the scope of psychology is that the "dynamism" of these regularities, their ultimate ground, is thought to lie in the propositions rather than in the psychological structures of the thinker.[54] Even if these latter structures were quite different, the assumption is that logical laws would still govern the thinking of the individual, as an ideal to be striven after, even if perhaps not always followed.

To the extent that the scientist's procedures are not completely bound by logical rule, however, it would seem that psychological considerations may have to be taken into account. The formulation of a hypothesis, for instance, is not a deductive process. It may be guided by analogies but by definition it goes "beyond the evidence," i.e., beyond what could be arrived at on the basis of formal logic alone. It is relevant, therefore, to inquire whether the pattern of *discovery*, say in science, can be at least partially accounted for in terms of psychological laws and theories.[55] To "account for" it here means to situate it as one human ability among others, to show if possible "how it works," an ambitious mechanistic metaphor but implying nothing much more than that the characteristic stages in discovery should be categorized in some general way. A good example of this sort of effort is Koestler's massive work, *The Act of Creation*,[56] which explores creativity in art, humor, and science, and suggests as an "explanation" of what occurs an ability on man's part to juxtapose hitherto unrelated matrices of thought. Whether this is an "explanation" or only an insightful *description* of what happens can be debated, but it is enough for us to note that insofar as it would "explain," it would do so by pointing to

[54] But many would disagree. Kant reduced logical (and many other sorts of) law to "psychological" structures, in a sense of 'psychological' admittedly very different from that of contemporary psychology. In an informal paper at the International Congress of Philosophy in Mexico City in 1963, Carnap argued that the grounds for accepting logical laws as "laws" (i.e., for making use of them in our thinking) are purely pragmatic and experimental—we find that they work. They are no more "psychological," then, than are physical laws. They do not express how our minds operate, but rather how the world is. In the lively debate which followed, Max Black urged that they are analytic features of the language we use, requiring no pragmatic or other sort of specific justification (Carnap's own earlier position). This sort of debate has been going on since Greek times, when Aristotle argued that logic and metaphysics presented different facets of the same basic structure, that of Being. It is not necessary for our discussion of the nature of PS that we should enter into this question in detail.

[55] One of the difficulties about answering this question is that there are so many different sorts of "psychology," ranging from depth psychology to behaviorist studies of animal behavior. Obviously, the implications of each of these for ES are likely to differ.

[56] New York: Macmillan, 1964.

a general human ability (possessed by some to a higher degree than others); it would then go on to break down this complex and puzzling ability into simpler and better understood sub-abilities. Contrast this with a *logical* explanation of the steps of a mental process, where rules would have to be given to justify each step of the process, or at least to estimate its inductive weight. To "explain" a process psychologically does not of itself justify the term of the process; at best, it only describes how we got there.

Michael Polanyi would broaden the scope of this sort of analysis to include not only discovery but also confirmation.[57] Or more exactly, he would assert that the conventionally sharp distinction drawn between these two rests upon some shaky empiricist assumptions. He then proceeds to give a philosophical-psychological account of scientific knowledge which sees it as an instance of the human ability to recognize pattern, to interpret tacit clues without necessarily being able to break down the process into well-defined compartments, "evidence" and "hypothesis," linked by explicit formal rules and analogies. There is (he claims) a "tacit structure" of knowing involved; to focus attention on one element in it may mean that the ability to see the whole as making sense may be lost. This brief summary does not do justice to a complex and admittedly controversial position;[58] it is introduced here only to illustrate the thesis that once one moves away from those limited parts of scientific activity which can be completely dissected in formal-logical terms, one has to take account of psychological factors side by side with the more properly logical-conceptual ones.

A complete epistemology of science (ES_2) cannot, therefore, leave psychological considerations out of account. Admittedly, such considerations are not relevant if we are merely interested in the validity of the scientific claims made, the extent to which they "explain" the data (ES_1). And this is the perspective in which the problem is most often discussed. But the wider perspective is a valid one, and the question of the methodology appropriate to it is deserving of more attention than it has so far received.

[57] *Personal Knowledge* (London: Routledge and Kegan Paul, 1958); *The Tacit Dimension* (New York: Doubleday, 1966).

[58] The main objection to the facet of it described here is that the methodological strands of scientific confirmation can, in fact, be separated off to a much greater extent than Polanyi allows. The steps by which an empirical "law" is built up, and the gradual H-D confirmation of a theory, are not just unstructurable personal acts of "seeing." The analogy with visual pattern recognition on which Polanyi frequently relies can easily be pressed too far.

Whether at this stage psychology can in *fact* offer much help on issues such as the nature of creative discovery is another matter. Such questions do not readily lend themselves to investigation in the prevailing behaviorist terms, and it is noteworthy that extensive empirical attempts in recent decades to correlate creativity with other more easily identifiable traits have been unsuccessful.[59] The "explanation" offered by psychology is, besides, of a very modest sort. It is obviously never going to reduce the creative act to specifiable rule; it can only search for some appropriate general categories in which to analyze it. Men have the ability to "see" a particular hypothesis as the best way of explaining the "facts." The distinction—one might almost say the tension—between the psychological and philosophical modes of approach to this ability soon becomes evident. Are we to rest content with describing it simply as an ability? Or ought we in each case where it operates attempt to specify the logical reasons why the hypothesis *is* the best one, or why it is confirmed by this piece of evidence?

These are the two principal ways of understanding recurrent patterns in the activities of the scientist. But how are these patterns to be *discovered?* This is where the historian comes in, because all these patterns are of themselves "historical," in the sense that they recur in time and can be documented by the ordinary methods of the historian. Does this not suggest that history ought to be added to logic and psychology as a third mode of recovering pattern in science, of "understanding science"? It is important to see why this is not the case. *HS* is *not* of itself a mode of understanding science, in the ordinary sense of discovering and explaining regularities in the practice of science. Its goal is to establish the singular, not the universal (as does epistemology or psychology). Insofar as it provides "understanding," it is an understanding of the past singular in its complexity and contingency, a different sense, therefore, of 'understanding.' To achieve it, the historian may make use of a variety of sciences: psychology, linguistics, sociology, as well as philosophy. But this does not mean that his own effort falls into the same methodological category as theirs. There is ultimately a quite fundamental division here. The historian is concerned with what happened just because it *did* happen. He may call upon universals of all sorts in his effort to establish what happened or why it happened. But his goal is not the assertion of a universal, a pattern, or the interlinking of such patterns. This is the task of the philosopher, the

---

[59] As a glance at such standard collections as *Scientific Creativity: Its Recognition and Development*, ed. C. W. Taylor and F. Barron (New York: Wiley, 1963), will quickly show.

sociologist, the economist, whose use of the materials of history does not commit them to the reconstruction of any specific set of historical events.

History is closely interwoven with these other fields. They are built up inductively from things that happen. But they are not concerned with the particularities of occurrences, only with their exemplification of a certain set of universals. The philosopher of science will discuss the nature of measurement, for instance, without adverting to any specific historic instance of it. The sociologist will assert a correlation between drug addiction and broken homes without giving historical details of the broken homes he investigated in making his generalization. Yet the philosopher and the sociologist have to begin from the activities of real people; they may not invent their material, they have to find it. This can easily be overlooked in the case of philosophy, because it is for the most part at such a high level of generality that specific reference to concrete instances, instances requiring the skill of the historian to establish or unravel them, is rarely found. When a philosopher speaks of the nature of discovery (say) in science, he will often suppose that what he is saying is so intuitively evident to anyone who has even a rudimentary acquaintanceship with science that reliance on specific instances, calling such instances in evidence so to speak, is simply unnecessary.

8. Logic and History

We have already seen that a philosopher who wishes to find an "external" warrant for a PS is likely to look either to a metaphysics (PSM) or to the properties of formal systems (PSL), whereas someone who relies on "internal" evidence (PSI) is likely to look for his evidence either to contemporary science or to episodes in HS. Of these approaches, the two that most strongly contrast with one another are the logical and the historical. It may be worth exploring these contrasts in more detail. It defines the ends of a spectrum of possible ways of relating HS to PS.

The logician and the historian approach the problem of relating two elements in a piece of scientific research in quite different ways. The logician seeks to discover a purely logical structure relating them, a structure of transparent intelligibility in its own right. This structure can be disentangled and its properties studied; it can be used to justify the move from one element to the other. What the historian seeks to establish is the fact that the elements occurred in a certain sequence, whether or not any formal structure can be discerned in their relationship. He may be able to "explain" the historical sequence by showing how it exemplifies some logical or psychological pattern. But his first concern is not with explanation but with a reconstruction of the past, no matter how opaque it seems.

In *PSL* in its most "external" form, science becomes the occasion for the logician to investigate certain formal structures that might not otherwise have come to his attention. His aim is to construct a theory of inference, a theory of confirmation, a semantics of scientific terms, or the like, in such a way that these can stand in their own right as formal systems. The important point is that *PSL*, so construed, does not rest upon an appeal to what is actually going on in science or to *HS*. When an exponent of *PSL* puts forward an inductive logic, he need not be claiming that this is what actually governs scientists in their evaluation of hypotheses. He may even be entirely indifferent to any reference from case studies in *HS*. It need not weaken his case to say that Newton did not follow the logical plan suggested. What the logician is saying is that in an abstractly described piece of scientific research, the logical relationships between the elements are of the following kind—whether or not the relationships he specifies describe any historical sequence or were grasped by the scientists involved in the research. In this external form, *PSL* is not an empirical study; it is basically of the same character as mathematics or any other formal discipline. It is only in a broad sense that it qualifies as "philosophy." It could be argued that it ought not be so described, any more than formal logic would nowadays be described as "philosophy." Nevertheless, this would be a mistake since *PSL* does illuminate the structures of empirical science, and does take its origin in them.[60]

---

[60] Two recent instances of *PSL* would be R. Carnap's *Philosophical Foundations of Physics* (New York: Basic Books, 1966), and Kyburg's *Philosophy of Science: A Formal Approach.* Carnap's book deals with probability theory, the logic of measurement, the logic of causal modalities, analyticity, correspondence rules. Kyburg organizes his book "around the concept of a formal system," and explicitly limits himself to *ES*. He leaves aside *OS* ("what does the fact that science exists at all tell us about the world?"), not because he thinks these unimportant, but because a formalist *ES* is, in his view, an indispensable starting point for any adequate discussion of them. His first chapters are characteristically headed "The Concept of a Formal System," "Quantities," "Scientific Terms," "Axioms," "Probability and Error," "Induction and Experiment." He makes extensive use of the predicate calculus and of the logic of relations, and indeed notes in his Preface that "the philosophy of science can be understood without knowing physics (though perhaps not without really understanding some science), but it cannot be understood without knowing some logic, an essential ingredient of every science." In the entire text (on the testimony of the careful index at the end), not a single scientist is mentioned, nor are there more than a few references to specific scientific works. Thus the weight of the book in no way rests on a reporting of scientific practice; an item from *HS* would not be relevant to any of the major points that Kyburg is making.

There is an obvious analogy here with the "rational mechanics" of Newton's successors. Newton developed a complex physical theory, which he himself on occasion liked to regard as a quasi-formal mathematical system, thus leaving aside all questions about the operational meaning of the concepts employed and of the empirical adequacy of the system as a means of understanding specific concrete problems, leaving aside in other words what would be regarded as the properly "physical" issues. This "rational mechanics," as it was called, was developed at the hands of Euler, Lagrange, Laplace, Hamilton, and many others. Its evolution was quite independent of any new empirical information, or any modifications in the empirical bearing of the concepts used. It was guided by purely mathematical principles; its criteria were those of pure mathematics, although its original impetus came from physics. What its proponents sought was a more elegant formal exposition, employing more economical and better defined concepts; this in turn would allow the hidden inconsistencies and vaguenesses of the earlier system to be eliminated. The construction of the well-known "Lagrangian" and "Hamiltonian" functions to help in formulating the state description of a Newtonian mechanical system was an instance of such a development. It was prompted not by any empirical inadequacy of the system but by a desire for conceptual improvement. In technique, the exponents of rational mechanics were mathematicians; its history has been similar to that of a branch of mathematics, even though its starting point was different. It is not, however, applied mathematics, a point that Clifford Truesdell has emphasized.[61] It does not simply apply a given mathematics to the formulation of a physical theory or to the solution of physical problems. Rather, new mathematical concepts have to be developed or old ones modified in response to the needs of the physicist, or as a further elaboration of a formal system first created by the physicist. Thus rational mechanics is not quite reducible either to physics or to mathematics. The analogy with *PSL*, with logic substituted for mathematics, is a fairly exact one; *PSL* likewise is not quite reducible either to philosophy or to logic.

If the logician, instead of considering general epistemological issues inherent in any part of science, turns to specific scientific theories with a view to formalizing them, he will have to take *HS* somewhat more seriously. If he aims to formalize Newtonian mechanics, he can scarcely do this without some reference to the documents. Yet this reference may serve

[61] *Essays in the History of Mechanics* (Heidelberg: Springer, 1968).

57

only as a starting point; he may settle for some convenient textbook account of Newtonian mechanics and focus on the logical issues involved in it, without pausing to ask whether the system he is analyzing is really that of Newton. If an objection is raised on this score, the logician is likely to be unmoved; his creative energies are concentrated on formal problems of structure, not on problems of historical interpretation. Thus his analysis of "Newtonian" mechanics is likely to identify this mechanics with a broad class of systems, independent of any particular historical text. Yet he may after all rightly claim that his analysis illuminates (at least to some degree) Newton's own work, its conceptual implications and its weaknesses. And he may well exhibit considerable historical sophistication in deciding how to formalize messy physical concepts like mass or force.

The value of such an approach to PS is that it uncovers logical structures, a thorough grasp of which is indispensable to the full understanding of science or of a particular scientific theory. Its limitations as PS have already been alluded to more than once, chief among them a remoteness from the actual workings of science, and a danger of escapism (from the point of view of PS not that of logic, of course) because of the allure of the free construct. The logical-empiricist school has naturally tended to PSL because the characteristically Humean epistemology of the empiricist made it difficult for him to take HS seriously as an independent source of philosophical insight. The variety of challenges to empiricism among philosophers of science in the past decade has brought with it a corresponding skepticism about the adequacy of logical reconstructionism as a program for PS. It has also opened the way for a much more thorough utilization of HS on the part of philosophers.

At the beginning of this section, it was noted that the logician and the historian represent the opposite ends of the intellectual spectrum in their approaches to science. Between them lie sociologists, economists, philosophers, theologians, etc. The historian tries to re-create the singular in all its individuality; he emphasizes context and distrusts generalization. The stuff of history is events, not concepts or propositions. For something to be part of history it is sufficient that it should simply have occurred. The historian does not begin with facts; he tries to establish them, and having established them, to understand them. The relationships on which this historical understanding rests are causal ones, broadly speaking. Being causal means that they are generalizable; they can in principle be conceptualized. If they are conceptualized, do they become logical? Can history take on the transparency of logic?

The answer (pace Hegel) is no, for two reasons principally. The conceptual-causal relations of the historian are not reducible to the formal rules of inference of the logician. These latter are not empirical; they are wholly independent of context. The causal patterns cited by the historian, and the concepts in terms of which these patterns are expressed, are learned empirically and are highly contextual in their application. The logician must go beyond the specific predicates of the scientist, the historian, the philosopher, to disengage (if he can) purely formal relations of inference holding between propositions employing unspecified predicate variables. But even if one were to allow a broader sense of 'logical' in which any conceptual-causal relation would be "logical," there is a further reason why history will not reduce to logic. It is impossible to re-create conceptually even the simplest historical event. There is no way to give an exhaustive listing of all the potentially relevant causal influences. Nor is it the case that all of these will have left a recoverable record. And the concepts in terms of which the event is described are at best only approximative and provisional. The work of the historian is by its nature a tentative one, then, always open to revision. Not only is the historical singular infinite in its complexity, but the evidence which would allow this complexity to be conceptually reconstructed is itself transient and soon irrecoverable. The boundaries of the historian's task are set by *matter*, in the Aristotelian sense of that term; whereas the logician (much more than the philosopher) can abstract from matter entirely.

To the extent, then, that a structure is logical, it has ceased to be historical. A deductive inference rests in no way for its validity upon experience or history. It is valid not because it has been followed innumerable times with success, but because it has the total transparency that makes reference to history unnecessary in its support. To the extent, on the other hand, that a particular discovery is nonformalizable, dependent for example upon a radical conceptual realignment, history will have to be called upon to warrant its reliability and significance. The tasks of the historian and of the logician are almost exactly complementary, therefore, and both are important if a full understanding of the genesis of that complex phenomenon known as "science" is to be achieved.

9. Can One Do History and Philosophy of Science Together?

This essay has made extensive use of dichotomies as a tool of methodological analysis. But it has also stressed that works of scholarship rarely fall into a single neat methodological category. One distinction, however, that might seem a reasonably sharp one is that between *HS* and *PS*. Ordinarily, it is easy to tell which of these genres a particular piece of research belongs to. Can they be validly blended in a single work? The answer would seem to be that they can. *PSI*, as we have seen, often involves a careful reading of the history of science as a *warrant* for the philosophical claims made. Such work accomplishes both a historical and a philosophical goal. The writer tries to illuminate the historical instance with all the relevant philosophical analysis he can produce so that, despite its singularity, he may understand it as best he can. He also uses the documented historical instance to make a further philosophical point; it serves not merely as illustration but as evidence for this point. This genre of "history and philosophy of science" (*HPS*) is a complex, even a risky, one, as we have already seen when discussing the work of Lakatos and Feyerabend above. There are obvious dangers involved in combining two methodologies so diverse (not to mention the dangers of infuriating two professional groups whose reflexes are so different!). A good piece of *HPS* will not blur the distinction between the historical and the philosophical points it is making; by making them at the same time, or at least in the same piece, there is no intention of claiming them to be ultimately identical. We have already seen that philosophy and historiography are at bottom irreducible to one another, no matter how closely they may be interlocked in practice. It is important to grasp in as precise a way as possible what the relation is between the historical and the philosophical motifs in such writing. The historical motif is prior and in a sense basic, for on the establishing of the analysis as *history* depends its *warrant* as philosophy. It is ultimately because something happened in a certain way that a point in *PS* can be made to rest partially upon it.

If someone, in order to make a general point about meaning variance, asserts for instance that Galileo's colleagues obtained telescopic results that differed from Galileo's, it is essential that he be correct in the historical claim if it is to serve as warrant and not merely as illustration. An illustration could be replaced by some other apposite instance if it proved historically inaccurate. But if the philosophical claim in any way *rests* upon the case histories cited, it is weakened if any of these are shown to be unreliable. The *HPS* writer may, of course, choose simply to cite his history from someone else and not attempt to bring any fresh support for the claim that it happened the way it is supposed to have. But even so, by making use of it for philosophical purposes, he will almost certainly have illuminated it, situated it, helped the reader to understand it better. And this, as we have already seen, is one of the two main functions of the historian. By making a series of points of general philosophical relevance in the context of Newton's dynamics, Dudley Shapere (to quote one recent example) has also illuminated Newton's own historical achievement.[62]

There is one particular category of philosophical problem where the HPS approach is seen at its best, and where the PSL methods of the logician prove inadequate. This is the investigation of the developmental aspects of science (S_2). If discovery in science were guided by logical laws, one could write a history of science as it had to occur. But, of course, science is not like this; central to it is human creativity, and there are the innumerable contingencies of influence and noninfluence. One can extract the partial logical structures of validation which are implicit in scientific research. But to see how change actually occurs in science, what factors are most often responsible for it, one has to have recourse to the historical record.

This is the approach taken by Mary Hesse, for example, in her Forces and Fields.[63] She traces some basic conceptual structures that have recurred in the analysis of continuous and discontinuous motion in mechanics. In particular, she emphasizes the complex philosophical problems that underlay many of the modifications of concept that occurred as mechanics attained a greater and greater precision. The resultant is good history of science; it also serves as the ground for a variety of epistemological and ontological assertions, notably the assertion of a generally realistic view of scientific constructs.

Thomas Kuhn's influential work, The Structure of Scientific Revolutions,[64] puts HS to even more striking philosophical use. He distinguishes

---

[62] "Philosophical Significance of Newton's Science," Texas Quarterly, 10 (Fall 1967), 201–215. Adolf Grünbaum argues this point cogently, and illustrates it from the historiography of relativity theory, in "The Bearing of Philosophy on History of Science," Science, 143 (1964), 1406–12.

[63] London: Nelson, 1961. One of the earliest and still one of the finest achievements of this genre was Pierre Duhem's ten-volume work, Le Système du Monde (Paris: Hermann, 1913–59). This is a history of mechanics, with special reference to celestial mechanics, but it also makes use of historical analyses to argue (in contrast to Mary Hesse) a generally positivist view of the nature of scientific theory. The prototype of HPS was the pioneering work of William Whewell, The Philosophy of the Inductive Sciences, Founded upon Their History (London, 1840).

[64] Chicago: University of Chicago Press, 1962.

between fundamental changes of "paradigm" in science ("revolutions" in his terminology), and theoretical developments that leave the basic "paradigm" unchanged. His main thesis is that changes of paradigm cannot be justified on empirical or even rational grounds, though *post factum* an effort will always be made to provide such rationalization. This is a bold claim; it denies the possibility of *any* sort of *PSL* applicable to significant advances in science. Indeed, it has seemed to many to call into question the entire set of formalist assumptions on which the logical analysis of science is based. Kuhn's *HPS* is thus the antithesis of *PSL*, which may help to explain the warmth it has generated. For us, the important thing about it is that *only* the history of science can serve as evidence in its support. It is a philosophical statement about the nature of S_2 and about the transformational characteristics of S_1. It could not be derived from a general theory of knowledge, nor could it rest upon a formal logic. Only a sensitive analysis of selected periods in *HS*, an analysis which leaves aside the preconceptions of later methodology, will suffice to tell whether it is correct or not.

Many criticisms have been leveled against the meaning-variance thesis as it variously appears in the work of Kuhn, Hanson, Feyerabend, Toulmin, and others. The main point of criticism is the tendency of these writers to exaggerate the nonformal elements in scientific change; like any other crusaders against an ancient dogma, they tend to underplay the evidence that counts in their opponents' favor. But their work has shown that the modifications of concept which lie at the root of scientific change cannot be accounted for along the deductivist lines traditionally favored by philosophers of science. If theories are regarded as quasi-axiomatic schemas, one can think of confirmation as a matter of checking inferences in the systematic way a *PSL* would demand. But how is one to justify the choice of the concepts in which both the theory and the evidence alleged in its support are expressed? On what grounds and by what means are these concepts altered? Variations of meaning of the sort that constantly occur in science cannot be accounted for in terms of a purely formal logic.

Yet it is somehow within these variations that the clue to scientific advance lies. It is not enough to lump all of them together under the label 'discovery,' treat it as irrational or irrelevant, and mark it off sharply from "validation," understood in an idealized *post factum* way that leaves aside the question of what actually *did* persuade people to adopt particular theories. Rather, one must begin with a careful analysis of the crucial moments of meaning change in HS, and try to see if patterns of any sort can be distinguished. If not, one may have to be content to point simply to an ability on the part of the well-trained scientist to discern a "good" theory without being able to specify what precisely makes it "good."

Kuhn and his colleagues are talking about the epistemology of science in the broader sense of science defined in section 1 (ES_2). The importance of HS to the resolution of their problem comes from the fact that on the one hand the knowing activity of the scientist is a temporal process, and on the other it is not usually subject to complete logical reduction (unless it is the derivation of a prediction from an already-given theory). If the knowledge processes of the scientist are part of what we wish to understand, we simply have to treat HS as our major source of evidence. But it is when we ask *ontological* questions about the import of the postulated theoretical structures of science (whether in general or in regard to specific structures) that the temporal dimension of science (and consequently the use of HS as a basic research tool) has to be taken most seriously. Only the history of science, it is clear, can serve to resolve these questions. What philosophers for a long time failed to see was that ontological questions necessarily involve the developmental aspects of science.[65] They cannot be answered (or more correctly, they will be wrongly answered) if one is content with examining a temporal "slice" of scientific work. What discloses the nature of the relation between the model and the modeled is not a logical structure of here-and-now predictions and verifications, but rather a dynamic pattern visible in the way models guide inquiry.

10. History as the Clue to Ontology

The realist-instrumentalist debate about the status of theoretical entities cannot be resolved (or, more exactly, is likely to be resolved in favor of the instrumentalist, on the good Occamist grounds that he is claiming less and achieving just as much) unless one takes into account the developmental aspect of science. And not just in an abstract way, but as a specific testimony to how theoretical entities have *in fact* guided research. The debated term 'real' can best be defined in this context by referring it to the object which gradually discloses itself through the progressive theoretical refinements offered by the scientist. The claim of a realist ontology of science is that the only way of explaining why the models of science function

---

[65] This point I first tried to make in a brief article in the 1955 *Proceedings* of the American Catholic Philosophical Association, "Realism and Modern Cosmology," 29, 137–150.

Ernan McMullin

so successfully in the overcoming of anomalies is that they approximate in some way the structure of the object. The resources that a "good" model seems to possess to meet the unexpected challenge from the data of the twilight world that lies over against the scientist can only come from its being a "fit" for that world. The long-lasting fertility of the good theory cannot be accounted for by simply alleging the endless creativity of the human mind in the face of anomaly. The model *guides*, and it guides in a way that a summary of the original "data" could never do, no matter how "creatively" made, unless there was a resonance between model and object.

But all of this needs to be documented in detail. In the limited space remaining to us it seems best to summarize a historical instance I have discussed more fully elsewhere.[66] In 1913, Bohr suggested his famous "planetary" model of the H-atom. He was guided by the results of Rutherford's scattering experiments which indicated that the atomic mass is concentrated in a very small nucleus, of positive charge proportional to the atomic number of the element. Spectroscopic data further suggested the idea of discrete energy levels within the atom; it was plausible to connect these with the negative charges required to keep the atom as a whole neutral. But a system composed of negative charges revolving around a positively charged center ought to radiate continuously, according to classical electrodynamics. The new quantum hypothesis gave the needed clue; by quantizing energy transfer, certain orbits were equivalently "canonized," thus fixing the permitted energy states of the given system.

The model was thus suggested by the Rutherford data, the Planck quantum hypothesis, and the energy-level principle already developed in spectroscopy. There was an analogy between it and the Copernican model of the solar system, though the differences between the two models (Coulomb instead of gravitational force, discrete orbits instead of continuously variable ones) are more evident than the similarities. In its original form, it immediately accounted for the Ritz "principle of combination" (1908) which summarized in one general formula the known data (Balmer, Lyman, and Paschen series) for the spectral frequencies of the radiation emitted by hydrogen:

$$\nu = R\left(\frac{1}{n^2} - \frac{1}{s^2}\right)$$

[66] "What Do Physical Models Tell Us?" *Logic, Methodology and Philosophy of Sciences*, vol. 3, ed. B. van Rootselaar and J. Staal (Amsterdam: North-Holland, 1968), pp. 385–396.

where $n = 1,2,3$ and s is an integer greater than n. Bohr was even able to derive a formula for R that made it equal to $(2\pi^2 me^4)/(h^3)$, thus allowing it to be calculated in terms of known quantities and giving a result that agreed (within three significant figures) with the value of R known from spectroscopic data.

The simple idea of a very light negatively charged particle revolving around a relatively heavy positively charged one thus accounted for all the basic facts about hydrogen with quite surprising ease and accuracy. But now the model *itself* suggested three further modifications, modifications that would be required if one were to suppose this to be a real structure, obeying the laws of physics as far as we know them. These were *not* required by the original data (the Ritz series and the Rutherford results); they were not in any sense contained in them. Nor were they directly implied by the original model, taken simply as a correlation. Only if the model is taken seriously as an approximation to a consistent physical reality is there any reason to suppose that theoretical modifications of the following sort would yield verifiable results.

In the original model, three simplifying physical assumptions had been made: first, the nucleus was assumed to be infinitely heavy (i.e., to be unaffected by the mass of the electron); second, the electron was restricted to circular orbits; third, the energy of the electron was calculated nonrelativistically. For *physical* consistency, the consequences of each of these assumptions had to be explored separately. When the finite mass of the nucleus was allowed for, R had to be multiplied by a factor of $(1 + m/M)$. This immediately explained why the lines for ionized helium (which is structurally similar to hydrogen) were not identical with the Ritz series: m/M for the two atoms is not the same. Calculation of the series for ionized helium immediately gave the correct results (the Pickering series already discovered in 1897). Second, elliptical orbits give the same energy levels as circles do, except in an electrical field, when splitting ought to occur. Since there is no physical reason why the special case of circularity should be favored, one would therefore expect this splitting. And there it was: the Stark effect, known since 1913, a fine structure of each H line, produced when the emitted atom is subjected to an intense electric field. The amount of the splitting and the polarizations produced were exactly predicted. Finally, a relativistic correction of the calculation of energy levels showed that *all* the H lines, even apart from the presence of fields, ought to show a very fine splitting. This had actually been observed, as

65

early as 1887, for the main Balmer line. It was soon verified for all the others, and in exactly the amounts predicted.

How is this striking series of successes to be accounted for? What relationship must be postulated between the model and the world the physicist is trying to understand? Since the model is the *only* possible mode of access we have to the world, there is no way of answering this question directly. But if we try to account for the career of the model (and the philosopher is forced to account for it somehow), there seems to be no satisfactory alternative to saying that the explanatory resources of the model are due to its having revealed, however imperfectly and incompletely, an "ontological" structure, i.e., a structure intrinsic to the world over against the observer, an anchor point in a network of causal relations stretching outward in the world. This is, of course, highly metaphysical and vague, as any discussion of ontology is forced to be because of the obvious limitations of language and proof in this domain. But if someone finds such a realism intolerably naive or hopelessly vague, he is still faced with the question: from where does the fertility of the model come, from the mind of the physicist, from the purely logical resources of the original construct—or from the object modeled and partially understood?

One further development of the Bohr model, rather different in logical type from the three already chronicled, is even more significant. In 1896, Zeeman had noticed a splitting of the spectral lines emitted by hydrogen in a magnetic field. The splitting was a very complex one, sometimes doublet, sometimes triplet, sometimes a baffling multiplet. It seemed that the Bohr model ought to explain it; after all, it had explained the apparently analogous effect of an electrical field. But all attempts to find some overlooked idealization or approximation, of the kind that had explained the Stark effect, failed. For fifteen years, much energy was expended on this problem. Then a number of people began to ask themselves: what if the orbital electron were to act as a tiny magnet? There was nothing in the original model to suggest this, but it was not inconsistent with the model. Since the electron is electrically charged, the easiest way to provide it with a magnetic field is to suppose it to be spinning. This is what Goudsmit and Uhlenbeck proposed (1926). Their postulate of electron spin allowed them to calculate the normal and anomalous Zeeman effects, not only for hydrogen but for other atoms as well. Here the theoretical modification was an explicit attempt to account for a set of unexplained data. This was done by adding a *new* complication to the original model, but one physically consistent with it. To give the electron a "spin" was to specify a parameter left undetermined in the first model. It was to explore one possible causal line in an only partially determined network. The frequent success of such efforts is what one would expect *if* a realist epistemology is the correct account of what it is that grounds scientific knowledge, makes it consistent and extensible outwards to an apparently unlimited degree.

In this section, we have departed from the neutrality of the earlier taxonomic enterprise to argue for a specific (and not especially popular) philosophical position. But our main point even here still remains (in the context of this paper) a taxonomic one. Whether or not one accepts a qualified realist view of physical models, one thing at least is clear. The only sort of evidence likely to be decisive—or even relevant—in this matter is that of the history of specific models. There is nothing in the logic of a model, considered as a purely formal structure, that would help one to an answer to the ontological question. The behavior of a model in time, the fact that it went this way instead of that way, is the best clue we have to its "real nature." In this respect the philosopher is not unlike the physicist himself who when investigating a fundamental particle follows its career in all sorts of different physical situations. In this way, he builds up a picture of what the capacities of his particle are. Likewise, the philosopher operating at a second level of inquiry chronicles the conceptual "behavior" of the double helix of DNA or of some other long-lived and productive model; only from its history can he learn how seriously he should take it as a clue to "real" structure. The model, when all is said and done, is *not* a physical particle or a Platonic entity; it is the creation of the physicist. Yet because its history does not seem to lie altogether within the grasp of its inventor, we are inclined to say that this history ought to be carefully looked into. The deep sources of history are what, after all, we mean by "reality."

Bayes's Theorem and the History of Science

0. Introduction

In his splendid introduction to this volume, Herbert Feigl rightly stresses the central importance of the distinction between the *context of discovery* and the *context of justification*. These terms were introduced by Hans Reichenbach to distinguish the social and psychological facts surrounding the discovery of a scientific hypothesis from the evidential considerations relevant to its justification.[1] The folklore of science is full of dramatic examples of the distinction; e.g., Kepler's mystical sense of celestial harmony[2] versus the confrontation of the postulated orbits with the observations of Tycho; Kekulé's drowsing vision of dancing snakes[3] versus the laboratory confirmation of the hexagonal structure of the benzine ring; Ramanujan's visitations in sleep by the Goddess of Namakkal[4] versus his waking demonstrations of the mathematical theorems. Each of these examples offers a fascinating insight into the personality of a working scientist, and each provides a vivid contrast between those psychological factors and the questions of evidence that must be taken into account in order to assess the truth or probability of the result. Moreover, as we all learned in our freshman logic courses, to confuse the source of a proposition with the evidence for it is to commit the genetic fallacy.

If one accepts the distinction between discovery and justification as viable, there is a strong temptation to maintain that this distinction marks the boundaries between history of science and philosophy of science. His-

AUTHOR'S NOTE: This work was supported in part by a research grant from the National Science Foundation.

[1] *Experience and Prediction* (Chicago: University of Chicago Press, 1938), section 1. I have offered an elementary discussion of the distinction in *Logic* (Englewood Cliffs, N.J.: Prentice-Hall, 1963), sections 1–3.

[2] A. Pannekoek, *A History of Astronomy* (New York: Interscience, 1961), p. 235.

[3] J. R. Partington, *A History of Chemistry* (London: Macmillan, 1964), IV, 553ff.

[4] G. H. Hardy et al., *Collected Papers of Srinivasa Ramanujan* (Cambridge: Cambridge University Press, 1927), p. xii.

tory is concerned with the *facts* surrounding the growth and development of science; philosophy is concerned with the *logical structure* of science, especially with the evidential relations between data and hypotheses or theories. As a matter of fact, Reichenbach described the transition from the context of discovery to the context of justification in terms of a *rational reconstruction*. On the one hand, the scientific innovator engages in thought processes that may be quite irrational or nonrational, manifesting no apparent logical structure: this is the road to discovery. On the other hand, when he wants to present his results to the community for judgment, he provides a reformulation in which the hypotheses and theories are shown in logical relation to the evidence that is offered in support of them: this is his rational reconstruction. The items in the context of discovery are *psychologically relevant* to the scientific conclusion; those in the context of justification are *logically relevant* to it. Since the philosopher of science is concerned with logical relations, not psychological ones, he is concerned with the rationally reconstructed theory, not with the actual process by which it came into being.

Views of the foregoing sort regarding the relations between the context of discovery and the context of justification have led to a conception of philosophy of science which might aptly be characterized as a "rational reconstructionist" or "logical reconstructionist" approach; this approach has been closely associated with the school of logical positivism, though by no means confined to it.[5] Critics of the reconstructionist view have suggested that it leaves the study of vital, living, growing science to the historian, while relegating philosophy of science to the dissection of scientific corpses—not the bodies of scientists, but of theories that have grown to the point of stagnation and ossification. According to such critics, the study of completed science is not the study of science at all. One cannot understand science unless he sees how it grows; to comprehend the logical structure of science, it is necessary to take account of scientific change and scientific revolution. Certain philosophers have claimed, consequently, that philosophy of science must deal with the logic of discovery as well as the logic of justification.[6] Philosophy of science, it has been said, cannot proceed apart from study of the history of science. Such arguments have

---

[5] Reichenbach, for example, was not a logical positivist; indeed, he was one of the earliest and most influential critics of that school.

[6] See, for example, N. R. Hanson, "Is There a Logic of Discovery," in *Current Issues in the Philosophy of Science*, ed. Herbert Feigl and Grover Maxwell (New York: Holt, Rinehart and Winston, 1961), pp. 20–35.

led to a challenge of the very distinction between discovery and justification.[7] Application of this distinction, it is claimed, has led to the reconstructionist approach, which separates philosophy of science from real science, and makes philosophy of science into an unrealistic and uninteresting form of empty symbol manipulation.

The foregoing remarks make it clear, I hope, that the distinction between the context of discovery and the context of justification is a major focal point for any fundamental discussion of the relations between history of science and philosophy of science. As the dispute seems to shape up, the reconstructionists rely heavily upon the viability of a sharp distinction, and they apparently conclude that there is no very significant relation between the two disciplines. Such marriages as occur between them —e.g., the International Union of History and Philosophy of Science, the National Science Foundation Panel for History and Philosophy of Science, and the Departments of History and Philosophy of Science at Melbourne and Indiana—are all marriages of convenience. The anti-reconstructionists, who find a basic organic unity between the two disciplines, seem to regard a rejection of the distinction between discovery and justification as a cornerstone of their view. Whatever approach one takes, it appears that the distinction between the context of discovery and the context of justification is the first order of business.

I must confess at this point, if it is not already apparent, that I am an unreconstructed reconstructionist, and I believe that the distinction between the context of discovery and the context of justification is viable, significant, and fundamental to the philosophy of science. I do not believe, however, that this view commits me to an intellectual divorce from my historical colleagues; in the balance of this essay I should like to explain why. I shall not be concerned to argue in favor of the distinction, but shall instead try (1) to clarify the distinction, and repudiate certain common misconceptions of it, (2) to show that a clear analysis of the nature of scientific confirmation is essential to an understanding of the distinction, and that a failure to deal adequately with the logic of confirmation can lead to serious historical misinterpretations, and (3) to argue that an adequate conception of the logic of confirmation leads to basic, and largely unnoticed, logical functions of historical information. In other words, I shall be attempting to show how certain aspects of the relations between history

---

[7] E.g., Thomas S. Kuhn, *The Structure of Scientific Revolutions* (Chicago: University of Chicago Press, 1962), p. 9.

and philosophy of science can be explicated within the reconstructionist framework. Some of my conclusions may appear idiosyncratic, but I shall take some pains along the way to argue that many of these views are widely shared.

1. The Distinction between Discovery and Justification

When one presents a distinction, it is natural to emphasize the differences between the two sorts of things, and to make the distinction appear more dichotomous than is actually intended. In the present instance, some commentators have apparently construed the distinction to imply that, first of all, a creative scientist goes through a great succession of irrational (or nonrational) processes, e.g., dreaming, being hit on the head, pacing the floor, or having dyspepsia, until a full-blown hypothesis is born. Only after these processes have terminated does the scientist go through the logical process of mustering and presenting his evidence so as to justify his hypothesis. Such a conception would, of course, be factually absurd; discovery and justification simply do not occur in that way. A more realistic account might go somewhat as follows. A scientist, searching for a hypothesis to explain some phenomenon, hits upon an idea, but soon casts it aside because he sees that it is inconsistent with some theory he accepts, or because it does not really explain the phenomenon in question. This phase undoubtedly involves considerable logical inference; it might, for instance, involve a mathematical calculation which shows that the explanation in question would not account for a result of the correct order of magnitude. After more searching around—in the meantime perhaps he attends a cocktail party and spends a restless night—he hits upon another idea, which also proves to be inadequate, but he sees that it can be improved by some modification or other. Again, by logical inference, he determines that his new hypothesis bears certain relations to the available evidence. He further realizes, however, that although his present hypothesis squares with the known facts, further modification would make it simpler and give it a wider explanatory range. Perhaps he devises and executes additional tests to check the applicability of his latest revision in new domains. And so it goes. What I am trying to suggest, by such science fiction, is that the processes of discovery and justification are intimately intertwined, with steps of one type alternating with steps of the other. There is no reason to conclude, from a distinction between the context of discovery and the context of justification, that the entire process of discovery must be com-

71

pleted before the process of justification can begin, and that the rational reconstruction can be undertaken only after the creative work has ended. Such conclusions are by no means warranted by the reconstructionist approach.

There is, moreover, no reason to suppose that the two contexts must be mutually exclusive. Not only may elements of the context of justification be temporally intercalated between elements of the context of discovery, but the two contexts may have items in common. The supposition that this cannot happen is perhaps the most widespread misunderstanding of the distinction between the two contexts. The most obvious example is the case in which a person or a machine discovers the answer to a problem by applying an algorithm, e.g., doing a sum, differentiating a polynomial, or finding a greatest common divisor. Empirical science also contains routine methods for finding answers to problems—which is to say, for discovering correct hypotheses. These are often the kinds of procedures that can be delegated to a technician or a machine, e.g., chemical analyses, ballistic testing, or determination of physical constants of a new compound. In such cases, the process of discovery and the process of justification may be nearly identical, though the fact that the machine blew a fuse, or the technician took a coffee break, could hardly qualify for inclusion in the latter context. Even though the two contexts are not mutually exclusive, the distinction does not vanish. The context of discovery consists of a number of items related to one another by psychological relevance, while the context of justification contains a number of items related to one another by (inductive and deductive) logical relevance. There is no reason at all why one and the same item cannot be both psychologically and logically relevant to some given hypothesis. Each context is a complex of entities all of which are interrelated in particular ways. The contexts are contrasted with one another, not on the ground that they can have no members in common, but rather on the basis of differences in the types of relations they incorporate. The fact that the two contexts can have items in common does not mean that the distinction is useless or perverse, for there are differences between logical and psychological relevance relations which are important for the understanding of science.

The problem of scientific discovery does not end with the thinking up of a hypothesis. One has also to discover evidence and the logical connections between the evidence and the hypothesis. The process of discovery is, therefore, involved in the very construction of the rational reconstruction. When the scientist publishes his hypothesis as acceptable, confirmed, or corroborated, along with the evidence and arguments upon which the claim is based, he is offering *his* rational reconstruction (the one he has *discovered*), and is presumably claiming that it is logically sound. This is a fact about the scientist; his evidence and arguments satisfy him. A critic—scientist or philosopher—might, of course, show that he has committed a logical or methodological error, and consequently, that his rational reconstruction is unsound. Such an occurrence would belong to the context of justification. However, even if the argument seems compelling to an entire scientific community, it may still be logically faulty. The convincing character of an argument is quite distinct from its validity; the former is a *psychological* characteristic, the latter is *logical*. Once more, even though there may be extensive overlap between the contexts of discovery and justification, it is important not to confuse them.

Considerations of the foregoing sort have led to serious controversy over the appropriate role of philosophy of science. On the other hand, it is sometimes claimed that philosophy of science must necessarily be a historically oriented empirical study of the methods scientists of the past and present have actually used, and of the canons they have accepted. On the other hand, it is sometimes maintained that such factual studies of the methods of science belong to the domain of the historian, and that the philosopher of science is concerned exclusively with logical and epistemological questions. Proponents of the latter view—which is essentially the reconstructionist approach—may appear quite open to the accusation that they are engaged in some sort of scholastic symbol mongering which has no connection whatever with actual science. To avoid this undesirable state of affairs, it may be suggested, we ought to break down the distinctions between history and philosophy, between psychology and logic, and, ultimately, between discovery and justification.

There is, I believe, a better alternative. While the philosopher of science may be basically concerned with abstract logical relations, he can hardly afford to ignore the actual methods that scientists have found acceptable. If a philosopher expounds a theory of the logical structure of science according to which almost all of modern physical science is methodologically unsound, it would be far more reasonable to conclude that the philosophical reasoning had gone astray than to suppose that modern science is logically misconceived. Just as certain empirical facts, such as geometrical diagrams or soap film experiments, may have great heuristic value for mathematics, so too may the historical facts of scientific development provide indispensable guidance for the formal studies of the logician. In spite of this, the philosopher of science is properly concerned with issues of logical correctness which cannot finally be answered by appeal to the history of science. One of the problems with which the philosopher of science might grapple is the question of what grounds we have for supposing scientific knowledge to be superior to other alleged types of knowledge, e.g., alchemy, astrology, or divination. The historian may be quick to reply that *he* has the means to answer that question, in terms of the relative success of physics, chemistry, and astronomy. It required the philosophical subtlety of David Hume to realize that such an answer involves a circular argument.[8] The philosopher of science, consequently, finds himself attempting to cope with problems on which the historical data may provide enormously useful guidance, but the solutions, if they are possible at all, must be logical, not historical, in character. The reason, ultimately, is that justification is a normative concept, while history provides only the facts.

I have been attempting to explain and defend the distinction between discovery and justification largely by answering objections to it, rather than by offering positive arguments. My attitude, roughly, is that it is such a plausible distinction to begin with, and its application yields such rich rewards in understanding, that it can well stand without any further justification. Like any useful tool, however, it must be wielded with some finesse; otherwise the damage it does may far outweigh its utility.

2. Bayes's Theorem and the Context of Justification

It would be a travesty to maintain, in any simpleminded way, that the historian of science is concerned only with matters of discovery, and not with matters of justification. In dealing with any significant case, say the replacement of an old theory by a new hypothesis, the historian will be deeply interested in such questions as whether, to what extent, and in what manner the old theory has been disconfirmed; and similarly, what evidence is offered in support of the new hypothesis, and how adequate it is. How strongly, he may ask, are factors such as national rivalry among scientists, esthetic disgust with certain types of theories, personal idiosyncrasies of influential figures, and other nonevidential factors operative? Since science aspires to provide objective knowledge of the world, it cannot be under-

[8] For further elaboration of this point see my *Foundations of Scientific Inference* (Pittsburgh: University of Pittsburgh Press, 1967), pp. 5–17.

stood historically without taking very seriously the role of evidence in scientific development and change. Such historical judgments—whether a particular historical development was or was not rationally justified on the basis of the evidence available at the time—depend crucially upon the historian's understanding of the logic of confirmation and disconfirmation. If the historian seriously misunderstands the logic of confirmation, he runs the risk of serious historical misevaluation. And to the possible rejoinder that any historian worth his salt has a sufficiently clear intuitive sense of what constitutes relevant scientific evidence and what does not, I must simply reply that I am not convinced.

Perhaps the most widely held picture of scientific confirmation is one that had great currency in the nineteenth century; it is known as the hypothetico-deductive (H-D) method. According to this view, a scientific hypothesis is tested by deducing observational consequences from it, and seeing whether these consequences actually do transpire. If a given consequence does occur, it constitutes a confirming instance for the hypothesis; if it does not occur, it is a disconfirming instance. There are two rather immediate difficulties with this characterization, and they are easily repaired. First, a scientific hypothesis, by itself, ordinarily does not have any observational consequences; it is usually necessary to supply some empirically determined initial conditions to make it possible validly to deduce any observational consequences at all. For example, from Kepler's law of planetary motion alone, it is impossible to deduce the position of Mars at some future time, but with initial conditions on the motion of Mars at some earlier time, a prediction of the position is possible. Similarly, from Hooke's law alone it is impossible to predict the elongation of a spring under a given weight, but with an empirically determined coefficient of elasticity, the prediction can be deduced. Second, it is frequently, if not always, necessary to make use of auxiliary hypotheses in order to connect the observations with the hypothesis that is being tested. For example, if a medical experimenter predicts that a certain bacillus will be found in the blood of a certain organism, he must conjoin to his medical hypothesis auxiliary hypotheses of optics which pertain to the operation of his microscope, for only in that way can he establish a deductive connection between what he observes under the microscope and the actual presence of the microorganism. With these additions, the H-D method can be schematized as follows:

Wesley C. Salmon

H (hypothesis being tested)
A (auxiliary hypotheses)
I (initial conditions)
––––––––––––––––––––––––––
O (observational consequence)

Since we are not primarily interested in epistemological problems about the reliability of the senses, let us assume for the purposes of the present discussion that the initial conditions I have been established as true by observation and, in addition, that we can ascertain by observation whether the observational consequence O is true or false. Let us assume, moreover, that for purposes of the present test of our hypothesis H, the auxiliary hypotheses A are accepted as unproblematic.[9] With these simplifying idealizations, we can say that H implies O; consequently if O turns out to be false, it follows that H must be false—this is the deductively valid *modus tollens*. Given the truth of O, however, nothing follows deductively about the truth of H. To infer the truth of H from the truth of O in these circumstances is obviously the elementary deductive fallacy of affirming the consequent. According to the H-D view the truth of O does, nevertheless, tend to confirm or lend probability to H. Presumably, if enough of the right kinds of observational consequences are deduced and found by observation to be true—i.e., if enough observational predictions are borne out by experience—the hypothesis can become quite highly confirmed. Scientific hypotheses can never be completely and irrefutably verified in this manner, but they can become sufficiently confirmed to be scientifically acceptable. According to this H-D conception, induction—the logical relation involved in the confirmation of scientific hypotheses—is a kind of inverse of deduction. The fact that a true observational prediction follows deductively from a given hypothesis (in conjunction with initial conditions and auxiliary hypotheses) means, according to the H-D view, that a relation of inductive support runs in the reverse direction from O to H.

The H-D account of scientific confirmation is, it seems to me, woefully inadequate. The situation is nicely expressed in a quip attributed to Morris R. Cohen: A logic text is a book that contains two parts; in the first (on deduction) the fallacies are explained, and in the second (on induction) they are committed. Quite clearly, we need a more satisfactory account of

---

[9] This is, of course, an unrealistic assumption, for as Pierre Duhem pointed out, in many cases the appropriate move upon encountering a disconfirming case is rejection or modification of an auxiliary hypothesis, rather than rejection of the principal hypothesis. This point does not affect the present discussion.

scientific confirmation. Automatically transforming a deductive fallacy into a correct inductive schema may offer an appealing way to account for scientific inference, but certainly our forms of inductive inference ought to have better credentials than that. The main shortcomings of the H-D method are strongly suggested by the fact that, given any finite body of observational evidence, there are infinitely many hypotheses which are confirmed by it in exactly the same manner; that is, there are infinitely many alternative hypotheses that could replace our hypothesis H in the schema above and still yield a valid deduction. This point is obvious if one considers the number of curves that can be drawn through a finite set of points on a graph. Hence, Hooke's law, which says that a certain function is a straight line, and Kepler's first law, which says that a planetary orbit is an ellipse, could each be replaced by infinitely many alternatives that would give rise to precisely the same observational consequences as Hooke's and Kepler's laws respectively. As it stands, the H-D method gives us no basis whatever for claiming that either of these laws is any better confirmed by the available evidence than is any one of the infinitude of alternatives. Clearly it stands in dire need of supplementation.

When we look around for a more adequate account of scientific confirmation, it is natural to see whether the mathematical calculus of probability can offer any resources. If we claim that the process of confirmation is one of lending probability to a hypothesis in the light of evidence, it is reasonable to see whether there are any theorems on probability that characterize confirmation. If so, such a theorem would provide some sort of valid schema for formal confirmation relations. Theorems do not, of course, come labeled for their specific applications, but Bayes's theorem does seem well suited for this role.[10]

In order to illuminate the use of Bayes's theorem, let us introduce a simple game. This game is played with two decks of cards made up as follows: deck I contains eight red and four black cards; deck II contains four red and eight black cards. A turn begins with the toss of an ordinary die; if the side six appears the player draws from deck I, and if any other side comes up he draws from deck II. The draw of a red card constitutes a win. There is a simple way to calculate the probability of a win in this

---

[10] I have discussed the Bayesian conception of confirmation at some length in *Foundations of Scientific Inference*, pp. 108–131 (Bayes's theorem is deduced within the formal calculus on pp. 58–62), and in "Inquiries into the Foundations of Science," in *Vistas in Science*, ed. David L. Arm (Albuquerque: University of New Mexico Press, 1968), pp. 1–24. The latter article is the less technical of the two.

game. Letting P(A,B) stand for the probability *from* A *to* B (i.e., the probability of B, given A), and letting A stand for tosses of the die, B for draws from deck I, and C for draws resulting in red, the following formula yields the desired probability:

(1) $P(A,C) = P(A,B)P(A \& B,C) + P(A,\sim B)P(A \& \sim B,C)$

where the ampersand stands for "and" and the tilde preceding a symbol negates it. Probability expressions appearing in the formula are P(A,C), probability of a red card on a play of this game; P(A,B), probability of drawing from deck I ($= 1/6$); P(A,$\sim$B), probability of drawing from deck II ($= 5/6$); P(A & B,C), probability of getting red on a draw from deck I ($= 2/3$); P(A & $\sim$B,C), probability of getting red on a draw from deck II ($= 1/3$). The probability of a win on any given play is 7/18.

Suppose, now, that a player has just drawn a red card, but you failed to notice from which deck he drew. We ask, what is the probability that it was drawn from deck I? The probability we wish to ascertain is P(A & C, B), the probability that a play which resulted in a red card was one on which the die turned up six, and the draw was made from deck I. Bayes's theorem

(2) $P(A \& C,B) = \dfrac{P(A,B)P(A \& B,C)}{P(A,B)P(A \& B,C) + P(A,\sim B)P(A \& \sim B,C)}$

supplies the answer. Substituting the available values on the right-hand side of the equation yields the value 2/7 for the desired probability. Note that although the probability of getting a red card if you draw from deck I is much greater than the probability of getting a red card if you draw from deck II, the probability that a given red draw came from deck I is much less than the probability that it came from deck II. This is because the vast majority of draws are made from deck II.

There is nothing controversial about either of the foregoing formulas, or about their application to games of the kind just described. The only difficulty concerns the legitimacy of extending the application of Bayes's theorem, formula (2), to the problem of confirmation of hypotheses. In order to see how that might go, let me redescribe the game, with some admitted stretching of usage. We can take the draw of a red card as an effect that can be produced in either of two ways, by throwing a six and drawing from deck I, or by tossing some other number and drawing from the other deck. There are, correspondingly, two causal hypotheses. When we ask for the probability that the red draw came from deck I, we are asking for the probability of the first of these hypotheses, given the evidence that a red card had been drawn. Looking now at the probability expressions that appear in Bayes's theorem, we have: $P(A,B)$, the prior probability of the first hypothesis; $P(A,\sim B)$, the prior probability that the first hypothesis does not hold; $P(A \& B,C)$, the probability of the effect (red card drawn) if the first hypothesis is correct; $P(A \& \sim B,C)$, the probability of the effect if the first hypothesis is incorrect; $P(A \& C,B)$, the posterior probability of the first hypothesis on the evidence that the effect has occurred. The probabilities $P(A \& B,C)$ and $P(A \& \sim B,C)$ are called likelihoods of the two hypotheses, but it is important to note clearly that they are not probabilities of hypotheses but, rather, probabilities of the effect. It is the posterior probability that we seek when we wish to determine the probability of the hypothesis in terms of the given evidence.

In order to apply Bayes's theorem, we must have three probabilities to plug into the right-hand side of (2). Since the two prior probabilities must add up to one, it is sufficient to know one of them, but the likelihoods are independent, so we just have both of them. Thus, in order to compute the posterior probability of our hypothesis, we need its prior probability, the probability that we would get the evidence we have if it is true, and the probability that we would get the evidence we have if it were false. None of these three is dispensable, except in a few obvious special cases.[11]

When the H-D schema was presented, we stipulated that the hypothesis being tested implied the evidence, so in that case $P(A \& B,C) = 1$. This value of one of the likelihoods does not determine a value for the posterior probability, and, indeed, the posterior probability can be arbitrarily small even in the case supplied by the H-D method. This fact shows the inadequacy of the H-D schema quite dramatically: even though the data confirm the hypothesis according to the H-D view, the posterior probability of the hypothesis in the light of the available evidence may be as small as you like—even zero in the limiting special case in which the prior probability of the hypothesis is zero.

If Bayes's theorem provides a correct formal schema for the logic of confirmation and disconfirmation of scientific hypotheses, it tells us that we need to take account of three factors in attempting to assess the degree to which a hypothesis is rendered probable by the evidence. Roughly, it says,

[11] Viz., if $P(A,B) = 0$ or $P(A \& B,C) = 0$, then $P(A \& C,B) = 0$; if $P(A,\sim B) = 0$ or $P(A \& \sim B,C) = 0$, then $P(A \& C,B) = 1$. Also, if $P(A,C) = 0$, the fraction becomes indeterminate, for, by (1), that is the denominator in (2).

we must consider how well our hypothesis explains the evidence we have (this is what the H-D schema requires), how well an alternative hypothesis might explain the same evidence, and the prior probability of the hypothesis. The philosophical obstacle that has always stood in the way of using Bayes's theorem to account for confirmation is the severe difficulty in understanding what a prior probability could be. I have argued elsewhere that it is essentially an assessment of what one might call the *plausibility* of the hypothesis, prior to, or apart from, the results of directly testing that hypothesis.[12] Without attempting to analyze what is meant by plausibility, I shall offer a few plausibility judgments of my own, just to illustrate the sort of thing I am talking about. For instance, I regard as quite implausible Velikovsky's hypotheses about the origin of Venus, any ESP theory that postulates transfer of information at a speed greater than that of light, and any teleological biological theory. Hypotheses of these kinds strike me as implausible because, in one way or another, they do not fit well with currently accepted scientific theory. I regard it as quite plausible that life originated on the face of the earth in accordance with straightforward physicochemical principles governing the formation of large "organic" molecules out of simpler inorganic ones. This does seem to fit well with what we know. You need not accept my plausibility judgments; you can supply your own. The only crucial issue is the existence of such prior probabilities for use in connection with Bayes's theorem.

Let us now return to the problems of the historian. I claim above that the analysis of the logic of confirmation could have a crucial bearing upon historical judgments. Having compared the H-D account of confirmation with the Bayesian analysis, we can see an obvious way in which this problem could arise. If a historian accepts the H-D analysis of confirmation, then there is no place for plausibility judgments in the logic of science—at least not in the context of justification. If such a historian finds plausibility considerations playing an important role historically in the judgments scientists render upon hypotheses, he will be forced to exclude them from the context of justification, and he may conclude that the course of scientific development is massively influenced by nonrational or nonevidential considerations. Such an "H-D historian" might well decide, along with the editors of *Harper's Magazine*, that it was scientific prejudice, not objective evaluation, that made the scientific community largely ignore Velikovsky's

[12] *Foundations of Scientific Inference* and "Inquiries into the Foundations of Science."

views.[13] He might similarly conclude that Einstein's commitment to the "principle of relativity" on the basis of plausibility arguments shows his views to have been based more upon preconceptions than upon objective evidence.[14] A "Bayesian historian," in contrast, will see these plausibility considerations as essential parts of the logic of confirmation, and he will place them squarely within the context of justification. The consequence is, I would say, that the historian of science who regards the H-D schema as a fully adequate characterization of the logical structure of scientific inference is in serious danger of erroneously excluding from the context of justification items that genuinely belong within it. The moral for the historian should be clear. There are considerations relating to the acceptance or rejection of scientific hypotheses which, on the H-D account, must be judged *evidentially* irrelevant to the truth or falsity of the hypothesis, but which are, nevertheless, used by scientists in making decisions about such acceptance or rejection. These same items, on the Bayesian account, become evidentially relevant. Hence, the judgment of whether scientists are making decisions on the basis of evidence, or on the basis of various psychological or social factors that are evidentially irrelevant, hinges crucially upon the question of whether the H-D or the Bayesian account of scientific inference is more nearly correct. It is entirely conceivable that one historian might attribute acceptance of a given hypothesis to nonrational considerations, while another might judge the same decision to have an entirely adequate rational basis. Which account is historically more satisfactory will depend mainly upon which account of scientific inference is more adequate. The historian can hardly be taken to be unconcerned with the context of justification, and with its differences from the context of discovery; indeed, if he is to do his job properly he must understand them very well.

3. The Status of Prior Probabilities

It would be rather easy, I imagine, for the historian, and others who are not intimately familiar with the technicalities of inductive logic and confirmation theory, to suppose that the H-D account of scientific inference is the correct one. This view is frequently expressed in the opening pages of introductory science texts, and in elementary logic books.[15] At the same

[13] This case is discussed in "Inquiries into the Foundations of Science."

[14] See Albert Einstein, "Autobiographical Notes," in *Albert Einstein: Philosopher-Scientist*, ed. Paul Arthur Schilpp (New York: Tudor, 1949), pp. 2–95.

[15] In my *Logic*, section 23, I have tried, without introducing any technicalities of

Wesley C. Salmon

time, it is important to raise the question of whether scientists in general —including the authors of the aforementioned introductory texts—actually comply with the H-D method in practice, or whether in fact they use something similar to the Bayesian approach sketched in the preceding section. I am strongly inclined to believe that the Bayesian schema comes closer than the H-D schema to capturing actual scientific practice, for it seems to me that scientists do make substantial use of plausibility considerations, even though they may feel somewhat embarrassed to admit it. I believe also that practicing scientists have excellent intuitions regarding what constitutes sound scientific methodology, but that they may not always be especially adept at fully articulating them. If we want the soundest guidance on the nature of scientific inference, we should look carefully at scientific practice, rather than the methodological pronouncements of scientists.

It is, moreover, the almost universal judgment of contemporary inductive logicians—the experts who concern themselves explicitly with the problems of confirmation of scientific hypotheses—that the simple H-D schema presented above is incomplete and inadequate. Acknowledging the well-known fact that there is very little agreement on which particular formulation among many available ones is most nearly a correct inductive logic, we can still see that among a wide variety of influential current approaches to the problems of confirmation, there is at least agreement in rejecting the H-D method. This is not the place to go into detailed discussions of the alternative theories, but I should like to mention five leading candidates, indicating how each demands something beyond what is contained in the H-D schema. In each case, I think, what needs to be added is closely akin to the plausibility considerations mentioned in the preceding section.

1. The most fully developed explicit confirmation theory available is Rudolf Carnap's theory of logical probability (degree of confirmation) contained in his monumental *Logical Foundations of Probability*.[16] In the systems of inductive logic he elaborated in that book, one begins with a formalized language and assigns a priori weights to all statements in that language, including, of course, all hypotheses. It is very easy to show that Carnap's theory of confirmation is thoroughly Bayesian, with the a priori

the probability calculus, to offer an introductory Bayesian account of the confirmation of hypotheses.

[16] Chicago: University of Chicago Press, 1950.

82

weights functioning precisely as the prior probabilities in Bayes's theorem. Although these systems had the awkward feature that general hypotheses all have prior probabilities, and consequently, posterior probabilities on any finite amount of evidence, equal to zero, Jaakko Hintikka has shown how this difficulty can be circumvented without fundamentally altering Carnap's conception of confirmation as a logical probability.[17]

2. Although not many exponents of the frequency theory of probability will agree that it even makes sense to talk about the probability of scientific hypotheses, those who do explicitly invoke Bayes's theorem for that purpose. Reichenbach is the leading figure in this school, although his treatment of the probability of hypotheses is unfortunately quite obscure in many important respects.[18] I have tried to clarify some of the basic points of misunderstanding.[19]

3. The important "Bayesian" approach to the foundations of statistics has become increasingly influential since the publication in 1954 of L. J. Savage's *The Foundations of Statistics*.[20] It has gained many adherents among philosophers as well as statisticians. This view is based upon a subjective interpretation of probability ("personal probability," as Savage prefers to say, in order to avoid confusion with earlier subjective interpretations), and it makes extensive use of Bayes's theorem. The prior probabilities are simply degrees of prior belief in the hypothesis, before the concrete evidence is available. The fact that the prior probabilities are so easily interpreted on this view means that Bayes's theorem is always available for use. Savage, himself, is not especially concerned with probabilities of general hypotheses, but those who are interested in such matters have a ready-made Bayesian theory of confirmation.[21] On this view, the prior probabilities are subjective plausibility judgments.

4. Nelson Goodman, whose influential *Fact, Fiction, and Forecast* poses and attempts to resolve "the new riddle of induction," clearly recognizes

---

[17] Jaakko Hintikka, "A Two-Dimensional Continuum of Inductive Methods," in *Aspects of Inductive Logic*, ed. Jaakko Hintikka and Patrick Suppes (Amsterdam: North-Holland, 1966), pp. 113–132.

[18] Hans Reichenbach, *The Theory of Probability* (Berkeley and Los Angeles: University of California Press, 1949), section 85.

[19] *Foundations of Scientific Inference*, pp. 115ff.

[20] New York: Wiley, 1954. An excellent exposition is found in Ward Edwards, Harold Lindman, and Leonard J. Savage, "Bayesian Statistical Inference for Psychological Research," *Psychological Review*, 70 (1963), 193–242.

[21] Sir Harold Jeffreys illustrates an explicitly Bayesian approach to the probability of hypotheses; see his *Scientific Inference* (Cambridge: Cambridge University Press, 1957), and *Theory of Probability* (Oxford: Clarendon Press, 1939).

that there is more to confirmation than mere confirming instances.[22] He attempts to circumvent the difficulties, which are essentially those connected with the H-D schema, by introducing the notion of "entrenchment" of terms that occur in hypotheses. Recognizing that a good deal of the experience of the human race becomes embedded in the languages we use, he brings this information to bear upon hypotheses that are candidates for confirmation. Although he never mentions Bayes's theorem or prior probabilities, the chapter in which he presents his solution can be read as a tract on the Bayesian approach to confirmation.

5. Sir Karl Popper rejects entirely the notions of confirmation and inductive logic.[23] His concept of *corroboration*, however, plays a central role in his theory of scientific methodology. Although corroboration is explicitly regarded as nonprobabilistic, it does offer a measure of how well a scientific hypothesis has stood up to tests. The measure of corroboration involves such factors as simplicity, content, and testability of hypotheses, as well as the seriousness of the attempts made to falsify them by experiment. Although Popper denies that a highly corroborated hypothesis is highly probable, the highly corroborated hypothesis does enjoy a certain status: it may be chosen over its less corroborated fellows for further testing, and if practical needs arise, it may be used for purposes of prediction. The important point, for the present discussion, is that Popper rejects the H-D schema, and introduces additional factors into his methodology that play a role somewhat analogous to our plausibility considerations.

The foregoing survey of major contemporary schools of thought on the logic of scientific confirmation strongly suggests not only that the naive H-D schema is *not universally accepted* nowadays by inductive logicians as an adequate characterization of the logic of scientific inference, but also that it is *not even a serious candidate* for that role. Given the wide popular acceptance of the H-D method, it seems entirely possible that significant numbers of historians of science may be accepting a view of confirmation that is known to be inadequate, and one which differs from the current serious contending views in ways that can have a profound influence upon historical judgments. It seems, therefore, that the branch of contempo-

---

[22] First edition (Cambridge, Mass.: Harvard University Press, 1955); second edition (Indianapolis: Bobbs-Merrill, 1965).

[23] *The Logic of Scientific Discovery* (New York: Basic Books, 1959). It is to be noted that, in spite of the title of his book, Popper accepts the distinction between discovery and justification, and explicitly declares that he is concerned with the latter but not the former.

rary philosophy of science that deals with inductive logic and confirmation theory may have some substantive material that is highly relevant to the professional activities of the historian of science.

It is fair to say, I believe, that one of the most basic points on which the leading contemporary theories of confirmation differ from one another is with regard to the nature of the prior probabilities. As already indicated, the logical theorist takes the prior probability as an a priori assessment of the hypothesis, the personalist takes the prior probability as a measure of subjective plausibility, the frequentist must look at the prior probability as some sort of success frequency for a certain type of hypothesis, Goodman would regard the prior probability as somehow based upon linguistic usage, and Popper (though he violently objects to regarding it as a prior probability) needs something like the potential explanatory value of the hypothesis. In addition, I should remark, N. R. Hanson held plausibility arguments to belong to the logic of discovery, but I have argued that, on his own analysis, they have an indispensable role in the logic of justification.[24]

This is not the place to go into a lengthy analysis of the virtues and shortcomings of the various views on the nature of prior probabilities.[25] Rather, I should like merely to point out a consequence of my view that is quite germane to the topic of the conference. If one adopts a frequency view of probability, and attempts to deal with the logic of confirmation by way of Bayes's theorem (as I do), then he is committed to regarding the prior probability as some sort of frequency—e.g., the frequency with which hypotheses relevantly similar to the one under consideration have enjoyed significant scientific success. Surely no one would claim that we have reliable statistics on such matters, or that we can come anywhere near assigning precise numerical values in a meaningful way. Fortunately, that turns out to be unnecessary; it is enough to have very, very rough estimates. But this approach does suggest that the question of the plausibility of a scientific hypothesis has something to do with our experience in dealing with scientific hypotheses of similar types. Thus, I should say, the reason I would place a rather low plausibility value on teleological hypotheses is closely related to our experience in the transitions from teleological to mechanical explanations in the physical and biological sciences and, to some extent, in the social sciences. To turn back toward teleological hypotheses

---

[24] *Foundations of Scientific Inference*, pp. 111–114, 118.
[25] This is done in the items mentioned in footnote 10.

would be to go against a great deal of scientific experience about what kinds of hypotheses work well scientifically. Similarly, when Watson and Crick were enraptured with the beauty of the double helix hypothesis for the structure of the DNA molecule, I believe their reaction was more than purely esthetic.[26] Experience indicated that hypotheses of that degree of simplicity tend to be successful, and they were inferring that it had not only beauty, but a good chance of being correct. Additional examples could easily be exhibited.

If I am right in claiming not only that prior probabilities constitute an indispensable ingredient in the confirmation of hypotheses and the context of justification, but also that our estimates of them are based upon empirical experience with scientific hypothesizing, then it is evident that the history of science plays a crucial, but largely unheralded, role in the current scientific enterprise. The history of science is, after all, a chronicle of our past experience with scientific hypothesizing and theorizing—with learning what sorts of hypotheses work and what sorts do not. Without the Bayesian analysis, one could say that the study of the history of science might have some (at least marginal) heuristic value for the scientist and philosopher of science, but on the Bayesian analysis, the data provided by the history of science constitute, *in addition*, an essential segment of the evidence relevant to the confirmation or disconfirmation of hypotheses. Philosophers of science and creative scientists ignore this fact at their peril.

[26] James D. Watson, *The Double Helix* (New York: New American Library, 1969). This book provides a fascinating account of the *discovery* of an important scientific hypothesis, and it illustrates many of the points I have been making. Perhaps if literary reviewers had had a clearer grasp of the distinction between the context of discovery and the context of justification they would have been less shocked at the emotions reported in the narrative.

Inference to Scientific Laws

The topic of inference to scientific laws is one to which, I believe, both philosophers of science and historians of science can contribute to their mutual benefit. This is by no means self-evident, and indeed has been denied by philosophers as well as historians. There is even a view that there is no such topic at all to discuss. The view, held by proponents of the hypothetico-deductive (H-D) picture of science, is that there are no inferences to laws, only *from* them. The scientist does not infer a law from the data. He invents it, guesses it, imagines it, and then derives consequences from it which he tests. For example, Popper, one of the foremost proponents of this view, speaks of theories, including laws, as "free creations of our own minds, the result of an almost poetic intuition," and he rejects the idea that they are inferred in any way from observations.[1] Again, in a recent work, Hempel writes: "The transition from data to theory requires creative imagination. Scientific hypotheses are not *derived* from observed facts, but invented in order to account for them. They constitute guesses at the connections that might obtain between the phenomena under study. . . ."[2] The physicist Feynman agrees. He writes: "In general we look for a new law by the following process. First we guess it. Then we compute the consequences of the guess to see what would be implied if this law that we guessed is right. Then we compare the result of the computation to nature . . . to see if it works. If it disagrees with experiment it is wrong. In that simple statement is the key to science."[3]

According to the H-D view scientists do not make inferences to laws,

AUTHOR'S NOTE: This paper contains in abbreviated form some material from chapters 6 and 7 of my *Law and Explanation in Science*, to be published.

[1] Karl Popper, *Conjectures and Refutations* (London: Routledge and Kegan Paul, 1965), p. 192.
[2] Carl G. Hempel, *Philosophy of Natural Science* (Englewood Cliffs, N.J.: Prentice-Hall, 1966), p. 15.
[3] Richard Feynman, *The Character of Physical Law* (Cambridge, Mass.: MIT Press, 1967), p. 156.

Peter Achinstein

only from them. How should this claim be construed? One way would be to say that the term "inference" is being used in a restricted sense to cover only deductive inferences from statements describing particular observations. On this construal the H-D theorist would be saying simply that scientists do not make deductive inferences to laws from nongeneral observation statements. The problem with this construal is that it does not do full justice to the claims of the H-D theorist. First, proponents of this view also deny that scientists arrive at hypotheses by way of inductive inferences. Popper, for example, writes that "induction, i.e. inference based on many observations, is a myth. It is neither a psychological fact, nor a fact of ordinary life, nor one of scientific procedure."[4] He holds that scientists and others do not in fact employ inductive inferences and that in principle they could not justifiably do so, since inductive reasoning is fallacious. While other H-D theorists do not accept Popper's view about the fallacy of inductive reasoning, they do share his view that as a matter of fact scientists do not arrive at hypotheses by inductive reasoning. Second, by saying that scientific hypotheses are invented, guessed, free creations of our own minds, H-D theorists appear to be contrasting these cases with those involving inferences. The claim seems to be that the scientist arrives at a law not by making an inference, not by engaging in a process of reasoning in which something is concluded on the basis of something else, but by making a conjecture. This conjecture may be a causal consequence of observations (among other things) but it is not inferred from them. Only after the conjecture is made do inferences occur.

What do we mean when we speak of a person as having inferred something? There are, I believe, two slightly different uses of "infer." In one use, when we say that A inferred that a proposition p is true or probable from the fact, or alleged fact, that q is true, we imply that A came to believe that p is true or probable. In another use, we imply simply that A believed that p is true or probable. In both uses we also imply that A's reason for believing that p is true or probable is the fact, or alleged fact, that q is true.[5] (If we know q to be false, we will say that A's reason for believing that p is true or probable is his belief that q is true.) But more must be involved in his reason than this. Suppose A comes to believe that God exists, and his reason for believing this is the alleged fact q, that if God does exist

[4] Popper, Conjectures and Refutations, p. 53.
[5] For a theory of inference according to which the second sense of "infer" above is basic and the first is derivative, see D. G. Brown, "The Nature of Inference," Philosophical Review, 64 (1955), 351–369.

and he does not believe it then God will punish him. Although A has come to believe that God exists and A's reason for believing this is the alleged fact that q is true, A has not *inferred* that God exists from this alleged fact. He has not done so because the alleged fact which constitutes A's reason for believing that God exists is not a fact which A believes makes it likely that God exists. It is, as we might say, a pragmatic or utilitarian reason rather than an "evidential" one, and it is only reasons of the latter sort that are involved in inferences. Accordingly, in what follows when I speak of A's reason for a certain belief I shall mean his "evidential" reason. However, if we want to spell out the conditions under which it is appropriate to say that A inferred that p is true or probable from the fact, or alleged fact, that q is true they are these: (1) A came to believe (or believed) that p is true or probable. (2) A's reason for believing this is the fact, or alleged fact, that q is true and the fact, or alleged fact, that q's being true makes it likely that p is true.

Reasoning is a broader concept. It includes, among other things, thinking about something and drawing certain conclusions, which means thinking leading to an inference (in either sense); it also includes examining a reason to see if it does support a belief. H-D theorists do not deny that there is reasoning in the case of laws. They simply assert that it is all of the second type: It involves only the examination of reasons to see if they do support a law, and so it takes place after the law is formulated. Their claim is that there is no reasoning which leads to laws in the first place; scientists do not infer laws, in the coming-to-believe sense of "infer" (which is the one that will be used in what follows[6]). It is this view that I want to reject.

In the early years of the nineteenth century Gay-Lussac performed numerous experiments in which gases were combined to form new compounds. In his experiments he noted, for example, that 100 parts (by volume) of oxygen combine with 200 parts of hydrogen to form water and that 100 parts of muriatic gas combine with 100 parts of ammonia gas to form ammonium chloride. After considering several other cases he wrote: "Thus it appears evident to me that gases always combine in the simplest proportions when they act on one another." (This is essentially Gay-Lussac's law.) Gay-Lussac came to believe that gases combine in simple ratios and his reason for believing this was the fact that various gases he observed combine in simple ratios. From his experiments he inferred that gases be-

---

[6] Those who, following Brown, believe that the second sense of "infer" is basic can understand the thesis under examination as saying that scientists do not come to *infer* laws from data. See *ibid.*, p. 355.

have in this way. It may be objected that Gay-Lussac in writing the words he did was not describing how he really came to believe the law. Quite possibly the law occurred to him after considering a single experiment with oxygen and hydrogen. It might also be objected that the law did not occur to Gay-Lussac simply on the basis of experimentation; a background in atomic theory was relevant as well. Let us grant that theoretical ideas played a part in the origin of Gay-Lussac's law. (I will go into this later.) Still there is an inference involved. It may be an inference from the result of one experiment together with theoretical assumptions. And the conclusion might not be the strong one that the law is definitely true, but a weaker one to the effect that the law may be true. Nevertheless, on the basis of at least certain experiments and theoretical assumptions, Gay-Lussac came to believe that (there is some likelihood that) gases combine in simple ratios by volume.

Why do proponents of the H-D account deny the existence of inferences to laws? There seem to be several strands in their thinking. First, they hold that there is no mechanical way to infer laws from data. As Hempel puts it, "There are . . . no generally applicable 'rules of induction' by which hypotheses can be mechanically derived or inferred from empirical data."[7] Second, they hold that laws and hypotheses generally are arrived at by an act of imagination. Third, they hold that the formulation of a hypothesis on the part of the scientist is a causal process in which many things are involved in addition to the observations he has made and the theories he holds, e.g., his personality, his training, and even his dreams.

We can grant these facts, but none of them establishes the nonexistence of inferences to laws. First, from the fact that laws cannot be inferred by mechanical reasoning from data it does not follow that they cannot be inferred. We might speak of someone as having engaged in mechanical reasoning leading to an inference from q to p if there are rules that permit p to be correctly inferred from q (rules that indicate that coming to believe p for the reason q is legitimate), if there are rules prescribing how the reasoning shall proceed (through what steps, or, if there are several possible ways, through what alternations of steps), and if the person actually came to believe p for the reason q solely by a conscious application of these rules. Given this characterization, it should be obvious that scientists do not engage in mechanical reasoning to laws. They do not infer laws from data solely by a conscious application of rules of the sort referred to above. But

[7] Hempel, *Philosophy of Natural Science*, p. 15.

it does not follow from this that they make no inferences to laws. A scientist may make an inference from data to law without engaging in reasoning that consists simply in a conscious application of rules that indicate what follows from what and what steps to take. Second, from the fact that to arrive at a law requires "creative imagination" or "poetic intuition" it does not follow that laws cannot be inferred. We might say that someone's inference involved imagination if the reasoning that led to it was not mechanical in the sense noted above. Many inferences that are made involve imagination; they are not mechanical but they are inferences nonetheless. Third, a scientist might not have made an inference from data to hypothesis if he had not had the personality, training, and dreams that he did have. He made the inference nonetheless, and the fact that he did make this particular inference instead of some other or none at all might be causally explained, in part at least, by reference to his personality, training, and dreams. The fact that there is a causal explanation of how a scientist came to believe a certain law for a certain reason does not preclude his having come to believe that law for that reason. It does not preclude his having made an inference to the law from observation and theory. Finally, there is no necessary incompatibility between conjecture and inference. A scientist may have made an inference to a hypothesis on the basis of certain data and the hypothesis may still be a conjecture, depending upon how strongly the data support the hypothesis.

What must now be determined is what sort of inferences are involved in the case of laws. This means characterizing the various types of reasons that scientists have for laws when they come to believe them and expressing these as "modes of inference." The latter will also indicate the kinds of reasons which scientists can offer in support of a law which they already believe and hence are not inferring. (Accordingly, when I speak of "modes of inference" I do not thereby imply that one who has or gives reasons that are in conformity with these modes is necessarily making an inference. He may be; but he may also be in the position of having or giving reasons for something that he or someone else has already inferred.)

The late Norwood Russell Hanson had some interesting views about the kinds of reasons scientists have for hypotheses as well as about why this topic should be of concern to both philosophers and historians of science. He held that one of the most important jobs of the historically minded philosopher of science or the philosophically minded historian is to provide a critical appraisal of the reasoning scientists actually employ. More-

91

over, he insisted, contrary to the H-D view, that scientists do make inferences to laws, and that these are neither deductive nor inductive but what, borrowing a term from Peirce, he called retroductive. The scientist begins by considering puzzling phenomena that have been observed. He makes an inference to a hypothesis which, if true, would explain the phenomena by organizing them into an "intelligible, systematic, conceptual pattern." The retroductive mode of inference Hanson characterizes as follows:

> Some surprising phenomenon P is observed.
> P would be explicable as a matter of course if [hypothesis] H were true.
> Hence there is reason to think that H is true.[8]

Now I believe that Hanson and Peirce (who expressed very similar views) are taking us in the right direction. But there are several problems with the retroductive account, as they express it. To begin with, the claim that the scientist always starts by considering simply observed phenomena is unacceptable. Maxwell developed his distribution law for molecular velocities by considering not observed phenomena but the unobserved molecular nature of a gas postulated by kinetic theory. Furthermore, even when the scientist develops a law by considering observed phenomena this, in the typical case, is not all he considers. Usually a theoretical background is also relevant. Gay-Lussac did not infer his law simply from his experiments with various gases. He inferred it from this together with, or in the light of, a theory he held about the molecular structure of gases. The retroductive inference, as characterized by Hanson and Peirce, neglects the background of theory which the scientist often has to begin with, and which may provide at least part of and in some cases the entire basis for an inference to a law.

The main problem, however, is that the Hanson-Peirce retroductive mode of inference is fallacious. From the fact that a hypothesis H, if true, would explain the data it does not in general follow that there is reason to think that H is true. To take a simple example, the hypothesis that I will be paid one million dollars when this paper is published would, if true, explain why I am writing the paper. But this provides no reason for thinking that I am about to become a millionaire. There are many "wild" hypotheses which if true would explain the data, but unless there is some other evidence in their favor this fact by itself lends no plausibility to them. This does not mean that there is nothing like retroductive reasoning in science

---

[8] N. R. Hanson, *Patterns of Discovery* (Cambridge: Cambridge University Press, 1958), p. 86.

(I mean *valid* reasoning). There is, but the Hanson-Peirce description of it is not adequate.

Finally, both Hanson and Peirce give the impression that there is only one mode of inference to laws—retroduction. Inductive and deductive reasoning exist, but they are employed only after retroduction to the law has occurred. This thesis I would also reject. There are several modes of inference to laws, including something akin to retroductive, but in addition, inductive, analogical, deductive, and possibly others. In what follows I want to try to formulate the retroductive or explanatory mode of inference in a more suitable way, and also say something about the (or one) inductive pattern of inference.

To get at the explanatory mode of inference we must keep in mind two things the Hanson-Peirce account neglects to mention. First, typically an inference to a hypothesis if made from observational data is made in the light of everything else we know, the background information, which includes accepted theories, principles, laws, etc. Second, when an inference is made to a hypothesis H on the grounds that it explains certain facts it is assumed that H offers a better explanation of these facts than other hypotheses with which H is incompatible. Frequently competing hypotheses are explicitly considered. One reason you reject the hypothesis that I am writing this paper because I will be paid a million dollars to do so, even though that hypothesis if true would explain why I am writing it, is that you can readily think of competing hypotheses which offer better explanations. In an explanatory inference one infers H from evidence E in the light of background information B on the grounds that H provides a good explanation of certain facts, an explanation that, given E and B, is better than that provided by competing hypotheses. We might put this mode of inference in the form of an argument, as follows, in which the premises indicate reasons for believing Conclusion 1 which, in turn, indicates a reason for believing Conclusion 2:

Premise 1: Evidence E is obtained in the light of background information B.

Premise 2: Hypothesis H_1 is capable of providing a set of answers S_1 to questions concerning facts F which may be part of E or B. Incompatible competing hypotheses H_2, . . . , H_n that it is reasonable to consider given E and B provide sets of answers S_2, . . . , S_n to the same questions concerning F.

Conclusion 1: H_1 offers a good explanation of F. (This conclusion is warranted provided that the answers supplied by H_1 reasonably satisfy criteria for good explanations, and do so on the whole better than competing hypotheses $H_2, \ldots, H_n$.[9])

Conclusion 2: H_1 (or H_1 is a plausible hypothesis).

Alternative hypotheses mentioned in Premise 2 might be explicitly considered by someone employing this mode of inference or they might not. In the former case we can say that the reasoner infers the explanatory power of H from the evidence and the fact that H offers such and such answers to questions while competing hypotheses offer such and such different ones. In the latter case he infers the explanatory power of H from the evidence and the fact that H offers such and such answers to questions, where he assumes that whatever competing hypotheses there are that are reasonable to consider, given E and B, do not supply answers that are more plausible.

There are various additional qualifications and embellishments on this, but I can't go into them here.[10] Let me turn to a different pattern of inference, which I will call inductive. One may legitimately infer that all F's are G's on the ground that all observed F's have been G's—provided that two crucial assumptions are made, which are similar to those in the explanatory case, and which are often omitted by champions of induction. One is that the background information is taken into account. The other concerns hypotheses that conflict with "All F's are G's" that might be deemed reasonable to consider as possible alternatives. The inference from "All observed F's are G's" to "All F's are G's" is warranted to the extent that such hypotheses are not plausible, given the inductive evidence and the background information. (The plausibility of a competing hypothesis would itself be inferred and the inference could be an explanatory one, an inductive one, or one of several other types I have not here discussed.) In an inductive inference one infers that all F's are G's from the inductive evidence, in the light of the background information and alternative incompatible hypotheses. Let me express this in the form of an argument, as follows:

Premise 1: Inductive evidence E is obtained in the light of background information B. The evidence is that F's have been examined and all those examined have been G.

Premise 2: $H_2, \ldots, H_n$ are hypotheses conflicting with "All F's are G's" that, given E and B, it is reasonable to consider as possible alternatives to "All F's are G's."

---

[9] For an account of these criteria, as well as an analysis of explanation, see my "Explanation," *American Philosophical Quarterly*, Monograph Series No. 3 (1969).

[10] They are discussed in my *Law and Explanation* (forthcoming), chapter 6.

Conclusion: All F's are G's, or it is plausible to assume that all F's are G's. (This conclusion is warranted to the extent that hypotheses in the set $H_2, \ldots, H_n$ are not plausible given E and B, i.e., to the extent that no inference is warranted from E and B to the conclusion that such a hypothesis is plausible.)

One who uses this mode of inference may explicitly consider alternative hypotheses or he may just assume that whatever competing hypotheses that might be deemed reasonable to consider as alternatives are not in fact plausible, given E and B.

No separate account of variety of instances is needed. We vary instances to rule out competing hypotheses. In determining whether all F's are G's, if we vary the F's with respect to some property H—if we observe both F's that are H and those that are not—we do so in order to rule out the competing hypothesis that it is only those F's that are also H's which are G's, and therefore that some F's are not G's. If no conflicting hypothesis known to those familiar with E and B is plausible, then the factor of variety has been taken into account. This is the best way to do so, since variety is always relativized to competing hypotheses. Otherwise the question of what sort of variety is relevant cannot be answered.

Here, then, are two of several possible modes of inference, rather simply expressed, and we shall want to ask whether and to what extent they are actually exemplified in scientific reasoning to laws. Before turning to this question, there are two issues that need brief mention. The first is whether, indeed, we have here two distinct modes of inference, whether, for example, inductive inference is really a special case of explanatory inference. This thesis has recently been propounded by Gilbert Harman who gives several arguments in favor of it which are somewhat involved and which I do not have time to discuss here.[11] Suffice it to say that I do not find the arguments convincing, but they are worthy of consideration, and I simply want to call attention to an interesting philosophical issue. The second issue concerns a distinction, somewhat notorious in the philosophy of science, between the context of discovery and the context of justification.

Those who embrace the H-D view described earlier hold that questions about the discovery of a hypothesis are empirical questions best left to the historian of science and to the psychologist. The philosopher of science,

[11] Gilbert H. Harman, "The Inference to the Best Explanation," *Philosophical Review*, 64 (1965), 88–95; also "Enumerative Induction as Inference to the Best Explanation," *Journal of Philosophy*, 65 (1968), 529–533. I discuss Harman's claim in *Law and Explanation*, chapter 6.

on this view, is and should be concerned only with the justification of hypotheses. It is only in this context that the scientist reasons and that his reasoning can be appraised. There is no logic of discovery, only a logic of justification. This claim has been denied in recent years, most strongly by Hanson. There is, he insisted, a logic of discovery, which is different from the logic of justification and which should be of interest to the philosopher of science. The logic of discovery is concerned with reasons for *suggesting* a hypothesis in the first place, the logic of justification with reasons for *accepting* a hypothesis once it has already been suggested.[12] Hanson does not claim that these two "logics" are mutually exclusive. Some reasons for suggesting H are, he admits, also reasons for accepting H, but many are not. What are reasons for accepting H? Repeated observations supporting H, new predictions from H which are confirmed by observations, derivability of H from established theories. What are reasons for suggesting H in the first place? The most important, according to Hanson, are explanatory ones: "Does this hypothesis look as if it might *explain* these facts?"[13] The retroductive inference pattern is meant to set out this particular mode of reasoning explicitly.

Now I cannot accept the idea that there are two "logics," that there are reasons for suggesting hypotheses in the first place and reasons for accepting them which on the whole are different. Take Hanson's retroductive or explanatory reasoning, which he contends falls under the logic of discovery. That a hypothesis offers a plausible explanation of certain facts can be a reason for suggesting it, but it can also be a reason for accepting it once it has been suggested. Or take deductive reasoning from an established theory, which, according to Hanson, falls under the logic of justification. The fact that a hypothesis follows deductively from an established theory may be a reason for accepting it, but it may also be a reason for suggesting it as plausible in the first place. Indeed, any reason for suggesting a hypothesis can also be a reason for accepting it, and conversely. There is no special logic of discovery as opposed to a logic of justification.

There is, however, a distinction worth making. If a scientist first came to be acquainted with a hypothesis in the course of reasoning to its truth or plausibility we might say that his reasoning occurred in a context of discovery. If the scientist had been acquainted with the hypothesis before his reasoning occurred and had engaged in the reasoning in the course of

---

[12] N. R. Hanson, "The Logic of Discovery," in Boruch Brody and Nicholas Capaldi, eds., *Science: Men, Methods, Goals* (New York: Benjamin, 1968), pp. 150–162.
[13] *Ibid.*, p. 153.

attempting to defend the hypothesis we might say that his reasoning took place in a context of justification. In the former case there is an inference since the scientist came to believe a hypothesis, and so we can also say that the scientist's inference occurred in a context of discovery. In the latter case there need not be an inference, since the scientist may already believe the hypothesis. However, in both cases we can speak of reasoning, and in both cases the type of reason the scientist has for his hypothesis can be the same. The present distinction is by no means exhaustive. A scientist might have known about hypothesis H before reasoning on the basis of the data that H is true or plausible, and he might not have reasoned that H is true or plausible in the course of attempting to defend H. He might simply have considered the data and reasoned that H is true. If so his reasoning would not fall into a context of discovery or a context of justification.

Can reasoning involving either of the modes of inference I have discussed occur in any of these contexts? The following argument might be given in favor of a negative answer: "Look at the modes of inference described earlier. Each contains in a premise a reference to the hypothesis whose plausibility is being asserted in the conclusion. Thus, the second premise of the explanatory mode of inference contains the statement that hypothesis H_1 is capable of providing a set of answers S_1 to questions concerning facts F. This means that someone who is reasoning to the plausibility of H_1 must already have known about H_1 before he reasoned that H_1 is plausible. So his reasoning cannot have taken place in a context of discovery, but only in a context of justification or perhaps in neither context. A similar point can be made about inductive reasoning, as formulated above."

My reply to this argument is to reject the assumption that these modes of inference require one to have known about H_1 before reasoning that H_1 is plausible. Suppose that while I ponder the evidence and background information the following thought occurs to me: a certain hypothesis H_1 is plausible because the evidence and background information are what they are and because that hypothesis provides such and such answers to questions concerning F while competitors offer such and such different answers. If this thought does occur to me, and if before this I had not had the belief that H_1 is plausible, then I have made an inference from the evidence and background information, and from the fact that H_1 and its competitors offer such and such explanations, to the fact that H_1 is plausible. I have come to believe that H_1 is plausible and my reason for this belief

97

has to do with the nature of the evidence and the nature of the explanations provided by H_1 and its competitors. It is quite possible for me to have had the thought described above without being acquainted with the hypothesis H_1 at some time prior to that thought. But this means that it is possible for my reasoning to H_1's plausibility to take place in a context of discovery. That is, it is possible for me to first come to be acquainted with H_1 in the course of making an inference to H_1's plausibility. On the other hand, of course, H_1 might have been known to me before the thought described above occurred to me, and I might have expressed this thought in the course of attempting to defend H_1. If so, my reasoning took place in a context of justification. It should be evident that what I have said here about explanatory reasoning can be extended to inductive reasoning as I have formulated it. Reasoning of either of these types can occur in a context of discovery, in a context of justification, or in neither of these contexts.

Now to the main question. Why should all this be of concern to philosophers and historians alike? I take it that one interest of at least some historians of science as well as of at least some philosophers is to understand the reasoning scientists have actually employed in the case of laws. The historian may be interested in this for its own sake, the philosopher also, or possibly as a means of generalizing to some broader truth about scientific method, or as in the case of Hanson, in order to provide a critical appraisal of the reasoning. Suppose we want to consider what reasoning Gay-Lussac employed in connection with his law of combining volumes. The historian of science M. P. Crosland in his generally informative paper entitled "The Origins of Gay-Lussac's Law of Combining Volumes of Gases" considers, as he puts it, "factors guiding Gay-Lussac to his law."[14] He mentions experiments that had been performed relating to the composition of the atmosphere, a theoretical interest on Gay-Lussac's part in questions of chemical affinity, certain experiments with boron trifluoride, and so on. But in this discussion there is perhaps a little too much general talk of ideas in the air, or, to use Crosland's terms, of ideas that "converged" in Gay-Lussac's work.[15] Did Gay-Lussac actually reason from these ideas? If so, what form or forms did his reasoning take? Did it proceed in a context of discovery, justification, or neither? By appeal to the modes of inference I described earlier, as well as others I did not, and also by appeal

[14] *Annals of Science*, 17 (1961), 1–26.
[15] *Ibid.*, p. 8.

to the distinction between reasoning occurring in a context of discovery and that occurring in a context of justification, I think that certain important questions can be raised that might otherwise be overlooked. With these categories and this distinction the historian and the philosopher can both contribute to the study of the origin of a law.

Gay-Lussac's "Memoir on the Combination of Gaseous Substances with Each Other" was published in 1809. Near the beginning of this paper he reports that from the fact, determined by himself and Humboldt, that water is composed of 100 parts by volume of oxygen and 200 of hydrogen he inferred that other gases might also combine in simple ratios. What he says seems to indicate that this inference was made in a context of discovery; prior to this the hypothesis had not been known by Gay-Lussac. The inference was not made solely from the observed fact about the composition of water, but from this in the light of theoretical and observational background information.

The theoretical aspect is indicated in the first two paragraphs of the paper. Gay-Lussac points out that in gases, by contrast to solids and liquids, the force of cohesion between molecules is slight, thus allowing regularities in expansion and contraction to exist in the case of gases by contrast to liquids and solids. Because of this theoretical assumption regarding the uniformity of gases the number of instances of gases combining in simple ratios by volume that Gay-Lussac needs in order to make a reasonable inference about all gases is considerably reduced. A number of other facts formed part of the background information and several of these are noted by Crosland in his article on Gay-Lussac's law. For example, the idea that combination by volume should be more regular than combination by weight was something Gay-Lussac inherited from Berthollet, whose assistant he was. Again, before the discovery of his law Gay-Lussac had been interested in the problem of the amount of acids and alkalis necessary to neutralize each other. Indeed, according to Crosland, among the most important experiments leading to the law were Gay-Lussac's experiments with fluoboric acid gas (boron trifluoride) and ammonia gas. If this is right then Gay-Lussac's reasoning, in the context of discovery, might better be described as involving an inference from simple combining ratios of hydrogen and oxygen and of fluoboric acid gas and ammonia gas to simple combining ratios of all gases, in the light of background information.

There are other facts Gay-Lussac cited in his paper that provided grounds for the law, but they did so more clearly in a context of justification; they are facts Gay-Lussac invoked after the law had occurred to him, which he appealed to in its support. For example, he writes: "We might even now conclude that gases combine with each other in very simple ratios; but I shall give some fresh proofs. According to the experiments of M. Berthollet, ammonia is composed of 100 of nitrogen, 300 of hydrogen, by volume. I have found that sulfuric acid is composed of 100 of sulphurous gas, 50 of oxygen gas." Gay-Lussac goes on to cite many other cases and concludes by saying: "Thus it appears evident to me that gases always combine in the simplest proportions when they act on one another; and we have seen in reality in all the preceding examples that the ratio of combination is 1 to 1, 1 to 2, or 1 to 3." Here, clearly, is reasoning based on the data Gay-Lussac had compiled, together with background information. When he was writing his paper it was evidently reasoning that occurred in a context of justification.

The question now is which modes of inference best describe Gay-Lussac's reasoning in the context of discovery and in the context of justification. What comes closest in both cases, I would claim, is the inductive mode. From the fact that hydrogen and oxygen combine in a simple ratio and that various acid gases do so as well when combined with ammonia, Gay-Lussac inferred, in the light of the background information, that it is plausible to think that all gases combine in simple ratios. The background information included the idea that gases, because of their molecular structure, should obey simple laws, and the idea, proposed by Berthollet, that combination by volume should be more regular than combination by weight. The first premise in an inductive argument would include the inductive evidence regarding the combination of various gases and background information of the sort just mentioned. The second premise would note any hypotheses that conflict with Gay-Lussac's law that he actually considered. Were there such hypotheses?

At the beginning of his paper Gay-Lussac cites Berthollet's hypothesis that compounds are formed in *variable* proportions. At the end of his paper Gay-Lussac agrees with Berthollet that in general this is so, but gases form a special case, and indeed there is reason to think that Berthollet was willing to treat gases as special. Accordingly, Gay-Lussac does consider a leading hypothesis of the day which might seem incompatible with his law but, he claims, is really not so. He does not, however, explicitly consider hypotheses that he recognizes to be inconsistent with his law. As Crosland emphasizes, one of Gay-Lussac's prime passions in life was to discover a law and he was very prone to generalize without careful consideration of alternatives. We can say that he inferred the plausibility of his law from the inductive evidence E and background information B, where he assumed that whatever competing hypotheses might be deemed reasonable to consider as alternatives are not in fact plausible, given E and B.

To really nail down the claim that Gay-Lussac was reasoning inductively —especially against the attacks of those who see all scientific reasoning as explanatory—I would have to consider the counterclaim that his reasoning better fits the explanatory or some other mode. I don't have time here to try to persuade you of this. Suffice it to say that I find no evidence that Gay-Lussac reasoned to his law on the ground that it provides explanations of why particular gases combine in simple ratios or in the particular ratios he cites. No doubt he believed that his law could be used in certain situations to provide such explanations, but there is no evidence that he reasoned to its plausibility on such grounds.

To classify Gay-Lussac's reasoning as inductive is in no way to minimize his achievement. Note that this is not what some philosophers call induction by simple enumeration. Gay-Lussac reasoned not simply from instances but from these in the light of a good deal of theoretical and experimental background information. Moreover, obtaining the experimental data in the first place and then detecting a regularity in his own experimental results as well as in those of others required considerable ability and imagination. Some people belittle inductive reasoning as trivial and uncharacteristic of sophisticated science, unlike explanatory reasoning. Coming to believe that all F's are G's may require thinking that is quite sophisticated, whether this involves considering the sorts of explanations a hypothesis and its competitors offer, or considering whether all observed F's are G's and the plausibility of competing hypotheses. Of course, once it has been established that all the F's observed have been G's and once it has been assumed that hypotheses incompatible with "All F's are G's" that are worthy of consideration are implausible, it does not require much intellectual effort to decide whether "All F's are G's" is a plausible hypothesis. But the analogous point holds true for explanatory reasoning. Once it has been established that H provides such and such an explanation of the facts, and once it has been assumed that competitors worthy of being considered are not as successful, it does not require much intellectual effort to decide whether H is plausible.

Peter Achinstein

Since I have mentioned explanatory reasoning, let me briefly cite one example, Avogadro's reasoning to his law that equal volumes of gases contain the same number of molecules. This is particularly interesting because Avogadro reasoned that his law is plausible on the ground that it affords a plausible explanation of why gases combine in simple ratios by volume, i.e., it affords a plausible explanation of Gay-Lussac's law. In his 1811 paper Avogadro begins with some simple assumptions of atomic theory. The fact that substances which combine to form compounds do so in fixed proportions by weight is explained by assuming that it is the molecules within the substances which combine and that the relative weights of the substances in the compound depend on the relative numbers of the molecules which combine to form that compound. In view of this fact, one reasonable explanation of why gases combine by volume in simple ratios is that equal volumes of gases contain equal numbers of molecules—the latter statement being Avogadro's law. We might begin to formulate Avogadro's reasoning, in conformity with the explanatory mode, as follows:

Premise 1: Gay-Lussac's law is accepted, as is much of atomic theory developed by Dalton.

Premise 2: Avogadro's law, when taken together with certain other assumptions from atomic theory, is capable of providing an answer to the question "Why does Gay-Lussac's law hold?" (The answer is given above.)

An explanatory inference will also include within the second premise mention of any alternative incompatible hypotheses that are being considered. Avogadro does consider a hypothesis that is the contradictory of his and would have to be implied by any competitor, viz. that the number of molecules contained in a given volume is different for different gases. He suggests that if this were so and we want to account for Gay-Lussac's law then we would need to invoke laws governing distances between molecules, something we are not required to do under the supposition of his law; moreover, these laws would be more complex than his law. So we can continue to formulate his reasoning as follows:

Premise 2 (cont.): Furthermore, if it is assumed that the number of molecules contained in a given volume is different for different gases then any answer to the question of why Gay-Lussac's law holds would need to invoke molecular laws governing distances between molecules, which we are not required to do if we assume Avogadro's law; these laws would be more complex than the latter.

Conclusion 1: It is plausible to suppose that Avogadro's law, when conjoined with certain assumptions from atomic theory, provides a good explanation of why Gay-Lussac's law holds. (Avogadro believed that this conclusion is warranted because he believed that his law supplies an answer that is good and indeed better than any that would be offered by competitors.)

Conclusion 2: Avogadro's law is plausible.

In what context did this explanatory reasoning take place? When Avogadro wrote his paper it occurred in a context of justification, since one of his aims in this paper was to defend his law. Avogadro's reasoning did not, however, occur originally in a context of discovery. The proposition that equal volumes of gases contain equal numbers of molecules had been considered earlier by Bernoulli and Dalton, and Avogadro was aware that Dalton, at least, had rejected the idea. Avogadro was aware of the proposition before he inferred that it is plausible.

So much by way of examples. Let me now try to bring together some of the main strands of this paper. My thesis is that the topic of inference to scientific laws is one to which both philosophers and historians can contribute to their mutual benefit. I considered and rejected the view, held by H-D theorists, that there is no such topic since there are no inferences to laws, only from them. I then examined the Hanson-Peirce view which champions the idea of inference to laws but which, I contend, oversimplifies its nature by suggesting that it is of one type, retroductive or explanatory, and also, I believe, oversimplifies the description of this type. I have been defending the idea that there are several modes of inference to laws that scientists actually employ; that these represent reasons scientists have for laws when they come to believe them as well as reasons scientists may offer for laws already inferred; that in explanatory reasoning one reasons not simply from observational data but from a theoretical background in addition, and sometimes from this alone; and that one considers as well alternative competing hypotheses, or at least assumes that these do not provide plausible explanations. I reject the idea that there is a logic of discovery as opposed to a logic of justification, and propose instead that we consider whether the reasoning of a scientist took place in a context of discovery, in a context of justification, or in neither context. The type of reasons the scientist has can be the same in all cases. If we can formulate various modes of inference—and I suggested how to begin at least to formulate two—we can ask whether, or to what extent, reasoning employed in particular instances by scientists conforms to these modes. And we can consider the context in which it took place. Doing these things should provide a better understanding of the origin of the law.

But who is to do what? How is the task to be divided? Those who don't want to draw a very sharp distinction between the historian and the philosopher of science may say that it doesn't matter. Others who see a sharper separation would urge the philosopher to formulate various patterns of inference, the historian to employ these in raising questions about particular cases, and the philosopher to utilize the results of the historian's exploration to modify and refine his formulations. At any rate, as I see it, there is need for contributions by both which are influenced by contributions by both.

COMMENT BY ARNOLD KOSLOW

Professor Achinstein, in his fine paper, tries to expand the horizons of much of current philosophy of science and to focus attention on certain new ways of understanding scientific behavior. His initial remarks indicate that the hypothetico-deductive account of science, as defended by Hempel for example, does not recognize that there are inferences *to* laws, only inferences *from* laws. However, when Hempel asserted that scientific hypotheses are not *derived* from observed facts, I think that what he intended was the rather simple, true statement that there is no deductive inference from singular statements to essentially general ones (modulo the general difficulty of distinguishing sharply between singular and general statements). The emphasis is upon "singular," since no one would deny that there could be inference to laws from other laws or from general theoretical statements. Although this observation is a relatively low-level logical one, it does have a point if we want to show what is wrong with the view that all laws are merely reports which summarize the results of experiments or observed facts. This simple logical point has not been challenged by any of Professor Achinstein's examples. The examples which are supposed to illustrate the inferences from singular statements S to a law L use statements such as "100 parts (by volume) of oxygen combine with 200 parts of hydrogen to form water," "100 parts of nitrogen combine with 300 parts of hydrogen to form ammonia," and these are not singular, but general statements.

Professor Achinstein claims that H-D accounts have failed to notice that there are many modes of inference to laws other than inductive and deductive ones. Before we consider one such candidate, the explanatory or retroductive mode of inference, it is perhaps worth asking whether Professor Achinstein has been fair to the literature. After all, in his sense of "inference" (making an inference seems to involve a process of reasoning in which something is concluded on the basis of something else), Carnap's principle of total evidence might qualify as a mode of inference which is neither inductive nor deductive.

Achinstein's explanatory mode of inference seems to be a two-stage affair. From evidence E ("obtained in the light of background information B"), together with the fact that of the class C of mutually incompatible hypotheses which are reasonable to consider, given E and B, hypothesis H_1 explains certain facts F better than any other hypothesis of that class, one concludes (according to this mode of inference) first that "H_1 offers a good explanation of F." A second conclusion is H_1 (or "H_1 is a plausible, reasonable hypothesis"). I have several questions about this schema.

I do not understand how this schema contributes to a solution of one of Hanson's problems. Hanson was concerned with explicating the predicate "H is reasonable to consider" or "R is a reason for considering, suggesting, or entertaining hypothesis H." But Professor Achinstein uses the expanded predicate "H is reasonable to consider, given evidence E and background information B," to determine the membership of the class C mentioned in our description of the schema of explanatory inference. Clearly then, the schema cannot yield an explication of the expanded Hanson predicate. Also, the second conclusion, that H_1 is reasonable, raises some minor problems. In the statement of his second premise, Achinstein grants that H_1 is a hypothesis which is reasonable to consider (given E and B); that is, H_1 belongs to the class C. If the second conclusion is that H_1 is reasonable to consider, then it is a correct but straightforward and unexciting conclusion from the second premise alone. If the conclusion is that H_1 is reasonable to believe, then the rule is an interesting rule for the acceptance of hypotheses and deserves closer examination.

The explanatory mode seems to depend upon the cogency of two conditions whose satisfaction is not at all obvious. The first (A) is that there is some way of ranking explanations of the same facts so that, given any two explanations, it is assumed that one of the two is the better explanation. In this way, saying H_1 provides an explanation of certain facts, which, given E and B, is better than that provided by competing hypotheses, makes sense. The second condition (B) which Achinstein's schema seems to appeal to is the possibility of competing explanations (of the same facts), where competition requires that the explanatory premises of the two competing explanations are incompatible.

It is certainly not obvious that any two explanations (of the same facts) can be ranked, and some of the difficulties connected with specific methods of ranking are by now notorious. For example, if particular explanations are ranked by the explanatory power of their premises, and this explanatory power in turn is ranked by comparing their associated sets of explanatory consequences (a method first suggested, I think, by Popper), it follows that not all explanations are comparable. The associated sets of explanatory consequences may only partially overlap, neither being included totally within the other.

The second point (B) is more serious. It is not obvious that there can be competing explanations based upon incompatible explanatory premises. On some accounts of scientific explanation, it is impossible to have such competition. If explanatory premises must be *true*, this point is obvious. If the premises must be *well confirmed* (by some agreed-upon body of evidence), there cannot be competing explanations. Further, if an explanation for (or by) a person O requires that O believe the premises, then here too, there cannot be competing explanations without requiring that people believe contradictory statements. It is therefore necessary either to give some analysis of explanation which allows for competing scientific explanations or to alter the schema of explanatory inference so that it refers to *potential* explanations rather than explanation simpliciter. The first limb of this disjunction is a task which confronts both philosophers and historians of science.[1] Historians of science quite often speak of competing

---

[1] The notion of competing explanations also raises problems for certain proposed theories of confirmation. B. Brody for example has recently argued ("Confirmation and Explanation," *Journal of Philosophy*, 65 (1968), 282–299) that certain adequacy conditions such as (1) if B is a logical consequence of A, then every statement which is evidence for B is evidence for A and (2) if B is a logical consequence of A, then every statement which is evidence for A is evidence for B, be replaced by (1') if A explains B, then every statement which is evidence for B is evidence for A, and (2') if A explains B, then every statement which is evidence for A is evidence for B. Conditions (1') and (2') have objectionable consequences: the explanans and explanandum must have exactly the same evidence. Further, if B and B* are any two explanations of, say, S, then B and B* must have exactly the same evidence. A third difficulty involves explanations which compete in that the explanans are incompatible. If H and H* are competing explanations of S, then every statement which is evidence for S is evidence both for H and H*. This is a consequence of (1') alone. If it is granted that any statement is evidence for itself, then the incompatible H and H* are evidence for each other. Finally, (1') and (2') together imply that there cannot be evidence which distinguishes between H and the incompatible H*—that is, evidence for H which is not evidence for H* or conversely. A host of unpalatable consequences seem generable in all their infinite variety.

explanations without revealing what concept of explanation underlies their description. As I have indicated above, some very standard accounts of explanation render the historian's description incoherent.

However, if the rule of explanatory inference refers to a class of relevant competing *potential* explanations, then it is unsatisfactory. For Achinstein's rule, so understood, states that if H_1 is the best of a class of relevant potential explanations of F (given evidence E and background information B), and evidence E is "obtained in the light of B," then H_1 offers a good explanation of F. But why should this be true? If we rank a number of relevant acts which are poor options, and A is the best of these, should we conclude that A is a good option? Should we conclude that H_1 offers a good explanation (or even a good potential explanation) of F, if it is only the best of a poor lot?

Professor Achinstein maintains, contrary to Hanson, that there is no separate logic of discovery as opposed to a logic of justification. There is no special Hanson predicate "R is a reason for considering, suggesting, or entertaining hypothesis H" which requires explication. Indeed, Professor Achinstein seems to identify that predicate with the more familiar one, "R is a reason for believing hypothesis H," when he states that ". . . any reason for suggesting a hypothesis can also be a reason for accepting it, and conversely." Let me suggest two counterexamples to the proposed identity. First, Einstein's special theory of relativity requires that all lawlike statements be Lorentz-invariant. That a statement is Lorentz-invariant is a reason for considering, suggesting, or entertaining it, but it is not a reason for believing it. Second, when Pauli delivered an advance report at Columbia University on, I believe, Heisenberg's axiomatic field theory, it was Bohr, so the story goes, who told Pauli that what everyone was looking for was a crazy theory, but the trouble was that Heisenberg's theory was not crazy enough. One might reconstruct this anecdote more academically by saying that "is sufficiently crazy" is a piquant way of referring to a specifiable feature which is a reason for considering, suggesting, or entertaining a theory, but not for believing it.

COMMENT BY PETER A. BOWMAN

Professor Achinstein claims that Hanson in his *Patterns of Discovery* does not do justice to the role of theory in retroductive inference, and Hanson would undoubtedly agree with him. For in a later treatment of the

Peter Achinstein

question "Is there a logic of scientific discovery?"[1] Hanson proposes the following revised characterization of such an inference:

(1) Some surprising, astonishing phenomena p_1, p_2, p_3 . . . are encountered.

(2) But p_1, p_2, p_3 . . . would not be surprising were a hypothesis of H's type to obtain. They would follow as a matter of course from something like H and would be explained by it.

(3) Therefore there is good reason for elaborating a hypothesis of the type of H; for proposing it as a possible hypothesis from whose assumption p_1, p_2, p_3 . . . might be explained.[2]

In a footnote to step (1), Hanson writes: "The astonishment may consist in the fact that p is at variance with accepted *theories*—for example, the discovery of discontinuous emission of radiation by hot black bodies, or the photoelectric effect, the Compton effect, and the continuous β-ray spectrum, or the orbital aberrations of Mercury, the refrangibility of white light, and the high velocities of Mars at 90 degrees. What is important here is *that* the phenomena are encountered as anomalous, not *why* they are so regarded."[3] Thus, it is not true for Hanson's later characterization that, as Professor Achinstein puts it, he "neglects the background of theory which the scientist often has to begin with . . ."

Moreover, in emending steps (2) and (3) to read "hypothesis of the type H" (rather than simply "hypothesis H"), Hanson is attending to the other aspect of theory which Professor Achinstein says he neglects; namely, "the background of theory . . . which may provide at least part of, and in some cases, the entire, basis for an inference to a law." But in the final analysis he leaves the crucial role of theory unexplicated, as I will now show. Hanson's historical example given along with the later characterization seems to suggest at least part of the "condition" (or "methodological rule" or "methodological demand") which the commentator on his paper, Paul Feyerabend, attributes to him and then shows to be problematic: "the explanans . . . must not be inconsistent with certain . . . theories . . . which are held either by the inventor, or by the (scientific) community in which he lives."[4] However, in his rejoinder Hanson denies "the very possibility of such a 'rule,' " saying, "at most, an argument making it plausible to explore one *kind* of hypothesis, rather than others, can be

---

[1] H. Feigl and G. Maxwell, eds., *Current Issues in the Philosophy of Science* (New York: Holt, Rinehart and Winston, 1961), pp. 20–35.
[2] *Ibid.*, p. 33.
[3] *Ibid.*
[4] *Ibid.*, p. 36.

entertained before experiment."[5] What role theory plays in this argument Hanson does not say.

Nonetheless, I can find no reason to suggest, as Professor Achinstein does on page 91, that Hanson would restrict this argument to what is characterized by the former as "the explanatory mode of inference." Admittedly, Hanson makes no general remarks to this point, but at one place in the paper under discussion[6] he shows that he wants to admit at least "analogical arguments, and those based on the recognition of formal symmetries."

REPLY BY PETER ACHINSTEIN

1. Koslow claims that when Hempel says (1) that scientists do not derive laws from observed facts, he means simply (2) that they do not (and cannot) make deductive inferences to laws, which are general statements, from singular observation statements. I agree that Hempel would certainly support claim (2), but it is by no means clear to me, as it seems to Koslow, that he equates (1) with (2). The fact that Hempel speaks of hypotheses as being freely invented, the fact that he rejects the idea of inductive rules of inference which can be used to get from data to a hypothesis, the fact that he conceives of the role of observation solely as one of testing hypotheses already invented, and the fact that he approvingly quotes Popper, who explicitly denies that scientists make inductive or deductive inferences to laws, suggest that for him the meaning of (1) is not exhausted by (2).

2. The second conclusion of my explanatory mode of inference is H_1 or that H_1 is a plausible hypothesis. Koslow suggests that this conclusion may be part of the second premise of this inference, which if so would trivialize the inference. Fortunately for me, premise 2 does not assert H_1 or that H_1 is a plausible hypothesis.

3. Koslow says that it is not obvious that any two explanations of the same facts can be ranked, and he indicates that if we compare explanations by comparing the "explanatory consequences" of hypotheses, then, since these may only partially overlap, the explanations will not be comparable. But why assume that this is the only or even the right way to compare explanations? Koslow seems to have in the back of his mind the deductive model of explanation and some formal or semiformal criteria for compar-

[5] *Ibid.*, pp. 40–41.
[6] *Ibid.*, pp. 26–27.

ing explanations. I have no such model in the back of my mind, though in the paper itself I do not discuss the nature of explanation or the manner of comparing hypotheses.[1]

4. Koslow contrasts explanations with potential explanations, where the former but not the latter are assumed to be either true or well confirmed. The hypotheses with which H_1 is compared would be offering potential explanations, to use Koslow's term. But, he objects, if H_1 is the best of the lot of potential explanations being considered it does not follow that H_1 offers a good explanation, since the best of a bad lot may not be good. I agree. However, this is not damaging to the explanatory mode of inference, since if we are to conclude that H_1 offers a good explanation it is required by this mode that "the answers supplied by H_1 reasonably satisfy criteria for good explanations," as well as that the answers supplied by H_1 satisfy these criteria on the whole better than competing hypotheses. The explanatory mode of inference does not allow us to infer that H_1 is a good explanation solely on the ground that it is better than the competitors under examination.

5. Koslow suggests two counterexamples to my anti-Hansonian thesis that any reason for suggesting a hypothesis can also be a reason for accepting it, and conversely: the fact that a statement is Lorentz-invariant can be a reason for suggesting it, but not for accepting it; the fact that a theory is "sufficiently crazy" can be a reason for suggesting it, but not for accepting it. I think my thesis does need some reworking. There are two senses of "can be reason": (a) can be a good reason that someone might have, (b) can be a reason, good or bad, that someone might have. In sense (b) my thesis seems true enough, though perhaps trivial. Any reason that someone might have for suggesting a hypothesis is a reason someone might have for accepting it. What about sense (a)? The valid point suggested by Koslow's examples seems to me to be that a reason for suggesting a hypothesis as plausible in the first place need not be as strong as one for accepting the hypothesis. R might be a good (enough) reason for suggesting H but not a good enough reason for accepting it. Accordingly, my general thesis might be expressed as follows: Any type of reason that can be a good reason for suggesting H in the first place can provide at least some reason, though not necessarily a conclusive one or one as good, in favor of accepting it. Now with reference to the types of reasons I discuss in the paper I

[1] But see my "Explanation," *American Philosophical Quarterly*, Monograph Series No. 3 (1969).

believe that a stronger thesis holds. Inductive and explanatory reasons can be good reasons for suggesting a hypothesis in the first place and equally good reasons for accepting a hypothesis already suggested. However, the stronger version does not hold generally. Koslow's second example does not satisfy the stronger version of my thesis though it does, I believe, satisfy the weaker one: If the fact that a statement is Lorentz-invariant is a good reason for suggesting it in the first place, then this fact provides at least some reason, though by no means a conclusive one, in favor of accepting it once it has been suggested. Koslow's second example, on the other hand, must be construed as hyperbole. If a theory really is crazy then the fact that it is provides no reason for suggesting it in the first place.

6. I have just one comment to make on Bowman's remarks. Bowman agrees with me that in *Patterns of Discovery* Hanson does not take adequate account of the role of theories in retroductive inferences. But, he claims, in a later paper Hanson rectifies this to some extent when, in connection with retroductive reasoning, he talks about (a) the observed phenomena as being at variance with accepted theories, and (b) a hypothesis of the *type* H rather than simply hypothesis H. I do not see how (b) brings in the notion of a theoretical background. Certainly (a) does, but it is only a theoretical background that is at variance with the phenomena observed, whereas in many cases the theoretical background, or at least a large part of it, that forms a basis for a retroductive inference is not at variance with the observed phenomena.

Science: Has Its Present Past a Future?

Before you study the history, study the historian . . . Before you study the historian, study his historical and social environment.

E. H. CARR, *What Is History?*

The fact is that the historian projects into history the interests and the scale of values of his own time . . .

ALEXANDRE KOYRE, *Scientific Change*

To judge from the pages of *Isis*, the atom has not yet been split. The *Annals of Science* apparently do not include secret research of any kind. Those advances in chemical and biological understanding which now allow plague and disease to be spread around the world at will find no place in the *Archives Internationales d'Histoire des Sciences*. *History of Science* (an annual avowedly devoted to the highlighting of "outstanding historical problems") displays no curiosity over the industrial and military pressures which have shaped and sustained the unparalleled scientific activity of the last three decades. The situation is no different if one turns to any of the dozen or so other journals now serving the history of science.

It is not only that the supremely important scientific revolution through which we have all lived seemingly invites no curiosity. *Perhaps more critical is the fact that historians of science have signally failed to make the pressures and perplexities of this revolution the springboard of their inquiries into other periods and problems.* As a newly established professional discipline, the history of science is undeniably born out of the tensions and aspirations engendered by modern science. How paradoxical then that such tensions and aspirations apparently hold no interest for the recently legitimized practitioners of this discipline. Yet it would be a mistake to

AUTHOR'S NOTE: I owe thanks to many colleagues for their generous response to, and thoughtful criticism of, earlier versions of this paper. Though my debts are far wider, I particular wish to acknowledge the help so freely afforded by Yehuda Elkana, Mary Hesse, Russell McCormmach, Bruce Mazlish, Everett Mendelsohn, Jerome Ravetz, and Charles Rosenberg. I am also grateful to the National Science Foundation for partial support of this work.

suppose that the present world does not deeply, and perhaps perversely, affect the historiographic assumptions of this new profession. After all, withdrawal is every bit as much a response as engagement.

The historiographic assumptions that have now characterized Western history of science for a generation are well known and easily accessible.[1] At the same time these assumptions are curiously devoid of historical analysis. To begin that analysis, and to explore the historical, sociological, and intellectual roots of prevailing presuppositions, is one purpose of this present essay. To consider newer and still neglected alternatives, another. Historians of science have so far seemed averse to any sustained debate over historiography and ideology. It is the writer's firm conviction that such debate, and the reordered priorities it will induce, are urgently needed.[2]

II

It is not necessary here to invoke the sacred names of Tannery, Duhem, and Sarton, still less of such "precursor" historians of science as Delambre, Whewell, and Kopp. Instead it seems more fruitful to concentrate on the 1930's. That was truly a golden decade for innovation and debate within the history of science, and the discipline still lives within its shadow. In the thirties previous analyses (often antiquarian in intent and amateur in ap-

---

[1] These assumptions are best displayed in the essays dating from 1943 to 1960, now collected in Alexandre Koyré, Metaphysics and Measurement: Essays in the Scientific Revolution (London: Chapman and Hall, 1968). See also such widely read texts as Herbert Butterfield, The Origins of Modern Science, 1300–1800 (London: Bell, 1957), Charles C. Gillispie, The Edge of Objectivity (Princeton, N.J.: Princeton University Press, 1960), and A. Rupert Hall, From Galileo to Newton, 1630–1720 (London: Collins, 1963).

[2] The early 1960's showed a still unfulfilled promise of critical thought on historiographic issues. At a 1961 symposium Henry Guerlac voiced a timely and unheeded warning of the distortions inherent in "the newer history of science with its strong flavour of idealism and super-rationalism." His remarks, and a spirited response by Koyré, appear in Scientific Change, ed. A. C. Crombie (London: Heinemann, 1963), pp. 797–812 and 847–857. A more extended defense of "the newer history of science" also appeared in 1963, in A. R. Hall's "Merton Revisited, or Science and Society in the Seventeenth Century," History of Science, vol. 2, pp. 1–16. The same year also saw publication of Joseph Agassi's supporting tract "Toward an Historiography of Science," History and Theory, supplement 2. Since that time there has been only limited sniping in the book review sections of learned journals. The solitary exception is the historiographic commentary embedded in T. S. Kuhn's survey of "The History of Science" in The International Encyclopedia of the Social Sciences, ed. David L. Sills (New York: Macmillan, 1968). Mention must also be made of A. R. Hall's "Can the History of Science Be History?" British Journal for the History of Science, 4 (1969), 207–220, which appeared as this present essay went to press. Hall's courteous and temperate response to this "friendly critic" well displays his own very different historiographic position.

proach) were replaced by what may now be seen as three types of proto-professionalism.[3]

The first of these was Marxian in form. In an age of economic depression and looming fascism it is not surprising that Marxian analyses held the center of the stage. Equally reasonably, England was the locus of both the major discussions and the major historical concern. Professor B. Hessen's classic 1931 paper on "The Social and Economic Roots of Newton's *Principia*" was read in London at the Second International Congress of the History of Science. The paper brilliantly exemplified the way in which economic determinism could contribute to rewriting the history of science. Spurred both by this example and by contemporary events, a loosely linked group of English scientists set to work to write those Marxist histories that would illuminate the previous growth and present condition of the scientific estate. J. D. Bernal, J. G. Crowther, Lancelot Hogben, and Joseph Needham come immediately to mind as members of this group.[4] Lacking younger disciples, for reasons we shall shortly explore, this English coterie has yet remained highly productive, if remarkably captive to its youthful ideas and pointedly ignored by present-day professional historians of science.

The basic position of the Marxist writers may be simply stated. In Hessen's words: "The method of production of material existence conditions the social, political and intellectual process of the life of society." Interpreted in a weak way, such a statement would be hard to quarrel with. But the Marxist interpretation was anything but weak. Hessen's own essay abounds with assertions like "The struggle of the university, and non-university science serving the needs of the rising bourgeoisie, was a reflection in the ideological realm of the class struggle between the bourgeoisie and feudalism." Or again, "Science flourished step by step with the devel-

---

[3] As Agassi neglects to his cost, and Kuhn is careful to stress, the writing of histories of particular sciences has an exceedingly long and involved history. No adequate account of this history exists: the best introduction is Kuhn's "History" article. Here it is unnecessary to reach back beyond the 1930's, in which "amateur" activity reached a new peak and the embryonic professional discipline received enduring prenatal impressions.

[4] Other associates included C. H. Waddington, J. B. S. Haldane, the scientist manqué C. P. Snow, the anthropologist V. Gordon Childe, and the classicist Benjamin Farrington. Neither a more complete enumeration of the group nor a listing of their works is possible here. No more is it possible to discuss the extent of, and shifts in, the Marxism of the various members. For this, the analytical study now being conducted by Mr. Gary Werskey of Harvard and Edinburgh universities must be awaited. However, some insight into the period, and the influence of Hessen, may be gained from such works as J. G. Crowther, *British Scientists of the Nineteenth Century* (London: K. Paul, Trench, Trubner, 1935).

opment and flourishing of the bourgeoisie. In order to develop its industry, the bourgeoisie needed science . . ."[5] Such simplistic sloganeering weakens the writing of this whole school, including Bernal's otherwise impressive *Science in History*. Published almost a quarter century after Hessen, *Science in History* represents a last, late flowering of the Marxist tradition. It bravely insists that "Greek science reflects the rise and decline of the money-dominated, slave-owning iron age society. The long interval of the Middle Ages marked the growth and instability of feudal subsistence economy with little use for science. It was not until the bonds of feudal order were broken by the rise of the bourgeoisie that science could advance. . . . The phases of the evolution of modern science mark the successive crises of capitalist economy,"[6] etc. Sufficient to say that by the time Bernal's work was published the relevance of such analyses had long been overtaken by changes in the political form and climate of the Western world.

The second type of protoprofessionalism was both more complex and less influential. Its *locus classicus* is R. K. Merton's 1938 monograph on "Science, Technology and Society in Seventeenth-Century England." Though indebted to Marxist canons and concerns in many ways, this monograph was quite separate in methodology and intent. An extensive historicosociological investigation, it stands unparalleled and unpursued to this present day. The reasons for this neglect have been partially analyzed by Professor Hall. His judgment that Merton's work "represented the culmination of an established tradition, not the beginning of a new one,"[7] is not the least provocative of his many sallies. Even so one must admit that few historians of science have assayed the sociological inquiries that Merton envisaged. Weber's "Protestant ethic," Hessen's economic determinism, and the sociologist's statistical method have as a combination proved singularly unappealing to the new profession, while sociology itself has been passing through a prolonged antihistorical phase.

[5] Quoted from p. 170 of *Science at the Cross Roads* (London, 1931?). This fascinating period piece, which owns neither editor nor publication date, contains 11 papers presented by the Soviet delegation to the Second International Congress of the History of Science and Technology, held in London in 1931. On Hessen, see the remarks in David Joravsky, *Soviet Marxism and Natural Science, 1917–1932* (New York: Columbia University Press, 1961), *passim*.

[6] John D. Bernal, *Science in History* (London: Watts, 1954), p. xi. The revisions Bernal has made in successive editions of this work would constitute an interesting study.

[7] "Merton Revisited," p. 1. Merton's study was published as pp. 360–632 of volume 4 of *Osiris*. Its long-overdue separate publication is now announced.

Arnold Thackray

The third influence of the thirties came out of a particular tradition in the history of philosophy, in its Gallic form. Distinct from Marxism and Merton as much in its fate as its approach, the work of Alexandre Koyré has captured the imagination of Western historians of science. If this assertion requires demonstration, it is only necessary to point to Professor Cohen's discussion of "the magistral influence" of Koyré, or to the remarks of Professors Clagett and Westfall. Clagett sees Koyré as having an "extraordinary influence on a generation of American scholars." In addition he stresses that "our students . . . were urged to take Koyré's studies as models," and notes "how dependent on his friendship and scholarship American historians of science had become [by the late 1950's]," and how they "sought him out repeatedly for advice and conversation." Going even further, Westfall asserts that "no single work has done more to shape the history of science as it is now practiced than Koyré's *Etudes galiléennes.*" Indeed Westfall himself admits being "molded by them to the extent that I cannot see them objectively."[8]

Koyré and his followers have sought to stress at the same time both the autonomy of, and the conceptual shifts in, the developing pattern of scientific thought. Their conviction has been that (to quote Professor Hall) "the intellectual change is one whose explanation must be sought in the history of the intellect." Why this position should have proved so enticing invites investigation.

As spokesman for the group, Hall attributes the appeal of Koyré to the way he made "peculiarly his own" the "analysis of the scientific revolution as a phenomenon of intellectual history." Even so, Hall feels constrained to add that "other factors" aided this appeal.[9] Since it is not self-evident that Koyré's stress on the intellectual history of science surpasses in cogency that of such earlier writers in the same tradition as Lovejoy, Metzger, Dampier-Whetham, Burtt, or Meyerson,[10] to name only the most obvious, it may be rewarding to pause and explicitly discuss those "other factors" which have contributed to Koyré's wide and enduring influence.

---

[8] For the remarks of Marshall Clagett and I. Bernard Cohen see *Isis*, 57 (1966), 157–166. In saying of the *Etudes Galiléennes* that "more than any other work it has been responsible for the new history of science," Cohen both anticipated Westfall and illustrated the remarkable degree of unanimity among American historians of science. For R. S. Westfall's remarks, see *Science*, 162 (1968), 553.

[9] "Merton Revisited," pp. 10, 11.

[10] See, for example, Arthur O. Lovejoy, *The Great Chain of Being: A Study of the History of an Idea* (Cambridge, Mass.: Harvard University Press, 1936), Hélène Metzger, *Newton, Stahl, Boerhaave et la Doctrine Chimique* (Paris: F. Alcan, 1930), William C. D. Dampier-Whetham, *A History of Science and Its Relations with Philosophy*

III

The history of science emerged as a professional discipline in the Western world in the 1950's, primarily in the United States. North America could boast perhaps five professional historians of science in 1950, twenty-five in 1960, and probably a hundred and twenty-five by the time this article appears in print.[11] The fifties was thus the crucial decade for defining standards, agreeing on methods, enrolling students, and creating a discipline. It was also the decade of the H bomb, the Cold War, Senator Joseph McCarthy, loyalty oaths, militant anticommunism, and the "silent generation" of students. There were therefore unusually complex political, ideological, social, and professional factors at work in the shaping of this new discipline.

The emerging profession of history of science had obvious and urgent need of effective tools, and of analytical methods possessing demonstrated worth. While Merton's work suggested one approach, it lay in no clearly developed tradition of historical writing about science. In contrast, the English Marxists did offer an articulated body of knowledge, but one that had failed to generate a body of contributions of comparable quality in the more pragmatic climate of America, even in the 1930's. Most of the English writers had also failed to refine, broaden, and develop their methods with the passage of time. The notable exception of Joseph Needham illustrates how content the majority were with traditional themes, an unsubtle use of Marxist ideology, and broadly popular rather than serious scholarly endeavor. Well attuned to the mood of the thirties, their vulgarizing work was poor preparation for the postwar world and the debut of a new profession. What their writing did generate was a stage setting for Koyré in the careful and restrained critique of the historian G. N. Clark, and the important original researches of his student Rupert Hall.[12]

and Religion (Cambridge: Cambridge University Press, 1929), Edwin A. Burtt, *The Metaphysical Foundations of Modern Physical Science: A Historical and Critical Essay* (New York: Harcourt, Brace, 1925), Emile Meyerson, *Identity and Reality*, trans. Kate Loewenberg (London: G. Allen and Unwin, 1930; original French edition, Paris, 1908).

[11] These figures are impressionistic estimates, but derive some support from D. J. de Solla Price, "Who's Who in the History of Science," *Technology and Society*, 5 (1969), 52–55, and "A Guide to Graduate Study and Research in the History of Science and Medicine," *Isis*, 58 (1967), 385–395. Throughout this present article, developments within the English-speaking world have been run together. A more careful differentiation of parallel and conflicting currents in Britain and the United States is obviously needed, but scarcely profitable within the limits of this paper.

[12] See G. N. Clark, *Science and Social Welfare in the Age of Newton* (Oxford: Clarendon Press, 1937), and A. Rupert Hall, *Ballistics in the Seventeenth Century* (Cam-

Arnold Thackray

Within a few years the movement to professionalism was dominant, and its favored means discovered. The sophisticated style, steadily widening output, impeccable scholarly credentials, effortless command of the textual sources, and uncompromising idealism of Koyré all carried exciting implications for the emerging profession. His work thus served as a more than welcome model around which to organize the history of science. Unashamed Marxism, the only viable alternative analytic tool, labored under all too obvious handicaps in both England and America.

More subtle factors were also important. A social history of science, whether patterned on Marx or Merton, might well demand primarily *historical* training of its professional practitioners, and see its role as contributing to historical debate. But a *discipline* is created and defined by exclusion and special expertise, however rhetorically desirable cultural bridges and a widened scholarly embrace may seem. The history of science has drawn its first generation of both professional practitioners and graduate students almost exclusively from the sciences, rather than from history. It would be interesting to explore the reasons for this, but here it is the fact itself that matters.

Ex-scientists could scarcely claim that their particular expertise (their *discipline*) lay in some special grasp of historical method, or some broader understanding of the historical context that escaped the history Ph.D. Matters were obviously different when it came to their ability to comprehend and critically evaluate the *scientific concepts* of past science. Here a scientific training was of obvious relevance. Who better to understand the physics of the seventeenth or the chemistry of the eighteenth century than the ex-physicist or chemist? Who better also to stress theoretical and idealistic aspects of the history of science than the former natural scientist, the new and zealous convert to conceptual studies. That the history of science was often taught to potential, and read by actual, scientists, merely served to reinforce already powerful trends.[18] The accident of timing which made

bridge: Cambridge University Press, 1952). Hall's book was the outgrowth of a 1949 Cambridge Ph.D. thesis. While Needham's work extended the Marxist analysis in one direction, highly original and important, though unfairly neglected, studies in another direction were undertaken in the United States by Edgar Zilsel: see, for example, his "The Genesis of the Concept of Scientific Progress," *Journal of the History of Ideas,* 6 (1945), 325–349.

[18] Revealing comments on the teaching situation and the current operating philosophies of historians of science may be found in the papers and discussions of section 8 of *Critical Problems in the History of Science,* ed. Marshall Clagett (Madison: University of Wisconsin Press, 1959) and section 26 of *Scientific Change,* ed. A. C. Crombie (London: Heinemann, 1963). See also the articles of A. C. Crombie and G. Buchdahl

the 1950's the decade of professionalization thus happily coincided with the widespread intellectual and social reaction against all things Marxian, and the growing awareness of the alternative explanatory model offered by Koyré.

Perhaps too, the "loss of innocence of science" played a part. After Hiroshima, the technological consequences of "pure" scientific research no longer seemed such unmixed blessings. The triumphant progress of Western science was no longer quite so triumphant and perhaps—disturbing thought—not even progress. One natural consequence was a renewal of interest in past and golden days, days in which science was no more (or less) than pure and autonomous thought, unstained by the pressures of technology and politics. How soothing too to hear that "the science of our epoch, like that of the Greeks, is essentially *theoria*, a search for the truth . . . an inherent and autonomous . . . development."[14]

Even Professor Hall would seem implicitly to admit the validity of this analysis. Though he argues that "the intellectual change is one whose explanation must be sought in the history of the intellect" when scientific research is at issue, this criterion apparently does not apply to the discipline of Clio. Discussing the current vogue for intellectual rather than social approaches to the history of science, he notes that "the historical evolution of this situation is of historical significance too." Does he therefore seek the explanation of this intellectual change "in the history of the intellect"? Not at all. Instead Hall argues that social approaches are at a discount because of "a certain revulsion from the treatment of scientists as puppets," a situation in which we are all "guiltily involved," and one we cannot review "without passion."[15] That an internalist should seek the explanation of the evolution of historical thought in such social and psychological terms, while vigorously insisting on intellectual autonomy for scientific ideas alone, is at least worth noting.

The foregoing analysis is not in any way intended to deny the obvious quality and intellectual power of the work associated with Professors Clag-

in *History of Science*, 1 (1962), 57–66. The alliance formed between the history and the philosophy of science, each small discipline seeking to professionalize and enlarge, also worked to aid the identification of *science* with *scientific thought*, so common throughout this period. This whole development deserves more detailed treatment, but here can only be summarily noted—though see the exchange between R. H. Shryock and H. Dingle in *Proceedings of the American Philosophical Society*, 99 (1955), 327–354, and H. Dingle, "History of Science and the Sociology of Science," *Scientific Monthly*, 82 (1956), 107–111.

[14] Koyré in *Scientific Change*, p. 856.
[15] "Merton Revisited," p. 15.

ett, Cohen, Crombie, Gillispie, Guerlac, Hall, Kuhn, Westfall, and their many students. They have not only created a discipline, but added immeasurably to our understanding of the science of the past. The prevailing standard of work in the history of science is indeed impressive. To realize this, it is only necessary to compare any recent issue of *Isis* with one from three decades back, or to contemplate the ever-escalating sophistication of research on Isaac Newton, over the last quarter century. The very necessary insistence on the intellectual validity *in its own terms* of previous scientific thought has been salutary. But it is not sufficient. As any analysis of recent research, any sustained conversation with younger scholars, any serious inquiry among present graduate students will reveal, the discipline of the history of science is now seriously distorted, and out of temper with the times.

The atomic and hydrogen bombs have long since lost their power to shock. For those who have come of age since 1960, even antiballistic missiles, chemical defoliants, and riot-control gases seem merely part of the accepted order, as inescapable as rain or sun. And from the computer, through the moon rocket, to the Xerox machine, the more peaceful technological abilities of modern science are everywhere displayed, and preach their silent sermon. Concepts such as "the free world" and "the Communist bloc" are now revealed for the simplistic slogans they always were, though peace on earth does not appear to be the inevitable outcome. Joseph McCarthy has been replaced by Eugene, who has vanished in his turn. Indicative of the shifting mood is the way that societies for social responsibility in science are now in vogue.[16] It is against this background that the future of the history of science invites fresh thought.

IV

Historians of science are fortunately peculiarly liberated from those degenerate scientisms which still dog some other areas of history. Prolonged exposure to the fads and fashions of past scientific thought has taught them that inductivism and positivism are at best philosophic and methodological prescriptions, not unyielding statements about the only way that truth may be attained. The desire to "tell it like it is" (or "wie es eigentlich

---

[16] Or consider the way responsible scientists are themselves calling for a new image of science: e.g., "Science can no longer be content to present itself as an activity independent of the rest of society, governed by its own rules and directed by the inner dynamic of its own processes." Robert S. Morison, "Science and Social Attitudes," *Science*, 165 (1969), 156.

gewesen," if you will) thus holds little lure. Lord Acton's vision of "ultimate history" has correspondingly small appeal. It is characteristic that in the recent and contrasting accounts of the nature of history given by two distinguished Cambridge historians, it should be E. H. Carr, the relativist, who draws heavily on modern work in the history and philosophy of science to support his position, and Geoffrey Elton, the old-style inductivist, who shows the traditional historian's disdain for this new discipline.[17]

If the historian of science knows too much about the relativity of scientific thought to be lured by any quasi-absolutist view of history, itself deriving from outmoded ideas about science, where may he turn instead? The intellectually, emotionally, and professionally satisfying response of the fifties and sixties lay in "the method of conceptual analysis, based on the model set before us by Koyré." Obviously one might continue and extend this approach. Koyré's own researches (and those of many of his disciples) focus exclusively on the period before 1700—and it is not, after all, self-evident that scientific change ended there. However, Koyré's method, with its stress on the interrelations of scientific, philosophical, and theological ideas, its demand for the complete mastery of a body of textual material, and its inevitable focus on the thoughts of great men, was perhaps mainly suited to the embryonic science of the sixteenth and seventeenth centuries. The few self-conscious attempts to extend its coverage into the nineteenth century have not been remarkable for their success.[18]

Koyré himself freely admitted that "history is always being renewed," and that "nothing changes more often and more quickly than the immutable past." He also acknowledged that his own stress on Platonism and idealism was "nothing else than a reaction against the attempts to interpret . . . modern science . . . as a promotion of arts and crafts, as an extension of technology, as an *ancilla praxi*."[19] It would therefore seem that to continue wholly preoccupied with conceptual analysis (or to return to Marxism, its polar opposite) would be to project into the seventies an approach more suited to the particular tensions of the 1930's. It is of course possible to argue that this approach is *the* method for the history of sci-

---

[17] See Edward H. Carr, *What Is History?* (London: Macmillan, 1961), and G. R. Elton, *The Practice of History* (London: Methuen, 1967), *passim*.

[18] The most interesting attempt is undoubtedly C. C. Gillispie's "Elements of Physical Idealism," in *Mélanges Alexandre Koyré* (Paris: Hermann, 1964), II, 206–224. On the proper method of approaching more recent science, see also the exchange between Professors T. S. Kuhn and L. P. Williams in *British Journal for the Philosophy of Science*, 18 (1967), 148–161.

[19] Quoted from *Scientific Change*, p. 852.

ence, and thus has a unique and timeless validity. But as the quotation above illustrates, Koyré himself would have had little patience with such an argument, whatever some more committed practitioners may feel.

What then are the possible alternatives? Admitting that conjecture and refutation lie at the base of all history, and that the discipline is (among other things) fundamentally committed to the search for a believable future, the present author wishes to appeal to the twin virtues of nonconformity and catholicity. The nonconformity lies in a very modest suggestion. It is that *concern about the future which reflects the present pressures and problems in science must be the basis for a far greater fraction of the historical enterprise.* We must face and accept the challenges posed by the multivarious tensions and aspirations which modern science has loosed into our world, and make these tensions the driving force of creative new research. Catholicity is also necessary, because as the tensions and aspirations are many-faceted, so must be the resulting history. This catholicity will embrace a variety of techniques, techniques that neither require nor allow of exhaustive enumeration here. Even so, four well-tried approaches from other areas of history would seem particularly relevant in the search for a comprehensible and usable past.

V

Perhaps above all an approach that embraces the sociology of knowledge is needed. A major weakness of the internalist position lies in its assumption that ideas can meaningfully be divorced from institutions. This is not to suggest that institutions make ideas, any more than incubators create eggs. However, the institutional context is often crucial in determining which ideas are adopted and flourish, and thus in creating regional and national styles in science. Dr. P. M. Rattansi's study of the physician-apothecary clash in Restoration England is one obvious example. Another is available in the writer's work on the political, and Professor Manuel's on the social, context in which the Royal Society became a vehicle for the creation of a British Newtonian tradition. Professor Kuhn's *Structure of Scientific Revolutions* suggests a variety of other ways in which the institutional background may be of the greatest significance, as do the essays of Joseph Ben-David. More generally still, the writings of Karl Mannheim explore the importance of a historical sociology of knowledge for any full understanding of our intellectual heritage. The work of Professor Namier suggests rather different ways to penetrate behind the outer label to the inner reality.[20] His techniques are now so well assimilated by the political historians that none would use the terms "Whig" and "Tory" as sufficient explanatory categories in themselves. Yet the denotation "Cartesian," "Newtonian," "Positivist," or "Romantic" is still too often made the complete explanation of a scientific group and its activities. The discussion of the social, political, and moral pressures covered by such labels has yet to begin. In this context it is revealing to note that Namier's archfoe, while unable to vanquish him on their common ground, yet enjoys a major reputation among historians of science for an internalist study of the evolution of scientific thought.[21]

If the sociology of knowledge is one area that the history of science has much both to learn from and to contribute to, role theory is another. From Znaniecki's pioneering study of *The Social Role of the Man of Knowledge* on, there is an extensive and important literature available. Though its insights have so far been ignored by historians of science, this literature has been ably exploited by historians of supposedly more traditional bent.[22]

The creation of modern science may be viewed in many complementary ways. One may if one wishes focus on a seventeenth-century conceptual revolution, or on the associated appearance of wholly new institutional forms. One might also usefully consider the emergence of, and shifts in, the social role of the man of science. The differences in ideology and social function between, say, Paracelsus, Robert Boyle, John Dalton, and R. B. Woodward are profound. Their investigation would illuminate some of the most critical phases in the growth of modern science. Again we are

[20] See, for example, P. M. Rattansi, "The Helmontian-Galenist Controversy in Restoration England," *Ambix*, 12 (1964), 1–23; A. Thackray, " 'The Business of Experimental Philosophy'—The Early Newtonian Group at the Royal Society," *Actes du XIIᵉ Congres International d'Histoire des Sciences* (Paris, in press); Frank Manuel, *A Portrait of Isaac Newton* (Cambridge, Mass.: Harvard University Press, 1968), chapter 13; J. Ben-David, "Scientific Productivity and Academic Organization in Nineteenth-Century Medicine," *American Sociological Review*, 25 (1960), 828–843; Karl Mannheim, *Ideology and Utopia: An Introduction to the Sociology of Knowledge*, trans. Louis Wirth and Edward Shils (New York: Harcourt, Brace, 1936); Lewis B. Namier, *The Structure of Politics at the Accession of George III* (London: Macmillan, 1929); A. Thackray, *Atoms and Powers: An Essay on Newtonian Matter-Theory and the Development of Chemistry* (Cambridge, Mass.: Harvard University Press, 1970).

[21] See Herbert Butterfield, *George III and the Historians* (London: Collins, 1957), and J. M. Price, "Sir Lewis Namier and His Critics," *Journal of British Studies*, 1 (1961), 71–93.

[22] Florian Znaniecki, *The Social Role of the Man of Knowledge* (New York: Columbia University Press, 1940); and, for example, Thomas C. Cochran, *Railroad Leaders, 1840–1890: The Business Mind in Action* (Cambridge, Mass.: Harvard University Press, 1953). See also Michael P. Banton, *Roles: An Introduction to the Study of Social Relations* (London: Tavistock, 1965).

confronted with a paradox. One of the most obvious and central themes in the history of science is the changing social role of the practitioner. From virtuoso to natural philosopher to scientist to, say, physical biochemist, these changes mirror profound shifts in the organization, content, and social function of scientific knowledge. Yet there is little literature that deals with the history of *science as a profession*.[23] Merton's pioneer work aside, almost nothing is known about the recruitment of men of science, the effect of the "Ph.D. machine," and the emergence and differential sociology of the various scientific disciplines. It is of course possible to argue that the evolution of, say, biochemical thought owes nothing to the competition and interaction between physicians, apothecaries, agricultural chemists, and "pure" research scientists. Possible but, one hopes, increasingly difficult.

A third neglected approach of crucial importance is through the interaction of material culture and intellectual forms—here, more specifically, of technology and science. The Marxist view of this interaction is encapsulated in Hessen's remark (quoting Engels) that "when after the dark night of the middle ages science again began to develop at a marvellous speed, industry was responsible." Not surprisingly such a dogmatic and extreme statement called forth its polar opposite. We have already noted Koyré's insistence that science "is essentially *theoria*, a search for the truth" and that it has "an inherent and autonomous development." It is time to move away from such barren antitheses. The division between "internal" and "external" history may accurately reflect much current work. However, to seek for "a demarcation of their respective fields of application with some degree of accuracy," as Professor Hall has urged,[24] would be to elevate a polemical division to the status of a methodological principle.

It would seem more fruitful to search for that *via media* which avoids the extreme formulations of either side. That material and intellectual cul-

---

[23] Though see E. Mendelsohn, "The Emergence of Science as a Profession in Nineteenth Century Europe," in *The Management of Scientists*, ed. Karl Hill (Boston: Beacon, 1964), pp. 3–48; J. Ben-David, "The Scientific Role: The Conditions of Its Establishment in Europe," *Minerva*, 4 (1965), 15–54, and "Social Factors in the Origins of a New Science: The Case of Psychology," *American Sociological Review*, 31 (1966), 451–465. Also useful to, but usually neglected by, the historian of science are the directly sociological discussions of modern science as a profession: e.g., Warren O. Hagstrom, *The Scientific Community* (New York: Basic Books, 1965), and Norman Storer, *The Social System of Science* (New York: Holt, Rinehart and Winston, 1966).

[24] "Merton Revisited," p. 15.

ture profoundly influence each other would seem a truism to the anthropologist. It is thus ironic that historians of man's greatest intellectual adventure so continually ignore this obvious theme. Studies of the first industrial revolution—I mean that of the eighteenth not the sixteenth century—offer a particularly apposite field. Between 1760 and 1880 the industrial bases of the Western world were transformed. The same period witnessed a clearly related event in the replacement of natural philosophy by positivistic science. Yet studies of the manifold and subtle interconnections of these two fundamental changes in Western culture have scarcely begun. Professor Guerlac's perceptive study of eighteenth-century French chemical technology reveals one possible approach. That it was promptly dubbed "un peu Marxiste" again reveals those fruitless antagonisms we must now renounce.[25]

My fourth field is the most irony-laden of all. If contemporary historians of science are agreed on one thing, it is that the success of science, its very intellectual power, its cutting *Edge of Objectivity* lies in the ability to quantify. One might naively suppose that those recruited from the ranks of science would, of all historians, prove the most eager to employ this demonstrated weapon. The reality is far otherwise. The impact of railroads on American history has been subject to masterly quantification. Every accepted and hallowed judgment of political history is now at the mercy of the computing behaviorist historians. Social mobility, family structure, the rise of the gentry: all are now moving out of the realm of easy and unsupported generalization into that of exact measurement.[26] The same cannot be said of any area within the history of science.

Quantification itself is not the *goal* of history, any more than of science. But in both cases it does allow a finer grasp, an easier handling, a more meaningful discussion of what can be known. The whole subject of quantification is now treated with considerable sophistication by general historians, as Professor Aydelotte's essay reveals.[27] Yet the matter is rarely

[25] See H. E. Guerlac, "Some French Antecedents of the Chemical Revolution," *Chymia*, 5 (1958), 73–112; and *Scientific Change*, p. 810.

[26] See, for example, R. W. Fogel, *Railroads and American Economic Growth: Essays in Econometric History* (Baltimore: Johns Hopkins Press, 1964); S. P. Hays, "The Social Analysis of American Political History," *Political Science Quarterly*, 80 (1965), 373–394; Stephan Thernstrom, *Poverty and Progress: Social Mobility in a Nineteenth Century City* (Cambridge, Mass.: Harvard University Press, 1964); *An Introduction to English Historical Demography from the Sixteenth to the Nineteenth Century*, ed. Edward A. Wrigley (London: Weidenfeld and Nicolson, 1966); Lawrence Stone, *The Crisis of the Aristocracy* (Oxford: Clarendon Press, 1965).

[27] "Quantification in History," *American Historical Review*, 71 (1966), 803–825.

broached by historians of science. Professor Price's somewhat differently conceived studies are the solitary exception that proves the rule. If we knew how many and who were the scientists of seventeenth-century England or nineteenth-century Germany, our discussions would be more informed. Education, religious affiliation, social class, regional location, occupation, and research interests are all susceptible of quantification and, with the aid of the digital computer, comparison and correlation. Such information would not of itself constitute the history of science. It would, however, remove much of the mist and murk that obscures current discussion. For example, we would be able accurately to specify the importance of Protestant Dissent and manufacturing interests among eighteenth-century English natural philosophers. Are Joseph Priestley and the Lunar Society really characteristic, or merely glamorous exceptions that obscure the rule? To answer this question would not explain the course of late-eighteenth-century science, but it would go far toward defining what any adequate explanation must encompass. At present, for want of hard statistics, we wallow in an impressionistic sea of unsupported generalizations.

VI

Enough of these laments. The argument is almost at an end. Lest it be misunderstood, the writer must again make clear that he values and is himself fundamentally indebted to that "method of conceptual analysis" advocated by Koyré. The untangling of past ideas will always be an important part of the task of the historian of science. The thesis of this paper is not that such activity is worthless. The point is rather that its present dominance is a reflection of past ideological positions no longer relevant, that it too narrowly restricts the range of discussion, and that it commands an unhealthy hegemony over the field. The urgent need is for a far greater diversity in methods of attack and forms of inquiry. This need parallels that for a greater concern with the major issues of today. War and peace, the use and misuse of research, the interrelations of science and technology, the effects of secrecy and military objectives, the reciprocal responsibilities of the scientist and society, the funding and control of science, problems of morale and status within the scientific enterprise: these must be the motivating pressures behind new historical research, aimed as much at the seventeenth and nineteenth as at the twentieth century.

The result of altered perspectives and priorities may not be that "history of systematized positive knowledge" for which Sarton yearned, nor that rethinking past thoughts of great men so popular at present. But in a world where the past, like so much else, is not what it used to be, fresh approaches may provide a firmer base from which to contemplate the present, and debate alternative futures. This search for a projectable past would now seem incumbent on the historian of science. To refuse the call and continue convinced that familiar problems and by now traditional methods are sufficient would be to condemn the discipline to the status of an esoteric luxury. Such a luxury would of course continue to be savored behind closed doors by the select few, while outside science and Western society continue what may yet prove the dance of death. The historian may rightly reject the call to direct social involvement. Equally, he should resist the lure of a past attractive only by its denial of present problems.

In conclusion, it may be appropriate to repeat Koyré's remark that "nothing changes more often and more quickly than the immutable past." The question before the historian of science is in what ways, and why, he can best aid that unceasing change.

COMMENT BY LAURENS LAUDAN

And, although in the hundred and twenty years or so, during which this ambition to imitate Science in its methods rather than its spirit has now dominated social studies, it has contributed scarcely anything to our understanding of social phenomena, not only does it continue to confuse and discredit the work of the social disciplines, but demands for further attempts in this direction are still presented to us as the latest revolutionary innovations which, if adopted, will secure rapid undreamed of progress.

HAYEK, *The Counter-Revolution of Science*

I am not quite sure how best to formulate my reactions to Professor Thackray's interesting and provocative paper. In substantial agreement with many of the specific proposals he makes, I am nonetheless rather uneasy about the general direction in which he takes the argument. Indeed, our differences at the general level are as great as our agreements over points of detail. By and large, we tend to agree at the level of truisms. Thackray wants pluralism and diversity in the writing of the history of science; so do I. He thinks it misleading to view the history of science entirely as the evolution of disembodied ideas. Again, I suspect we all nod our heads in enthusiastic support. He thinks historians of science should know something about how to handle statistical data. I believe he would find few living historians of science to play the role of *advocatus diaboli* against him. He insists that the historian projects his own interests into history. That point has surely been commonplace for every historian since the

Arnold Thackray

turn-of-the-century debates between Meyer, Weber, and Rickert![1] Like Agassi and Kuhn before him, Thackray, in parts of his paper, has chosen to do battle against an already vanquished enemy,[2] and as a result he often seems (mixing the metaphor) to be preaching to the converted. But Thackray's paper does not consist entirely of innocuous reflections on the historian's craft. On the contrary, there are two general theses in his paper which are far from tautological, and it is those which I should like to discuss briefly here.

Where I disagree most fundamentally with Thackray is on the question of the *value* of, and the urgency for, historiographical debate of the kind he hopes to provoke and intensify. Let me explain why. A historian always begins with a particular problem or set of problems. They may range from "Why was Faraday uneasy about action at a distance?" to "What factors influenced the growth of agricultural chemistry from 1800 to 1850?" The nature of the particular problem will usually indicate that certain factors are likely to be operative in the given situation, and that certain strategies are more likely to lead to a coherent solution to the historical puzzle. Clearly some questions and problems are amenable to quantitative analysis or a role-theoretical approach, while the resolution of others will depend almost completely on textual exegesis. More complex problems will require a more subtle combination of several such approaches. What we should avoid is dissipating our limited energies needlessly in pompous and protracted debates about the *general* nature of the history of science. Unless one is prepared to defend the highly dubious thesis that all scientific developments depend on the same sort of influences and pressures, then it is clearly foolish to argue that all (or even most) historical problems can be analyzed in the same way or in terms of the same categories of narration. If Thackray thinks the acceptance of Newtonian theory in England was a function of a power struggle within the Royal Society, then let us discuss that particular claim on its own merits. If a disciple of Hessen wants to treat Newton as a pawn of the English mercantile class, then we can discuss that argument. If, like Professor Frank Manuel, someone wishes to "explain" Newtonian action at a distance in terms of Newton's Freudian frustrations, we may even try to take that claim seriously. But I am convinced that an abstract discussion about historiography is going to leave

---

[1] See especially E. Meyer, *Zur Theorie und Methodik der Geschichte* (Berlin, 1902), pp. 37ff, or M. Weber's *Gesammelte Aufsätze zur Wissenschaftslehre* (Leipzig, 1922).
[2] Where Agassi's straw man was the "inductivist," and Kuhn's the "internalist," Thackray invents the "intellectualist."

all the interesting questions unresolved. It is also very likely to be counter-productive, for the validity of an intellectual, sociological, role-theoretical or psychological approach can only be determined by, and for, individual cases.

For the same reasons that it has generally been useless (and sometimes harmful) for the members of a scientific profession to worry themselves over general questions of method rather than getting on with their work,[3] we as historians must not allow ourselves to be diverted from the task at hand by a heated debate, however enticing, "over historiography and ideology." In the final analysis, the only important testimony to the soundness of a sociological or an intellectual approach to the history of science is that it gives plausible answers to interesting and important questions. It is a sad fact about the sociology of science that, Merton's work notwithstanding, it has failed to produce a signal contribution to the field which could be ranked with the books of (say) Koyré, Burtt, or Duhem. In lieu of giving us the genuine product, sociologically oriented historians of science have lately tried to move the debate into the arena of the philosophy and logic of history. What they have not been able to establish convincingly in the particular case, they have sought to establish as a *general* characteristic of the evolution of science.

If all this seems philistinic, I plead guilty to being a philistine in these matters. My excuse is that we are in a field with very many interesting problems and with far too few bodies (and perhaps still fewer minds) to solve them. It would be nothing short of retrograde for us to declare a moratorium on our researches in order to go through the motions of an insoluble general debate about the kinds of forces affecting scientific change. So, while Thackray views with dismay the fact that historians of science have not taken up the challenge to historiographical debate issued by Agassi, Kuhn, and Hall, I interpret their disinclination as a reassuring sign of widespread common sense among the practitioners of the subject.

There is a second aspect of Thackray's essay which I want to touch on briefly. Thackray claims that the historian's researches ought to be governed by considerations of contemporary relevance. We live, he insists, in a real world with genuine social problems growing out of science and its products. We have before us, to continue the paraphrase, two choices: either we can pursue historical investigations relevant to those problems or

[3] Witness the stalemates in psychology in the 1920's and 1930's whenever methodological questions became uppermost.

we can ignore them altogether. I find this argument nothing short of invidious, and embodying the kind of rhetoric and fuzzy-mindedness typical of the new right (otherwise known as "the new left"). Thackray's sloganistic "withdrawal is every bit as much a response as engagement" is but another version of the simplistic chant "if you're not part of the solution . . . you're part of the problem."

I am not so bold as Professor Thackray to attempt to speculate on the psychological and political motives which drew our colleagues Gillispie, Guerlac, Cohen, and others into the history of scientific ideas. Perhaps, as he suggests, they were afraid of McCarthy and the House Un-American Activities Committee. The point is that I do not really care very much what their motives might have been. A treatise on "the genetic fallacy" is not required to show that the history they have written stands more or less on its own, and that its importance and validity clearly must be assessed independently of whatever lurking traumas and fears might have been imbedded in their psyches. Fortunately, history is sufficiently empirical that *we generally can* (in spite of Professors Carr and Thackray) *assess the history without assessing the historian.* By the same token, the sociological history of science must needs be assessed by historical standards. All the good intentions in the world, leavened with heaps of relevance, will not justify such history unless it is sound history, for relevance is not a historical desideratum. We decide whether Merton's work is cogent and valid, not by asking what light it sheds on the National Science Foundation, but by asking what light it sheds on the early Royal Society.

There are several classic difficulties with the nebulous demand for relevance, even assuming that it is right to take that demand seriously. How, for instance, is the historian to interpret this demand? It might well happen that a study of the early years of the British Association is more "relevant" to modern problems than a history of the National Research Council since the last war. For all we know, a study of the financial support given to French scientific expeditions in the eighteenth century will be of greater relevance than an investigation of the science-policy-shaping practices of the National Science Foundation. How does one know in advance whether any given investigation is going to be (or is even likely to be) more relevant than another? Thackray's formula, identifying relevance with temporal proximity to the present day, is just too simpleminded to be taken seriously. Moreover, overt preoccupation with relevance normally tends to make for bad history, because this puts constraints on the historian, which should not be there, and which cause him to cast about for contemporary analogies, however farfetched. I think there is general agreement, for instance, that Professor Santillana's classic *The Crime of Galileo* was weakened by Santillana's penchant for casting Galileo in the role of Oppenheimer, fighting a seventeenth-century McCarthy. To write the book as he did certainly made it more relevant (for the 1950's). But it made it much less interesting and less reliable as a piece of history. This is but one of many cases one could cite to illustrate the Orwellian consequences of Thackray's plea for a "search for a . . . *usable* past."

Of course, it would be a piece of good fortune if one could manage to write sound history which was relevant as well. Nor is this necessarily as difficult as having one's cake and eating it too. But even if we concede that relevance, though not necessary, is nice when you can get it, I am not sure why this has any bearing on the historiographical dispute between the intellectualist historians of science and their detractors. Indeed, I believe there is an important sense of relevance according to which the intellectual history of science is crucially relevant, and it is a serious oversight on Thackray's part to have missed, or at least ignored, the kind of relevance to which I refer.

Anyone familiar with contemporary science will concede that there are many problems in the science of our day. Some of them, as Thackray observes, are related to the obviously moral difficulties which science poses for modern man. Others, no less significant in magnitude, are problems of a conceptual character. The scandals in quantum mechanics are well known, and the recent work of Dicke and others in relativity theory similarly suggests that the choice between classical mechanics and relativity theory is not as clear-cut as it once seemed. It is not implausible to think, as Jammer, Feyerabend, and Hanson (among others) have suggested, that a careful analysis of the history of science—especially the history of physics —may well shed considerable light on many of these thorny conceptual problems, perhaps even indicating possible ways out of certain theoretical impasses in which contemporary scientists find themselves. Thus, insofar as it is possible that historical studies will illuminate the intellectual difficulties in which modern scientists find themselves, the history of science is clearly relevant to modern concerns.

If the historian feels strongly about moral and social issues of the day (as one certainly hopes he will in times like these), then he has every obligation to let his views be known, by direct social involvement if necessary. But he is under no moral obligation to prostitute his intellectual interests by making them subordinate to a sense of social mission and an exaggerated sense of the importance and significance of the research he does produce. It would be a disaster if the profession as a whole were to accept the sophistical argument which concludes that *socially* relevant research is a moral imperative. It is neither immoral nor escapist to believe that there are important historical problems about medieval or seventeenth-century science, and to believe that those problems are completely irrelevant to modern *social* concerns.

Thackray suggests that a view like the one I am defending renders history of science an "esoteric luxury." The fact is that history of science (like most aspects of intellectual life) *is* a luxury; and I think Thackray would be hard-pressed to show how it could be otherwise. Indeed, it is in part precisely because it is a luxury that we must work to preserve and to cultivate it. To abandon intellectual luxuries is, after all, to compromise civilization. To act as if history of science is (or might become) more than a luxury is dangerous in the extreme; for it is likely to arouse expectations which, if not satisfied, will turn against the subject and undermine it. If there are those who attack the history of science because it is a luxury, we should face them here and now.

I have said more than I intended to say. By way of summary conclusion, let me reiterate my conviction that the debate about historiography of science, precisely because it is concerned with the general, is going to be hopelessly inconclusive. What we should be doing (and what we should be encouraging our students to do) is tackling significant but specific issues concerned with the evolution of science, bringing to bear as many techniques as we can master and as the problems require.

REPLY BY ARNOLD THACKRAY

An adequate reply to Professor Laudan is not possible here, for "art is long, and time is fleeting." Even so, I would like to correct some trivial misunderstandings and make two brief clarifications, in the hope of keeping the issues sharp.

I do *not* want to debate "the *general* nature of the history of science" (though, as I trust is apparent, I do hope for widespread discussion of ways of studying past science); I was *not* offering psychological and political explanations of individual actions (though I was essaying a first delineation of factors affecting the developing pattern of a newly emerging discipline); I did *not* (I trust) identify relevance with "temporal proximity to the present day" (though I do think historians should study recent science, and that relevance has the same benefits and dangers here as in the seventeenth century).

More generally, let me challenge Professor Laudan's belief that historians simply seek "plausible answers to interesting and important questions." Unless he is an inductivist, which I doubt, Professor Laudan presumably admits some general criteria for deciding which questions are interesting and important. Similarly, those "particular problems" with which his historians so innocently begin are not God-given but selected from the unending range of possible problems by the implicit or explicit application of general criteria. It is of course a well-worn debating technique to hide one's own preferred assumptions behind the label "objective" (or even "obvious") while attacking one's opponent not for his facts or his logic but for his temerity in starting from general propositions. This present discussion deserves better, especially from so able a philosopher of science.

Finally, let me deny that I asked for a history "governed by" relevance. Yet when one searches in vain among the scholarly articles of the past decade for material which illuminates our present social dilemmas, it does not seem entirely rash and extreme to ask for some changes in emphasis. I began by quoting Koyré, because he clearly realized that "the interests and the scale of values of his own time" shaped his work. Free discussion and critical debate will make present interests and values our ready servants: a conspiracy of silence serves only to perpetuate the unacknowledged rule of the interests and values of yesteryear.

━━━━━ MARY HESSE ━━━━━

Hermeticism and Historiography: An Apology for the Internal History of Science

In her pioneering study of the natural magic and hermetic tradition in relation to sixteenth- and seventeenth-century science, Frances Yates specifically disclaims the intention of contributing to the history of science proper: "with the history of genuine science leading up to Galileo's mechanics this book has nothing whatever to do. That story belongs to the history of science proper . . . The phenomenon of Galileo derives from the continuous development in Middle Ages and Renaissance of the rational traditions of Greek science."[1] More recently, however, she has made bolder claims for the relevance of the hermetic tradition:

I would thus urge that the history of science in this period, instead of being read solely forwards for its premonitions of what was to come, should also be read backwards, seeking its connections with what had gone before. A history of science may emerge from such efforts which will be exaggerated and partly wrong. But then the history of science from the solely forward-looking point of view has also been exaggerated and partly wrong, misinterpreting the old thinkers by picking out from the context of their thought as a whole only what seems to point in the direction of modern developments. Only in the perhaps fairly distant future will a proper balance be established in which the two types of inquiry, both of which are essential, will each contribute their quota to a new assessment.[2]

These two quotations serve well to introduce some issues in the historiography of science which deserve attention from philosophers. My interest

AUTHOR'S NOTE: This paper originated from discussions in the Research Seminar on "Science and History" at King's College, Cambridge, organized by Dr. P. M. Rattansi and Dr. R. M. Young. I am glad to express my indebtedness to them and to other participants in the seminar, especially Dr. C. Webster and Miss F. A. Yates, but I would stress that they are in no way responsible for the views expressed here.
[1] *Giordano Bruno and the Hermetic Tradition* (London: Routledge and Kegan Paul, 1964), p. 447.
[2] "The Hermetic Tradition in Renaissance Science," in *Art, Science, and History in the Renaissance*, ed. C. S. Singleton (Baltimore: Johns Hopkins Press, 1968), p. 270.

in them was first aroused by a practical demarcation problem in the history of science. As the discipline has emerged from a certain scientific parochialism to take its place in general history, it is inevitable that some tension has arisen between the so-called "internal" and "external" approaches to the science of the past. On the one hand historians in the tradition of Cassirer, Collingwood, and Koyré have tended to regard the history of science as the history of rational thought about nature, evolving according to its own inner logic, and requiring for its understanding only the attempt on the part of the historian to "think the scientist's thoughts after him," what Miss Yates calls the "continuous development . . . of the rational tradition." This type of historiography of science has produced such masterpieces as Duhem's *Le Systéme du Monde*, Burtt's *Metaphysical Foundations of Modern Physical Science*, and Koyré's *Etudes Galiléennes*, and the bulk of papers which have appeared in the specialist journals in the last few decades. On the other hand, there is the view of science as an irreducibly social and cultural phenomenon, subject alike to rational and irrational influences, to magic as well as mathematics, religious sectarianism as well as logic, politics and economics as well as philosophy, and which is itself one of the major causative influences upon the general historical scene and inseparable from it.

These views are not in themselves incompatible, and no one would wish to deny that there is truth in both of them. But two further types of consideration have tended to bring them into conflict. The first is the claim frequently made by philosophers of the history of science that some particular view of the nature of science is implicit in every study of its history, and the second is the claim made or implied by some proponents of the integration of science with general history that the notion of its internal history as a history of pure concepts independent of "nonrational" factors is a delusion. The passages quoted from Miss Yates are themselves examples of both these claims, since they imply some specification of the "rational tradition" which is contrasted with other factors influencing scientific development, and suggest additionally that the history of this tradition has been distorted by being written from a forward-looking point of view "picking out . . . only what seems to point in the direction of modern developments." Both these elements in Miss Yates's analysis depend on an implicit philosophical position with regard to the nature of science, and the second calls in question the autonomy of internal history.

A more explicit statement of the effect of a philosophy of science on the historiography of science is given by A. R. Hall. Discussing Merton's influential sociological interpretation "Science, Technology and Society in Seventeenth-Century England,"[3] he describes Merton's view that long-time changes in science are primarily to be ascribed to social factors, and contrasts this with the implicit view of the intellectual historians that new intellectual attitudes are not "generated by or dependent upon anything external to science . . . the history of science is strictly analogous to the history of philosophy." Hall goes on:

Profoundly different historical points of view are involved. . . . To suppose that it is not worth while to take sides or that the determination of the historian's own attitude to the issue is not significant is to jeopardise the existence of the historiography of science as more than narration and chronicle. For example: how is the historian to conceive of science, before he undertakes to trace its development; is he to conceive it as above all a deep intellectual enterprise whose object is it [sic] to gain some comprehension of the cosmos in terms which are, in the last resort, philosophical? Or as an instruction-book for a bag of tricks by which men master natural resources and each other? . . . I have deliberately given an exaggerated emphasis to these rhetorical questions in order to indicate the violent imbalance between two points of view that one simply cannot ignore nor amalgamate.[4]

The second claim, that the notion of an independent internal history of science is a delusion, has been pervasive in recent literature, and is even beginning to act as a subtle disincentive to young scholars against working in the more traditional areas of the history of scientific ideas. Attempts to integrate the external and internal approaches abound, but detailed and critical analyses of the claim that internal history is inadequate are more difficult to find. Two recent brief statements of it may be taken as typical. The first occurs in Christopher Hill's reply to debates centering round his analysis of the role of puritanism and capitalism in the scientific revolution: "I am impenitent in my conviction that it is right to try to see society as a whole, and wrong to consider men's work and thought as though they existed in separate self-contained compartments."[5] The second occurs in Robert Kargon's Preface to his *Atomism in England from Hariot to Newton*:

. . . most historians of atomism . . . deal with their subject as if it existed, so to speak, in a void. In these works, atomism is treated as an ideologi-

[3] *Osiris*, 4 (1938), 360.
[4] "Merton Revisited," *History of Science*, 2 (1963), 1.
[5] "Debate: Puritanism, Capitalism and the Scientific Revolution," *Past and Present*, no. 29 (1964), 97.

cal development of a few major figures. Absent are truly *historical* relations between men and ideas; all stress is placed upon internal philosophical and scientific developments . . . Atomism becomes a concept developed by philosophical titans and not *real men*, facing *real problems*—social, political, theological, and personal, as well as scientific.[6]

These various examples are enough to show that the issues between internal and external history of science are also issues involving the relations of philosophy and history of science. Those who see a philosophy of science in every history of science may look to it to provide the definition of "genuine science" which serves to demarcate internal history, and if possible to guarantee its autonomy. Or, if their philosophy of science is a pragmatic and instrumental one, they may use it to demonstrate the noninde-pendence of internal history, for if science is an epiphenomenon of society or technology, then necessarily autonomous internal history is a delusion. The two claims tend, however, to work in opposite directions with respect to the relations between the history and the philosophy of science. The first imposes on historians the duty of being self-conscious and critical about their implicit philosophy, while the second results in increasing affiliation between history of science and general history rather than either science or philosophy, and consequently loosens the tie between the history of science and the philosophy of science.

These implications of the two claims are my excuse for embarking upon a discussion of internal and external history on this occasion, although I fully realize that it is presumptuous of a philosopher who does little more than keep up with trends in history of science to attempt to pass judgment

[6] Oxford: Clarendon Press, 1966, p. vii. Other comments on the relation of internal and external history are more guarded on the question of internal autonomy. Reviewing L. Pearce Williams's *Michael Faraday*, T. S. Kuhn notes that his "predominant concern" is "with Faraday's scientific ideas and their philosophical background," and comments that the "extrascientific events in Faraday's life" might have been exploited to give a more plausible picture of Faraday the man, without suggesting that Williams has failed to reveal an intrinsic connection between these extrascientific events and the scientific ideas themselves (*British Journal for the Philosophy of Science*, 18 (1967), 148). In even more tentative vein, Henry Guerlac deplores excessive introspection, on grounds of public relations with historians: ". . . if Syracuse does little to explain Archimedes, perhaps Greek culture as a whole may do so at least in part. And certainly, for the general historian, Archimedes does something to explain Syracuse. . . . if we concentrate exclusively on what has been called the internal history of science, on the filiation and unfolding of scientific ideas and technics, we may end up writing for ourselves alone, or for ourselves and the philosophers of science." A. C. Crombie, ed., *Scientific Change* (London: Heinemann, 1963), p. 876. See also A. W. Thackray's article "Science: Has Its Present Past a Future?" for a stimulating characterization of internal history as itself a product of social withdrawal on the part of twentieth-century historians of science.

on a domestic debate among the historians. However, one of the functions of philosophy in relation to history of science may be to suggest a few conceptual guidelines through what is a very complex and many-sided question.

To make that question manageable, I shall consider in particular in the last sections of this paper certain aspects of seventeenth-century science, where various new interpretations, of which Miss Yates's is one, are currently in the field alongside the internalist view we have inherited from Burtt, Koyré, Dijksterhuis, Butterfield, and their successors. What that internalist view is can be sufficiently indicated by reference to the chief characters of the story: Copernicus, Kepler, Galileo, Descartes, Boyle, and Newton, who are pictured as engaged, with supporting cast, in metaphysical, theoretical, and experimental argument whose internal rational structure is relatively independent of personal biographies, cultures, and politics. But first I shall consider some more general philosophical points arising from the relationship of external and internal history.

II

Out of the complex of issues already raised it is useful first to make a distinction between the historical *occasions* upon which scientific developments take place, and the *character* of those developments themselves. That there should be any activity describable as scientific obviously depends on a certain stability in some part of society, a certain degree of literacy, and a certain desire for intellectual pursuits. All these conditions are closely dependent on social environment. But in themselves they do not necessarily produce any conflict of interests between the external and internal approaches to science, or threaten the internal autonomy of scientific ideas. These more intimate relationships are involved in the question of how far external or nonrational factors influence or determine the character of the science done and the scientific conclusions reached, and it is this question that I shall be exclusively concerned with here.

A further distinction in the kinds of "external" factors involved is suggested by comparing the comments quoted in the last section from Hall and Yates. Hall contrasts internal and external in the traditional terms of intellectual and social; Miss Yates on the other hand implies a more subtle classification, in making the distinction not so much in terms of thought in general as opposed to social pressures, as between a particular kind of thought, the "rational tradition," and a variety of other kinds of mental or ideological influence. It is tempting to express Hall's distinction in terms of the current philosophical distinction between *reasons* and *causes*: social, political, and psychological factors act as causes (or partial causes) of the contemplation and acceptance of particular kinds of scientific theories, while what have traditionally been called internal factors provide reasons, and would serve to define Miss Yates's "rational tradition." But between these two Miss Yates suggests a third: the influences coming from an alternative tradition of thought or ideology, which can hardly be called mere unconscious causes, and which she would yet wish to distinguish from rational factors. But if the distinction between the occasion and the character of scientific development is kept in mind, the distinction between Hall's and Yates's types of "external" factor is seen to be more apparent than real. If some social, political, or psychological factor is to influence the character of scientific theory, there must be some sense in which it becomes an object of thought, even if perhaps it has to be called an "unconscious reason," or a rationalization, or even a bad, though conscious, reason. Some examples may help to illuminate this point.

Miss Yates's ideological tradition of natural magic and hermeticism undoubtedly provides in principle material for intellectual factors influencing the character of science which are by no means unconscious or merely causal, and which their proponents would regard as reasons, though Miss Yates seems to regard them not just as bad reasons, but even as *irrational* (not "genuine science"). On the other hand, some factors which look purely causal and unconscious at first sight may on more careful inspection reveal intellectual and rational components. For example it is suggested that familiarity with practical machines was a partial determinant of corpuscular mechanism, and that the existence of hydraulic and heat engines played a similar role in early nineteenth-century thermodynamics. If these were factors in determining the character of their contemporary science, they were not merely causal jogs on the mental processes of scientists, but owed something to the rational consideration that machines of various kinds might be macroscopic analogues for more fundamental elementary processes. There are, however, other cases where we are more disposed to regard the external factors as unconscious and therefore more nearly causal. It is interesting that the examples that come to mind are also highly controversial as historical explanations, for it is more difficult to establish unconscious causality upon scientific ideas than intellectual influence. Hill, for example, has suggested that a possible cause for Harvey's change of mind about the primacy of the heart and the primacy of the blood may have been the actual transition from monarchy, analogous to the heart, to republic, analogous to the blood.[7] And F. E. Manuel hints that Newton may have been more receptive to the idea of attraction because he may as a child have felt himself deprived of natural affection on the remarriage of his mother.[8] We may regard Hill's suggestion as bad history, and the influence suggested, if it had been good history, as bad science, but it is not at all clear that, for Harvey, this may not have been in principle a perfectly respectable analogical argument. Only Manuel's suggestion seems to preserve pure unconscious causality.

All of which is to say that the classification into internal and external influences on the character of science is by no means as simple as it looks at first sight. We seem to require some more fundamental framework in terms of which to discuss it; therefore let us return to Hall's suggestion that the historian's assessment of the various kinds of influence will depend upon his conception of what science is, and that external and internal history are committed to incompatible conceptions. He seems to have in mind the conflicts that might arise from the claim that sociological or psychological interpretations such as those just mentioned are *sufficient* explanations of theoretical development. Such a claim would have, of course, much wider ramifications on the intellectual scene than merely in the historiography of science. Let us, however, ask a more limited question, namely, is our approach to the historical question of interaction between the various factors that influence science predetermined by some prior philosophical analysis of the structure of science? In the course of this discussion some light may be shed on the obscure notion of the "rational tradition" and on the distinction between external and internal factors.

It is certainly the case that philosophical analyses of the structure of science have an effect on its historiography, even when historians disclaim any concern with philosophy of science. J. Agassi has engagingly described

---

[7] C. Hill, "William Harvey and the Idea of Monarchy," *Past and Present*, no. 27 (1964), 54. C. Webster opposes to Hill's view the internal explanation that it was Harvey's "failure to substantiate the Aristotelian idea of the primacy of the heart in embryology which led him to doubt other facets of the heart's primacy." "Harvey's De Generatione: Its Origins and Relevance to the Theory of Circulation," *British Journal for the History of Science*, 3 (1967), 274.

[8] *A Portrait of Isaac Newton* (Cambridge, Mass.: Harvard University Press, 1968), pp. 83–85.

the unhappy effects of *inductivism* in history of science.[9] An inductivist historian has implicit Baconian philosophical allegiances, concentrates on describing "hard facts" and experiments, reconstructs past arguments to fit an inductive structure, and judges past theoretical conceptions as true or false, significant or fit for ridicule, depending upon what are now acceptable theories. This is the prime example of what Miss Yates calls history read according to premonitions of what was to come, and if it is the type of historical tradition she has in mind, she is fully justified in asking for a corrective, although not necessarily in looking for the corrective in "nonrational" factors. Inductivism also produces the type of history which most obviously leads to conflict with externalist interpretations, for if the development of science consists of exhaustive inventories of hard facts, and careful generalizations to laws and theories which are fully and uniquely warranted by the facts, then although the particular facts selected for study might depend partly on external factors, the nature of the conclusions would not. The story of science would indeed unroll inexorably and cumulatively according to its own internal logic, and any externalist claims to provide additional causal explanations would be otiose or false.

A more plausible version of inductivism which leads to the same consequences might be called *naive realism*. This holds that theories properly arrived at are simply and perennially true, without specifying that they are necessarily arrived at by strictly inductive means. They might, for example, be justified by consilience of concepts in Whewell's sense, or they might be the product of perennial metaphysical principles as in seventeenth-century rationalist science. A quasi-Kantian view of science as flowing wholly from a priori categories is another variation on the same theme. Any such view, if correct, would justify autonomous internal history, since all imply that the character of science is determined solely by intellectual factors.

But no philosopher and few historians would now subscribe to any of these views of science, and certainly none of them can be equated with the views held by twentieth-century intellectual historians. Their view seems to come nearer to what Agassi calls *conventionalist* history, in which the theoretical system of each period is seen from its own standpoint as creating its own intellectual world in which facts themselves are interpreted wholly according to criteria of internal coherence. An extreme form of conventionalism (which is the only form considered by Agassi) implies that

[9] "Towards an Historiography of Science," *History and Theory* (The Hague), 2 (1963).

scientific theories not only are insulated from external social factors, but are also immune to intellectual and empirical constraints making for modification and evolution of their closed logical systems. But conventionalism in this form would hardly be adopted by any philosopher or historian if it implies that theoretical systems are immune to change, for it is universally agreed that both facts and alternative theories make for change, and most internal historians have taken full account of the resulting problem situations and conflicts which Agassi enjoins them to notice.

In a less extreme form of conventionalism, however, it might be held that changes of world view, insofar as they can be accounted for historically at all, are *wholly* functions of intellectual factors, and this seems indeed to be the philosophy of science which Hall ascribes to the intellectual historians in the passage quoted above. Such a view would provide an alternative source for the "rational tradition" of science to that described by Miss Yates as "forward-looking" history written in the light of modern science. For what is "rational" in a particular period may be different from what is now regarded as "scientific," but it may be recoverable as rational for that period when a historian sympathetically immerses himself in the literature of the period, and learns to think in terms of its own rules of argument and to use its criteria of truth. This is the approach now frequently adopted by internal historians, but it seems to need more specification than that just given to it, for in this sense the anthropologist might even claim to find a rational tradition among the Azandi witches. What seems to be closer to the actual philosophy of intellectual historians such as Cassirer, Collingwood, and Koyré, however, is the view that what counts as rational at any period is a timeless characteristic which shows itself to the historian who follows the rationally intelligible thought of the scientists of the past, and which transcends the cultural peculiarities of particular historical periods. Collingwood's version of this view, however, sometimes seems to include the recognition that it is the historian's own thought that structures the thought of the past—the timeless rationality of history is his own; thus Collingwood comes nearer to Miss Yates's "forward-looking" history.[10]

It is tempting to characterize this last approach as *neo-Hegelian*, not only because several of its practitioners stand recognizably in a Hegelian tradition of historiography, but because it shares with Hegel's philosophy

[10] See the quotation from an unpublished manuscript in the Editor's Preface, *The Idea of History* (Oxford: Clarendon Press, 1946), p. xii.

of history the emphasis on history as the history of thought, developing by the exercise of human reason according to implicit and autonomous logical constraints. Just as Collingwood remarks that Hegel's philosophy of history is most successful when applied to the history of philosophy,[11] it might be held that it is also the appropriate approach to history of science seen as a sequence of intellectual structures developing according to internal rational criteria. On the other hand the epithet "neo-Hegelian" must be used with care, for there are at least two important respects in which Hegelian historiography differs from that of the internal historians of science. In the first place the dictum that all history is the history of thought applies to *all* history, not just to the history of intellectual endeavors such as philosophy and science. In other words, a Hegelian history of the social and political factors influencing science could be written as easily as a history of scientific ideas, for all factors, internal and external, would be seen as the products of human reason operating upon the relationships between man and the natural world. Hence, strictly speaking, Hegelian history is neutral on the issue of the intellectual autonomy of science, although this freedom of interpretation has hardly been exploited by the historians of science who stand nearest to the Hegelian tradition.

Secondly, a strictly Hegelian history of science would be more restrictive in its characterization of the "internal logic" of science than the internal historians would generally be prepared to accept. For the Hegelian logic is dialectical, and would commit internal history to the view that science proceeds by revolutionary alternations of thesis, antithesis, and synthesis, rather than by accumulation or evolution. Traces of such a view may indeed be found, for example in the title of Koyré's *From the Closed World to the Infinite Universe*, but there is little evidence in general that internal historians have structured their work according to a dialectical logic in any but the loosest sense of being prepared to accept the occurrence from time to time of radical conceptual revolutions. Koyré's own effective restriction of the source of new conceptual frameworks to *intellectual* systems already found in the history of thought (the seventeenth-century revolution as a revival of Archimedes and atomism, for example) seems rather to be a *historical* judgment about the relative importance of ideas and techniques in the seventeenth century than a product of any a priori structuring of scientific development according to a particular view of its logic.

Again, if we consider the *deductivist* analysis of science which has been

[11] *Ibid.*, p. 120.

almost universally accepted by philosophers of science until recently, it is clear that it entails no claim to internal autonomy nor any necessary conflict between the standpoints of external and internal history. This view has been characterized by a radical distinction between the sociology and psychology of science and its logic, or as it is sometimes expressed, the contexts of *discovery* and of *justification*. How a hypothesis is arrived at is not a question for philosophy of science; it is a matter of the individual or group psychology of scientists, or of historical investigation of external pressures upon science as a social phenomenon. The question for philosophy or logic is solely the question whether the hypotheses thus "nonrationally" thrown up are viable in the light of facts, that is, whether they satisfy the formal conditions of confirmation and falsifiability adumbrated by deductivist philosophers. Although this view places a heavy straitjacket on the philosophy of science, it appears to exert no restraint at all upon its history, much less to cause any possible conflict with the external approach. It allows historians to take seriously as scientific whatever theories were contemplated in the past, arrived at by whatever external or internal influences, and however apparently bizarre, just so long as these theories were treated according to the deductivists' logical criteria. The use of the terms "logical" and even "rational" in this analysis is indeed far narrower than in the intellectual historian's "internal logic of science," or his view of the history of science as the history of man's rational thought about nature. For deductivism characterizes all influences leading to discovery as nonlogical or even nonrational, and leaves the whole context of discovery to the efforts of the historian without offering him any criteria of distinction between kinds of influence on discovery. And since for given evidence, a theory satisfying the deductive criteria is never unique, even the kinds of concepts adopted are open to nonlogical influences. Hence within deductivism as a view of science it would even seem impossible to make the distinction between intellectual and social influences on discovery, and a fortiori no general conflict could be generated between them, and no general claim to internal autonomy sustained. Perhaps the only external view which would in principle conflict with deductivism would be one in which every deductive as well as nondeductive argument was interpreted indifferently as a psychological or sociological epiphenomenon, but surely no such approach to history has ever been seriously practiced, even among Freudian or Marxist historians.

What has in fact happened is not that a deductivist approach to internal history has conflicted with an external history, but that all forms of historical investigation, internal as well as external, have led to radical questioning of the deductive view of science itself. For they have revealed the impossibility of drawing any sharp line between the contexts of justification and of discovery, between the "rational" arguments as defined by deductivism and the psychological and cultural processes which determine what kinds of theory are contemplated, and even between the "hard facts" which must be respected as tests of theory and the way these facts were interpreted in a given cultural environment.[12] It is no accident that the current attacks upon all these entrenched dichotomies of deductivism come either from historians of science (explicitly and recently from Kuhn, but also implicitly from Duhem) or from philosophers who are deeply immersed in history of science and conduct their discussions by means of detailed case histories (Popper in his later writings, Buchdahl, Feyerabend, Hanson, Harré, Lakatos, Toulmin). Some of these writers have moved to a position similar to that described above as conventionalism, with stress on the role of intellectual and even in some cases inductive factors[13] in scientific development, but without any implication that external causes of change are excluded, or that there is any intrinsic conflict between the approaches of internal and external history.[14]

I have used the comments of various historians on the relations between internal and external history to suggest several general conclusions which may now be drawn together:

[12] Cf. T. S. Kuhn, The Structure of Scientific Revolutions (Chicago: University of Chicago Press, 1962), p. 8.

[13] Particularly in N. R. Hanson, Patterns of Discovery (Cambridge: Cambridge University Press, 1958); R. Harré, Theories and Things (London: Sheed and Ward, 1961); M. B. Hesse, Models and Analogies in Science (Notre Dame, Ind.: University of Notre Dame Press, 1966).

[14] Kuhn has given an interesting analysis of the external-internal distinction from an external sociological rather than an internal intellectual viewpoint, including an attempt to explain the apparent autonomy of internal history. The account depends on his analysis of the pre-paradigm and paradigm stages of a science; in the mature paradigm stages, the scientific community is a sociological entity whose structure and practices embody the paradigm of rationality for that group: ". . . the practitioners of a mature science are effectively insulated from the cultural milieu in which they live their extra-professional lives. That quite special, though still incomplete, insulation is the presumptive reason why the internal approach to the history of science, conceived as autonomous and self-contained, has seemed so nearly successful." "Science: The History of Science," International Encyclopedia of the Social Sciences, ed. David L. Sills (New York: Macmillan, 1968), p. 81. This analysis, however, still evades the question whether the scientific group is to be defined in terms of rationality or rationality in terms of the scientific group.

Mary Hesse

1. The notion of internal history as generally understood involves essentially the conception of a "rational tradition" and of "intellectual factors."

2. The suggestion that the rational tradition should be defined in terms of a particular philosophical analysis of science is too simple a view of the relations between philosophy and history of science. Of the accounts discussed, only inductivism or the various forms of a priorism would provide a complete specification of rationality with respect to science; conventionalism and deductivism leave open many questions about the development of theories and about intellectual influences, questions which can only be answered historically; and what I have guardedly called neo-Hegelianism seems to presuppose an autonomous rational tradition, but little indication is given by its practitioners of how this is to be recognized, except perhaps by the subjective experience of the historian.

3. Among analyses currently adopted by philosophers of science, namely deductivism and various species of conventionalism, there is much room for feedback between historical investigations and philosophical accounts of scientific structure.

4. None of these views exclude the possibility that external factors may be partial causes of particular scientific developments, including particular concepts and theories, none claim a priori autonomy of internal history, and none entail any necessary conflict between external and internal interpretations.

5. Internal history is not necessarily history read forwards in the light of what is to come, picking out particular precursors either of modern theories or of modern analyses of rational method, for it may be practiced by "thinking men's thoughts after them" according to their own theories and criteria of rationality. But this method involves some judgment by the historian about what in the past is to count as "rationality."

No general specification of "rationality" seems, however, to be forthcoming from the philosophical analyses discussed, and therefore we cannot expect that there will be any general answer to the question whether there is a relatively autonomous internal history. All such questions must be asked in particular cases, which is to say that they are essentially historical questions. I therefore turn now to some specific recent examples in seventeenth-century historiography with these questions in mind. I wish to stress particularly at this point that what I am attempting here is not to make firsthand historical judgments, for which I have no competence, but rather to ask how far the evidence cited by historians themselves supports any attack upon the received internal tradition.

III

The significance of recent studies of the hermetic and natural magic tradition in seventeenth-century science is precisely that they seem to erode the intuitive distinctions in terms of which external history and internal history have been understood, by pointing to a set of factors conditioning the character as well as the occasion of science, which if not intellectual are certainly mental, and which are also closely related to important sociological considerations. They have therefore raised sharply the question of the nature of internal history in general, and in particular the autonomy and adequacy of the received tradition of the mechanical philosophy. The suggestion does indeed arise that this tradition is an unacceptable distortion, since it fails to account for a large element of thought about the natural world that even a neo-Hegelian view would regard as relevant. Hence it is sometimes made to appear that there are fundamental objections to the notion of internal history in general. My philosophical discussion has already indicated that such general objections are misplaced; I shall now suggest that they are not supported by the evidence as cited by the historians of this particular case either.

First it should be remarked that interest in hermeticism and magic is no external imposition upon the received history of the mechanical philosophy, for many historians had previously noted odd features of this story which seemed to intrude unexplained in the internal development of concept, theory, and fact. To take just one example, in a set of essays entitled *The Making of Modern Science* in 1960, Charles Raven, alone of all the contributors who concentrated on the mathematical and physical sciences, remarked: "Immensely important as they have been in establishing a reign of law and an urban and materialistic society, [Copernicus, Kepler, Galileo, and Newton] neither initiated the emergence nor gave rise to the transformation of modern man. The heliocentric cosmology was less disturbing than the rejection of spontaneous generation, or creation as an act rather than a continuing process, and of witchcraft, astrology, and magic."[15] "Witchcraft, astrology, and magic" point better to the general climate of thought and belief in which modern science arose than do the

---

[15] "Living Things in the Frame of Nature," in *The Making of Modern Science*, ed. A. R. Hall (Leicester: Leicester University Press, 1960).

antiseptic details of the Democritan-Archimedean tradition. And they remind us of the context in which it is natural to find references to the writings of Hermes, Orpheus, Moses, and other pseudo-priscine authors scattered through Copernicus, Gilbert, Kepler, Wallis, and Newton, to find More immersed in the Cabala, Glanvill writing in 1666 a work entitled *A Philosophical Endeavour towards the Defence of the Being of Witches and Apparitions*, and Boyle seriously discussing the effectiveness of amulets, sympathetic powers, and the magnetic cure of wounds.

The hermetic writings were a group of Gnostic texts actually dating from the second and third centuries A.D., but believed in the sixteenth century to be contemporary with or earlier than Moses, and originating in Egypt. They consequently carried all the ancient authority so much revered in the Renaissance; they were quite non-Aristotelian in spirit and hence reinforced any antischolastic tendencies of Renaissance thought; and since they were in fact written in the Christian era, they contained some elements of Judaism and Christianity which were regarded as prophetic and so enhanced still further their authority. P. M. Rattansi epitomizes well the main tenets of hermeticism in contrast with the careful distinctions maintained in medieval scholasticism between the natural and the marvelous, the magical, and the miraculous:

For Hermeticism, by contrast, man was a *magus* or operator who, by reaching back to a secret tradition of knowledge which gave a truer insight into the basic forces in the universe than the qualitative physics of Aristotle, could command these forces for human ends. Nature was linked by correspondences, by secret ties of sympathy and antipathy, and by stellar influences; the pervasive nature of the Neo-Platonic World-Soul made everything including matter, alive and sentient. Knowledge of these links laid the basis for a 'natural magical' control of nature. The techniques of manipulation were understood mainly in magical terms (incantations, amulets and images, music, numerologies).[16]

It is of this tradition that he says on the previous page: "It was not completely vanquished by the rise of the mechanical philosophy. Without taking full account of that tradition, it is impossible . . . to attain a full picture of the 'new science.' "

There is here, as in other unguarded comments on the hermetic tradition,[17] more than a hint of the notion that by adding to the picture all in-

---

[16] "The Intellectual Origins of the Royal Society," *Notes and Records of the Royal Society*, 23 (1968), 132.

[17] Cf. F. A. Yates: "If we want the truth about the history of thought, we must omit nothing" (*Giordano Bruno*, p. 204). Fortunately few historians attempt the impossible

fluences that fed into it, of all conceivable degrees of relevance, we get nearer to some form of complete description or complete understanding of the "whole picture." But the view of history as complete description, or "telling it like it was," is an error analogous to the error of inductivism in science. It presupposes that history is a search for hard facts, which are relatively independent of each other, and that the full picture is attained by accumulating as many of these as possible. Even the claim to get nearer the full picture by adding more factors should be treated with caution. Throwing more light on a picture may distort what has already been seen, and certainly judgments of relative significance are required if the picture is not to become flat and overcrowded. The historian's task is not to spell out in tedious detail every minor writing or trivial biography of forgotten figures, or every twist and turn of the social scene which had some bearing on the personnel and institutions of science. Even Miss Yates visibly and sensibly wilts before some of the details of her magico-cabalist authors. In the present context, the historian's task is rather, first, to follow up the loose ends of the received internal tradition where necessary to explain oddities and answer unanswered questions and, second, to investigate such other factors related to the science of the period as have their own intrinsic interest, perhaps because they are opened up from the side of general history, or because they have an importance or fascination of their own. It is then a historical question whether the received internal tradition should be jettisoned or modified, and whether some relatively autonomous understanding of the nature of "rationality" in seventeenth-century science emerges.

In order to discuss this it is useful to distinguish among the elements found by historians in the hermetic complex three themes whose relation to the history of science ought to be considered separately. These are (1) the social and political affiliations of certain religious sects, and the schools of Paracelsian and Helmontian doctors and chemists, (2) the full-scale hermetic and natural magic tradition as a way of thought and life in such writers as Paracelsus himself, Bruno, and Fludd, and (3) the doctrines of extended spirits and powers of matter which persisted even in later seventeenth-century science in opposition to corpuscular mechanism.

Of the first of these factors it does not seem to be anywhere claimed that they provided more than the occasion and the motivation for certain de-

task of carrying out such injunctions in practice. Their history is infinitely more sensitive than some of their throw-away methodological remarks about it.

velopments connected with the new science.[18] The enthusiastic fervor, utopianism, and even revolutionary character of certain religious sects do indeed throw much light on the institutional expressions of science. Such sectarian figures as Hartlib, Dury, and Comenius helped to encourage Baconian allegiances in the early Royal Society, and the anti-establishment circles in which they, and some other founding fathers of the Society, moved go some way to explain the suspicion with which the Society was viewed by Royalists and Churchmen in the Restoration period. But none of this seems to impinge essentially on the internal tradition of the mechanical philosophy. Similarly, when Rattansi explores the reasons for the popularity of Paracelsian and Helmontian medicine in England in the mid-century, he concludes: "Paracelsus and his chemical doctrines were brought into prominence because of factors that do not belong purely to the internal history of chemistry . . . but must be referred to the larger social and political environment."[19] In other words the Paracelsian iatro-chemists stand to some extent *outside* an "internal history of chemistry" which is presupposed here to be independent of the social environment which brought them into prominence. Again, in the debates which have followed the related theses of Merton and Hill regarding the influence of puritanism on seventeenth-century science, several commentators have remarked that the argument suffered from too little conceptual clarity about what was to count as "science" (and indeed as "puritanism"). Far from suggesting a restructuring of the internal tradition, these debates presupposed its existence, and the disputants were counseled to look at what had been achieved in internal history in order to acquire some internal specification of what "science" is.[20]

The case with the second and third elements of the hermetic complex is different, because here it is not a question of interacting social factors, but

[18] Apart from the classic studies of R. K. Merton, and C. Hill's *Intellectual Origins of the English Revolution* (Oxford: Oxford University Press, 1965), recent partly "external" studies of these matters are to be found in M. Purver, *The Royal Society: Concept and Creation* (Cambridge, Mass.: MIT Press, 1967); R. H. Kargon, *Atomism in England from Hariot to Newton* (Oxford: Clarendon Press, 1966); P. M. Rattansi, "The Helmontian-Galenist Controversy in Restoration England," *Ambix*, 12 (1964), 1; and C. Webster, "English Medical Reformers of the Puritan Revolution: A Background to the 'Society of Chymical Physitians,'" *Ambix*, 14 (1967), 16.

[19] "Paracelsus and the Puritan Revolution," *Ambix*, 11 (1963), 31.

[20] Cf. A. R. Hall, "The Scholar and the Craftsman in the Scientific Revolution," in *Critical Problems in the History of Science*, ed. M. Clagett (Madison: University of Wisconsin Press, 1959); H. F. Kearney, "Puritanism, Capitalism and the Scientific Revolution," *Past and Present*, no. 28 (1964), 81, and "Puritanism and Science: Problems of Definition," *Past and Present*, no. 31 (1965), 104.

of intellectual factors which might be held to be necessary ingredients of the history of science seen as the history of thought. Their close relation to the social factors just discussed has obscured the fact that the real challenge to the received internal tradition comes not so much from "external" factors concerned with social and political pressures as from within history seen as "thinking men's thoughts after them." Miss Yates excludes them from the "rational tradition," and yet they are undoubtedly in men's heads, and presupposed in much of their literature. Can an internal history which has neglected them be defended?

There are two ways in which it might be defended. The first is to peer into the internal tradition to see whether any implicit specification of "rational science" for the seventeenth century can be detected which will serve to distinguish and perhaps isolate the activities described in this tradition from hermetic thought. The second is to look for the explicit comments of each tradition on the other. Both methods are illuminating, but the first alone is indecisive, since it reveals a notion of "rationality" that is at best a loosely clustered family resemblance concept, some elements of which seem almost as remote from our views of rationality as do the tenets of hermeticism. If we are to heed warnings not to read back our methods into the past (a sin only less grievous than inductivism), it is difficult to see how to draw the line between what legitimately concerns internal history and what does not.

IV

Close attention to the intellectual context of seventeenth-century science has revealed a multiplicity of ways in which contemporary writers themselves saw the new philosophy. No simple inductivist or deductivist account or any mixture of these is sufficient to do justice to the complexity of either their theory of rationality or its practice. For striking illustration of this we need go no further than the first chapter of Professor Sabra's book *Theories of Light from Descartes to Newton*.[21] In Descartes's optics alone he finds no fewer than six interpretations of rational method:

1. The method of the *Discourse*, which is supposed to be but is not the method of the *Dioptric*.

2. The method explicitly described in the *Dioptric*, which consists of arguing from possibly false assumptions to observation; that is, it is a spe-

[21] A. I. Sabra, *Theories of Light from Descartes to Newton* (London: Oldbourne, 1967).

cies of "saving the phenomena" which is not concerned with the true nature of light.

3. One method actually used in the *Dioptric*, of deduction of phenomena from claimed metaphysical principles.

4. Another method actually used in the *Dioptric*, of circular argument from effects to causes to predictions of further effects in which supposed causes are held "sufficiently proved" by the truth of their consequences.

5. Descartes's own specification of this circular argument in terms of the ancient mathematical method of analysis and synthesis: in order to solve a problem, look for that from which a solution would result, and again what that would result from, until we come to something known or a first principle. Then by synthesis retrace our steps deductively until we reach the solution.

6. But Descartes knows that in many problems knowledge of the general principle does not give sufficient specific information about the original problem, and so he proceeds by analogy with other cases whose solutions are known.

Thus within one topic Descartes almost runs the whole gamut of "methods" proposed in seventeenth-century science. And to this list may be added several interpretations of Baconian method, both in Bacon's own work and in the tensions between the inductive and hypothetical methods in the early Royal Society, culminating in Newton's Baconian claim to "deduction from the phenomena."

In addition to all these contemporary attempts to characterize scientific method, it is clear in many expositions and apologies for the new philosophy that a certain kind of content as well as certain kinds of arguments is regarded as necessary. Science is to be mathematical, mechanist, and hence nonanimist and nonteleological; in other words it is to deal with facts interpreted in a certain way. This is seen, for example, in various discussions of the nature of the primary qualities, most consistently and pervasively in Descartes. But his is not the only system in which the conceptual substance as well as the method of the new philosophy is essential to its specification. Even Bacon, who officially rejects any anticipation of the nature of the most general forms until after the inductive ladder has been ascended step by step, seems forced by his own method to specify in advance of what kinds the general forms must be, and vacillates at various times between a quasi-atomism, a continuum theory based on density and rarity as the primary qualities, and even the Paracelsian principles sulphur and mercury. Hooke, similarly, in attempting a reconstruction of Bacon's method in his "philosophical algebra," presupposes a fundamental theory of primary mechanical vibrations, which he elaborates not inductively but hypothetically in his explanations of elasticity and of light.[22] Method and metaphysics are inseparable in seventeenth-century science; hence it is useless to seek for any perennial paradigm of rationality in the study of seventeenth-century method.

Even "concern for the facts" cannot be taken easily as a common ingredient of the new philosophy. On the one hand Galileo expresses disinterest in the actual form of projectile trajectories, and Descartes in the actual behavior of colliding particles, once the mathematical principles are established; and on the other hand Bacon, Mersenne, More, Boyle, Digby, Glanvill, and a host of lesser figures accept as factual some instances of magic, sorcery, telepathy, and sympathetic cures, and sometimes try to rationalize them in terms of mechanical explanations. These attempts were highly artificial, and with the subsequent success of the mechanical philosophy the alleged facts themselves dropped out of the purview of science, perhaps prematurely. That mathematical and mechanical science become to some extent constitutive of facts is seen clearly in Mersenne, for whom this framework determines what is to count as a natural phenomenon—all else is to be rigorously excluded as the sphere of the supernatural and miraculous.[23]

But when all this has been said, our intuition remains that however varied may be the explicit and implicit methodologies of seventeenth-century science, they are still worlds away from hermeticism. This intuition is in fact confirmed by several examples, cited by our group of historians, of intellectual dispute between adherents of the two traditions. In fact it soon becomes clear in their work that the hermetic tradition does not provide merely an extra factor to be noted and added to the picture, but rather its importance is that it provides the occasion for some conscious self-definition of the new science in the course of vigorous repudiation of the hermetics and all their works.

This can be illustrated from several examples of dispute. In an exchange

---

[22] On method and mechanism in Bacon and Hooke, see my "Francis Bacon," in A Critical History of Western Philosophy, ed. D. J. O'Connor (London: Collier-Macmillan, 1964), p. 141; and "Hooke's Philosophical Algebra," Isis, 57 (1966), 67; cf. also R. Harré, Matter and Method (London: Macmillan, 1964).

[23] See R. Lenoble, Mersenne; ou La naissance du mécanisme (Paris: J. Vrin, 1943), p. 7.

of polemics with the English Rosicrucian doctor Robert Fludd, Kepler dissociates himself from the interpretation of mathematics found in the hermetic writers.[24] Kepler does indeed himself believe in a mathematical harmony of the cosmos as the image or analogue of God and the soul, but his geometry is Euclidean, his conclusions require proof, and they must correspond with facts (that is, the kind of facts Kepler inherited in Brahe's planetary observations). According to Fludd, on the other hand, Kepler merely "excogitates the exterior movement. . . . I contemplate the internal and essential impulses." Fludd *complains* that geometry is dominated by Euclid, while arithmetic is full of "definitions, principles and discussions of theoretical operations . . . addition, subtraction, multiplication, division, golden numbers, fractions, square roots and the extraction of cubes." There is, he goes on, no "arcane arithmetic," no understanding of the significance of the number 4, deriving from the sacred name of God.[25]

In less measured tones than Kepler, Mersenne devotes himself to combating the arrogance and impiety of the terrible magicians.[26] Their arbitrary numerologies do not even agree among themselves; they do not understand that words are mere *flatus voces*, merely conventional signs or sounds, not images or causes. The proportion of the planetary distances may exhibit harmony, but whether it does or not is a matter of fact, not of cosmic analogies. Moreover, astrology, magic, and the Cabala are not just harmless games, they reduce human freedom to cosmic determinism and hence are morally reprehensible. Although some alleged examples of sorcery may be facts, use of sorcery is morally detestable; the magicians are guilty of arrogance and impiety in their claim that the human intellect is divinely inspired and is the measure of things. When Fludd replies to this onslaught with equal violence, Mersenne requests Gassendi to take up the cause, and he, slightly reluctantly but for friendship's sake, drops what he

[24] The documents have been presented by W. Pauli, "The Influence of Archetypal Ideas on the Scientific Theories of Kepler," in C. G. Jung and W. Pauli, *The Interpretation of Nature and the Psyche*, English translation (London: Routledge and Kegan Paul, 1955), p. 151. See also A. G. Debus, "Renaissance Chemistry and the Work of Robert Fludd," *Ambix*, 14 (1967), 42, and "Mathematics and Nature in the Chemical Texts of the Renaissance," *Ambix*, 15 (1968), 1.

[25] Debus, "Mathematics and Nature in the Chemical Texts of the Renaissance," p. 17; Pauli, "The Influence of Archetypal Ideas on the Scientific Theories of Kepler," pp. 196, 102ff.

[26] Lenoble, *Mersenne*; Yates, *Giordano Bruno*, chapter XXII.

is doing in order to study Fludd's writings.[27] That is a measure of the externality of the hermetics at this period to the new philosophy.

Another such polemical exchange is Seth Ward's Vindiciae Academiarum, written in reply to an attack upon the academic activities of the University of Oxford by John Webster.[28] Webster berates Oxford for its neglect of the new science, citing indifferently as representatives of that science Bacon, Copernicus, Galileo, Paracelsus, Boehme, Fludd, and the Rosy Cross. Ward replies with careful distinctions between the true natural language or universal character "where every word were a definition and contained the nature of the thing," and "that which the Caballists and Rosycrucians have vainly sought for in the Hebrew." Hieroglyphics and cryptography were invented for concealment, grammar and language for explication. Magic is a "cheat and imposture . . . with the pretence of specificall vertues, and occult celestiall signatures and taking [credulous men] off from observation and experiment. . . . The discoveries of the symphonies of nature, and the rules of applying agent and materiall causes to produce effects, is the true naturall magick." Both Mersenne and Ward take Aristotle for an ally against the magicians: it is not Aristotle, rational though wrongheaded, who is the enemy of the new philosophy, but "the windy impostures of magick and astrology, of signatures and physiognomy."[29]

Rattansi characterizes the situation accurately when he contrasts "the emotionally-charged and mystical flavour of Hermeticism, its rejection of corrupted reason and praise of 'experience' (which meant mystical illumination as well as manual operations), and its search for knowledge in arbitrary scriptural interpretation," with "a sober and disenchanted system of natural knowledge, harmonized with traditional religion," and goes on: "To move from one to the other was to change one conceptual scheme for ordering natural knowledge to another, with an accompanying shift in the choice of problems, methods, and explanatory models."[30] The change of

---

[27] G. Sortais, La Philosophie Moderne depuis Bacon jusqu'à Leibniz, vol. 2 (Paris: P. Lethielleux, 1920), p. 43.
[28] S. Ward, Vindiciae Academiarum (Oxford, 1654); J. Webster, The Examination of Academies (London, 1654).
[29] Ward, Vindiciae Academiarum, pp. 22, 34, 36; Lenoble, Mersenne, p. 146.
[30] Rattansi, "Intellectual Origins," p. 139. Rattansi also notes several examples of conversions from hermeticism to the mechanical philosophy in the 1650's, including Boyle, Barrow, and Charleton (p. 136). But these seem to have been conversions from certain animist concepts in their theories of matter rather than from what I have called the full-blooded hermeticism of Paracelsus, Bruno, and Fludd. I shall discuss the significance of these theoretical conceptions in the next section. Again, the picture becomes

sensibility is also a contemporary view. For example Glanvill: "among the Egyptians and Arabians, the Paracelsians, and some other moderns, chemistry was very phantastic, unintelligible, and delusive . . . the Royal Society have refined it from its dross, and made it honest, sober, and intelligible . . ."[31] And Sprat's plea for a "close, naked, natural way of speaking" is directed as much at the "Egyptians" as at the Aristotelians.[32]

In view of all this, any suggestion of a confluence of hermeticism and mechanism into the melting pot of the new science would be a mistake. In all that constituted its essence as a way of thought and life, hermeticism was not only vanquished by the mid-century, but had provided the occasion for the new philosophy to mark out its own relative independence of all such traditions. The style of argument required in the polemics is itself significantly different from that adopted in domestic scientific disputes. It involves rhetoric and ridicule, and appeals to theological and moral principle, and sometimes political and pragmatic test. Thus, Allen Debus shows too great a tolerance and fails to highlight the element of conflict in a study of the work of Robert Fludd when he concludes: "I do not believe that it is sound to dismiss the work of these men as valueless for the rise of modern science, as has often been done, simply because they were not right in our terms. The work of Robert Fludd had been taken quite seriously in the second quarter of the seventeenth century and it had resulted in a major confrontation between the supporters of the mystical neo-Platonic universe and representatives of what we would call a more modern outlook."[33] That the hermetic tradition had a large popular following in the mid-century and even later does not amount to "influence" upon the development of rational science, much less call in question its autonomy, nor does the fact that it provoked leading protagonists of that science show that internal history which neglects it is necessarily distorted.

The error of cumulative history is one to which conventionalists are par-

<hr>

unduly blurred when "hermeticism" is said to be one of the traditions of "scientific inquiry" that must be taken account of (p. 140).

[31] J. Glanvill, *Plus Ultra* (London, 1668), p. 12.

[32] T. Sprat, *The History of the Royal Society of London* (London, 1667), section XX; see also section III.

[33] Debus, "Renaissance Chemistry and the Work of Robert Fludd," p. 58. See also his "Mathematics and Nature in the Chemical Texts of the Renaissance," p. 2, where he speaks of neglect by historians of science of "non-modern views of the role mathematics should play." But it is inconceivable that historians should not make judgments about that "non-modern role," and it has yet to be shown that its neglect has seriously distorted our view of the internal tradition. Miss Yates's comments on Bruno's mathematics are less conventionalist: "Bruno is not at all in the line of the advance of mathe-

ticularly prone, and its effects on historiography are the opposite from those of inductivism. Instead of selecting only those factors which lead to modern developments, the conventionalist is tempted to try to select as many factors as possible, to improve the total internal coherence of his story, as if to know all is to understand all. In his indictment of the inductivist, or whig, interpretation of history, Butterfield allowed that history must be selective, but required that judgments of importance should be the judgments of the period, not our judgments.[34] But how is one to follow this advice in the history of seventeenth-century science? Are we to take Mersenne's judgments of importance or Fludd's? To reply that the seventeenth century itself clearly accepted Mersenne's view rather than Fludd's is not sufficient, for this would be like relying on the popular verdict of Athens upon Socratic philosophy to dictate our judgments of intrinsic importance. So long as we select science as our subject matter, we are bound to write forward-looking history in the limited sense that we regard as important what we recognize as our own rationality, having some historical continuity with our own science. This does not imply, as we have seen, that we impose our own theories or even our own views of method on the science of the past. And if it seems in danger of becoming a circular definition of internal history as that which is continuous with our science according to our internal history, the only cure is to look more closely at the record to see whether the relative autonomy of internal history can be maintained in spite of possible disturbing factors. Pursuit of the hermetic hare has surely so far shown that in regard to the seventeenth century it can.

V

When all this has been said, however, it remains true that some of the language, the problems, and the concepts of natural magic persist in the new science, though in sterilized form. The oddity of the conceptual background of the internal history remains when it is read merely in its own terms, but it is now possible to treat the strange concepts as providing alternative theories, to be tested and argued for in accordance with recognized patterns of rationality. Gilbert, for example, adopts the explanatory analogy of "soul" for the magnet, but only after rejecting it for the attrac-

matical and mechanical science. Rather he is a reactionary who would push the Copernican diagram or a compass invention back towards 'mathesis.' " *Giordano Bruno,* p. 324.

[34] H. Butterfield, *The Whig Interpretation of History* (London: Bell, 1931), p. 24.

tion of electrified bodies on experimental grounds, and because for him "soul" is the only available principle of order and harmony which are so obviously exhibited by magnets. He does indeed quote favorably "Hermes, Zoroaster, Orpheus" as recognizing a world soul, but only to parallel his own conclusions, which come from experiments, not from philosophical speculations and ancient books: "we do not at all quote the ancients and the Greeks as our supporters."[35] Again, Bacon sometimes speaks of the powers of matter in terms of "desire," "aversion," "instinct," and in his later works inanimate bodies have "perceptions," but such expressions are usually found in the context of discussions of Democritan atomism, in which Bacon is fully aware of the difficulties of accounting for such phenomena as cohesion, impenetrability, and magnetism in terms of passive matter and motion alone.[36] Even so, his quasi-animism cost him the approval of Mersenne, who placed the line between old and new "between Campanella and Bacon on one side, and Galileo and Descartes on the other."[37]

Examples could be multiplied; a few more will have to suffice. When Henry More recants his allegiance to Cartesianism, this is not only in fear of its theological implications, but also because he sees that it will not work in detail even in mechanical examples.[38] In this he is followed by Leibniz, who introduced into mechanical philosophy that thoroughly legitimized offspring of sixteenth-century animism, the *vis viva* of mechanical systems. Even Gassendi is not a pure Democritan atomist. For example, he holds that attraction is a principle of motion equally fundamental with impulse, and explains it in terms of an attracting body emitting material rays which "grasp" the object attracted. There is no mechanical account of how this action takes place; in fact motion toward a magnet is ascribed to a quasi-soul in the object which is stimulated into motion by the magnetic emanations.[39]

Preeminently, of course, as has most recently been emphasized by J. E.

[35] W. Gilbert, *De Magnete* (London, 1600), Preface and p. 309. I have discussed Gilbert's animism in relation to his experimental method in "Gilbert and the Historians," *British Journal for the Philosophy of Science*, 11 (1960), 1, 130.

[36] For Bacon's relation to the natural magic tradition, see P. Rossi, *Francis Bacon, from Magic to Science*, trans. S. Rabinovitch (London, Routledge and Kegan Paul, 1968).

[37] Quoted in Lenoble, *Mersenne*, p. 12; see also *ibid.*, p. 30.

[38] This has been brought out in an unpublished M.Litt. thesis of P. M. L. Moir, "The Natural Philosophy of Henry More" (Cambridge, 1967).

[39] P. Gassendi, *Syntagma Philosophicum, Opera*, II (Lyons, 1658), p. 450.

McGuire,[40] Newton provides many examples of appeal to ancient wisdom and exploitation of nonmechanical concepts. Some of this is present in fragmentary form in the published work, but the attempt to understand its real significance in Newton's own thinking has demanded study of manuscripts neglected until very recently, and still mainly unpublished. About the light thrown by them on Newton's thought, three points may very briefly be made in the present context.

1. Those of Newton's ideas which were unavailable to his eighteenth-century successors were not historically significant for the development of eighteenth-century Newtonian science, which can therefore be studied independently of them. When such disciples as Desaguliers, s'Gravesande, Maclaurin, Musschenbroek, and Pemberton came to write epitomes of the *Principia*, these are indeed sometimes set in a theological framework, but it is one which is far removed from the explicit hermeticism of Newton's unpublished writings. The point is so obvious as to be hardly worth making that the public Newton is necessarily more significant for internal history than the private.

2. What I have called the sterilization of hermetic and Neoplatonic influences has gone very far in Newton even in his unpublished manuscripts. He does indeed suggest that the ancients attributed the power of gravity to atoms, and derived the proportion of the weights and distances of the planets from the inverse-square law for musical strings in unison, but, he goes on, they asserted this "without telling us the means *unless in figures* [my italics]: as by calling God harmony representing him and matter by the God Pan and his pipe."[41] Hence Newton's view is that we have to rediscover these things in order to decode the ancient myths which will tell us nothing on their own, although having decoded them we may indeed take comfort from the correspondence of our discoveries and theirs. Again, Newton seems to have had a conception of a continuum of being ranging from inert matter through immaterial, quasi-spiritual forces and principles to God, and it is certainly possible to find in this traces of the Dionysian celestial hierarchy of spirits ascending and descending between earth and heaven, mediated through the whole history of the concept of the Great Chain of Being. But now the terrible power of the spirits is exorcised, the

[40] J. E. McGuire, "Transmutation and Immutability: Newton's Doctrine of Physical Qualities," *Ambix*, 14 (1967), 69; "Force, Active Principles, and Newton's Invisible Realm," *Ambix*, 15 (1968), 154; "The Origin of Newton's Doctrine of Essential Qualities," *Centaurus*, 12 (1968), 233; and J. E. McGuire and P. M. Rattansi, "Newton and the 'Pipes of Pan,' " *Notes and Records of the Royal Society*, 21 (1966), 108.
[41] Quoted in McGuire and Rattansi, "Newton and the 'Pipes of Pan,' " p. 118.

laws of God are found throughout the hierarchy, and they are accessible not through mystic communion of the mind and the spirits, but through patient experimental investigation. Newton may use the image of microcosm and macrocosm, but the analogy is now that of man and created nature on the one hand, and the mind of God on the other, not that of Renaissance man bestriding the universe, his mind the measure of all things.

3. The details of Newton's philosophy of nature are not required for an understanding of the internal history of physics and astronomy any more than is Bruno's magical interpretation of the circles of Copernicus.[42] But this philosophy has its historical importance in answering a different kind of question, namely, how has one preeminent scientific thinker sought to reconcile a particular theology with a particular science? The same question might also be asked about Bruno, and has indeed been asked by Miss Yates, but just because Newton's conceptions are nearer to our own than Bruno's (and this is not a temporal matter, for Aristotle's are nearer to our own than Bruno's), Newton's proposed solution is more historically significant for us. The fortunate accident that large amounts of Newton's manuscript material survive does not so much enable us to complete the picture of early modern science, or even of Newton's own biography, as enable us to ask and answer a new set of questions which have for us their own intrinsic interest.

COMMENT BY ARNOLD THACKRAY

Dr. Hesse has presented us with an important but profoundly puzzling paper. Its importance lies in her pioneering discussion of the implications carried by the notions of "internal" and "external" history. With a masterly touch she exposes the—apparently insuperable—philosophical problems attending any attempt to construct a general defense of an autonomous internal history of science. I do not wish to comment on this first part of her paper, but simply to stand in awe of such commanding expertise. Things are different when I turn to the second, and puzzling, part of the argument.

Here, if I understand correctly, she wishes to defend on particular grounds and for a particular historical period that internalist view for which she has just denied any possibility of a general and abstract defense. But her particular defense *necessarily* demands appeal to just those general

---

[42] For which see Yates, *Giordano Bruno*, chapter XIII.

principles she has exposed as insufficient. Indeed, unless Dr. Hesse is resorting to that inductivism she herself deplores, I for one cannot understand how she finds it possible to defend a particular case without resort to general principles (however heavily disguised). Hence my puzzlement.

Dr. Hesse argues that for intellectual historians such as Collingwood (and presumably Koyré) "the timeless rationality of history" which they seek to impose is not in fact timeless but rather their own particular rationality. She admits that this approach is arbitrary and unsatisfactory. She also concedes that "no general specification of rationality seems . . . to be forthcoming." Yet Dr. Hesse then proceeds as if the general specification existed and the approach were adequate!

Her argument depends heavily on "the internal tradition of the mechanical philosophy"—as for instance in the statement that "the historian's task is . . . first, to follow up the loose ends of the received internal tradition." Now it is by no means self-evident (save possibly to a determined and modern-minded physicist) what such phraseology implies—the early Gilbert? the later Robert Boyle? Newton in his more theological moments? Again, Dr. Hesse's description *assumes* but does not prove that no rational(!) seventeenth-century figure curious about nature would adopt any other mode of approach. Similarly, she assumes rather than demonstrates the validity of the traditional internalist position when she writes that "so long as we select science as our subject matter, we are bound to write forward-looking history. . . ." If this were true (I do not think it is), one might reasonably inquire why any historian worth his salt should waste his time on such a necessarily anachronistic pursuit. Similar internalist assumptions are evident in the passages about how "our intuition" shows the difference between seventeenth-century science and hermeticism, about how science is to be "mathematical, mechanist" (alas for botany, zoology, geology, chemistry, etc.), and about how it is legitimate "to investigate such other factors related to the science of the period as have their own intrinsic interest" (according to whose timeless criteria of intrinsic interest?).

Though Dr. Hesse's fascinating paper raises a host of further questions, there is not time to pursue them. Therefore let me briefly reemphasize my unease.

Were Dr. Hesse merely saying "chaqu'un à son gout," her own particular preference for traditional internal history would be entirely unexceptionable. But in fact she wishes to be prescriptive, as her dismissal of

161

hermeticism indicates. It would therefore seem incumbent on Dr. Hesse *either* to show the prescriptive validity of those principles from which her assumptions derive (a process she herself has so excellently shown the difficulties of), or to allow less traditional and less determinedly "forward-looking" historians unhindered license peacefully to pursue their own hermetic hares.

Was Copernicus a Hermetist?

1. Internal-External History of Science

If a history of science is to deserve the name, it must be "internal." A fully qualified historian of science knows the discipline whose history he undertakes to write. He masters his subject not only in the chronological period which chiefly interests him but also in the earlier and later stages. Like an alert pedestrian trying to cross a busy two-way thoroughfare, he looks not only straight ahead but also to the right and to the left.

A historian of science, however, is more than a harried pedestrian. He understands not only his chosen subject but also its social setting, insofar as that background affected the science. To that extent he is a historian, and to that extent his history of science will be "external." By the nature of his craft the historian of science is perforce a hybrid creature. He is in part historian, in part scientist. His product is both internal and external, both scientific and historical.

If a history of science endeavored to be exclusively internal, it would inevitably miss the social forces which affect the development of science. On the other hand, any history of science which attempted to be exclusively external would ignore the inner self-correcting dynamic of science. A satisfactory history of science combines comprehension of the scientific subject matter with understanding of the historical period. Its narrative records positive achievements and illuminating failures. It pursues the ramifications of ideas, sound and unsound. It scrutinizes the societal pressures impinging on the thought and activity of scientists, while at the same time discarding supposititious farfetched and spurious connections.

2. The Case of Archimedes

We have been told that "Syracuse does little to explain Archimedes."[1] His discovery that a segment of a parabola equals four-thirds of the triangle

[1] A. C. Crombie, ed., *Scientific Change* (New York: Basic Books, 1963), pp. 855, 876.

having the same base and altitude as the segment[2] was an achievement in pure geometry, not connected with or explained by his residence in Syracuse. But it was the king of Syracuse who "persuaded Archimedes to make for him offensive and defensive machines for every type of siege"[3] in which Syracuse was embroiled, and with Syracuse Archimedes, who owed to this external pressure both intellectual stimulation and violent death at the hands of an enemy soldier.

3. Copernicus and Hermetism

We have also been told that

the core of the [Renaissance Neoplatonist] movement was Hermetic, involving a view of the cosmos as a network of magical forces with which man can operate. The Renaissance magus had his roots in the Hermetic core of Renaissance Neo-Platonism, and it is the Renaissance magus, I believe, who exemplifies that changed attitude of man to the cosmos which was the necessary preliminary to the rise of science.

The Renaissance magus was the immediate ancestor of the seventeenth-century scientist. "Neo-Platonism" . . . was indeed the body of thought which . . . prepared the way for the emergence of science.

The emergence of modern science should perhaps be regarded as proceeding in two phases, the first being the Hermetic or magical phase of the Renaissance with its basis in an animist philosophy, the second being the development in the seventeenth century of the first or classical period of modern science.

. . . revived Platonism with the accompanying Pythagoro-Platonic interest in number, the expansion of theories of harmony under the combined pressures of Pythagoro-Platonism, Hermetism, and Cabalism, the intensification of interest in astrology with which genuine astronomical research was bound up, and . . . the expansion of alchemy in new forms, it is, I think, impossible to deny that these were the Renaissance forces which turned men's minds in the direction out of which the scientific revolution was to come.[4]

It may be illuminating to view the scientific revolution as in two phases, the first phase consisting of an animistic universe operated by magic, the second phase of a mathematical universe operated by mechanics.[5]

[2] Archimedes, *Quadrature of the Parabola*, Proposition 17.
[3] Plutarch, *Life of Marcellus*, section 14.
[4] Frances A. Yates, "The Hermetic Tradition in Renaissance Science," in Charles S. Singleton, ed., *Art, Science, and History in the Renaissance* (Baltimore: Johns Hopkins Press, 1967), pp. 255, 258, 271, 273.
[5] Frances A. Yates, *Giordano Bruno and the Hermetic Tradition* (Chicago: University of Chicago Press, 1964), p. 452.

In the first phase of the scientific revolution Nicholas Copernicus published his *Revolutions* in 1543. We are told that

it is . . . in the atmosphere of the religion of the world that the Copernican revolution is introduced.

That religion of the world which runs as an undercurrent in much of Greek thought, particularly in Platonism and Stoicism, becomes in Hermetism actually a religion.

Egypt, and its magical religion, becomes identified with the Hermetic religion of the world.

Nor does Copernicus fail to adduce the authority of *prisci theologi* (though he does not actually use this expression), amongst them Pythagoras and Philolaus to support the hypothesis of earth-movement.[6]

Copernicus never adduced the authority of Pythagoras,[7] and he cited Philolaus not as a theologian but as an earth-moving astronomer (*Revolutions*, I, 5):

That the earth rotates, that it also travels with several motions, and that it is one of the heavenly bodies are said to have been the opinions of Philolaus the Pythagorean. He was no ordinary mathematician, inasmuch as Plato did not delay going to Italy for the sake of visiting him, as Plato's biographers report.[8]

Although Copernicus does not use the expression *prisci theologi*, he does discuss the *prisci philosophi*, the ancient philosophers who contended that the earth occupied the center of the universe.[9] He also mentions the *prisci mathematici*, the ancient mathematicians or astronomers who maintained that the earth was motionless.[10] Whatever theology may have been professed by these ancient philosophers, mathematicians, and astronomers did not concern Copernicus, who does not adduce their authority as much as he analyzes their shortcomings.

We are further told about Copernicus that "at the crucial moment, just after the diagram showing the new sun-centred system . . . comes a reference to Hermes Trismegistus on the sun":

At rest, however, in the middle of everything is the sun. For in this most beautiful temple, who would place this lamp in another or better position than that from which it can light up the whole thing at the same time? For, the sun is not inappropriately called by some people the lantern of the

[6] *Ibid.*, pp. 153, 4–5, 6, 153–154.
[7] Edward Rosen, "Was Copernicus a Pythagorean?" *Isis*, 53 (1962), 504–508.
[8] Only Diogenes Laertius so reports (*Lives of the Philosophers*, Plato, chapter 6).
[9] Copernicus, *Revolutions*, I, 7.
[10] *Ibid.*, V, 2.

universe, its mind by others, and its ruler by still others. The Thrice Greatest [labels it a] visible god . . . (Copernicus, *Revolutions*, I, 10).[11]

Where the foregoing quotation is cut off, Copernicus continues: "and Sophocles' *Electra*, the all-seeing." But Sophocles calls the sun all-seeing in his *Oedipus at Colonus*, not in his *Electra*.[12] Evidently Copernicus did not verify his quotation from Sophocles.

We were told just above that Copernicus makes "a reference to Hermes Trismegistus on the sun." But Copernicus does not mention the name Hermes, and his version of the accompanying epithet is "Trimegistus," as the manuscript written with his own hand clearly shows.[13]

We are told that in the passage quoted above from Copernicus's *Revolutions* (I, 10) "the main echo is surely of the words of Hermes Trismegistus in the *Asclepius*."[14] Hermes' words read as follows:

The sun illuminates the other stars not so much by the power of its light, as by its divinity and holiness, and you should hold him, O Asclepius, to be the second god, governing all things and spreading his light on all the living beings of the world, both those which have a soul and those which have not.[15]

The foregoing words of Hermes Trismegistus in the *Asclepius* do not call the sun a visible god, as Copernicus said that "Trimegistus" did. Yet we hear that Copernicus "quoted, near his diagram of the new system, Hermes Trismegistus in the *Asclepius* on the sun as the visible god."[16]

On the other hand, the expression "visible god" does occur in an ancient theologian whom Copernicus chides as follows: "Lactantius, otherwise an illustrious writer but hardly a mathematician, speaks quite childishly about the earth's shape when he mocks those who declared that the earth has the form of a globe" (*Revolutions*, Dedication-Preface). These puerilities concerning the earth's form were uttered by Lactantius in his *Divine Institutes* (III, 24). In that same work (IV, 6) Lactantius quoted Hermes as saying (in Greek) that "the second god was created visible." This visible second god was misequated by Lactantius with Jesus, although Hermes

---

[11] Yates, *Giordano Bruno*, p. 154.

[12] Edward Rosen, "Copernicus' Quotation from Sophocles," in *Didascaliae, Studies in Honor of Anselm M. Albareda*, ed. Sesto Prete (New York: Rosenthal, 1961), pp. 369–379.

[13] *Nikolaus Kopernikus Gesamtausgabe*, I (Munich and Berlin: Oldenbourg, 1944), fol. 10r, line 6.

[14] Yates, *Giordano Bruno*, p. 154.

[15] *Ibid.*, pp. 152–153.

[16] *Ibid.*, p. 238; Frances A. Yates, *The Art of Memory* (London: Routledge, 1966), p. 153.

obviously means the perceptible universe.[17] When Copernicus says that "many of the philosophers have called it a visible god," perhaps he is echoing Lactantius's Hermes, among others, but here (*Revolutions*, I, Introduction) Copernicus's visible god (*visibilem deum*) is the universe, not the sun.

Copernicus uses the Latin words *visibilem deum* and not the Greek *theon horaton*, as quoted by Lactantius from the original Greek text of the *Asclepius*, which was available to Lactantius, but since his time has been lost. The *Asclepius* has survived in a Latin translation, which renders our passage as *qui videri . . . possit*.[18] Since this Latin translation, which used to be misattributed to Lucius Apuleius, has neither *visibilem* nor *deum*, it did not provide the model for Copernicus's visible god, whether this was the universe, according to many of the philosophers, or the sun, according to "Trimegistus."

The numerous Greek passages inserted by Lactantius in his *Divine Institutes* must have annoyed readers unfamiliar with that language. For when Lactantius later wrote the *Epitome* of his *Divine Institutes*, he eliminated the Greek quotations, including ours, which he replaced by his own Latin translation. This contains the expression *deum visibilem* in chapter 37 (42),[19] by contrast with *qui videri . . . possit* in Pseudo-Apuleius. However, in Copernicus's time Lactantius's *Epitome* was printed from a defective manuscript lacking chapter 37 (42). Hence Copernicus never actually laid eyes on *deum visibilem* in Lactantius's *Epitome*. Nor did that expression occur in Pseudo-Apuleius's translation of the *Asclepius*. Had Copernicus ever handled a copy of Hermes, with his knowledge of Greek he would not have fumbled the epithet "Trismegistus." As in the case of his miscitation of Sophocles, he may have relied on an imperfect recollection of something he had once heard said by somebody, presumably one of his professors with access to the complete manuscript of Lactantius's *Epitome* on which our modern editions are based,[20] or to one of the manu-

---

[17] Walter Scott, *Hermetica* (Oxford: Clarendon Press, 1924–36), I, 299; III, 19–20, 47–48; *Corpus hermeticum*, ed. A. D. Nock and A.-J. Festugière (Paris, 1945–54), II, 305, 365.

[18] Scott, *Hermetica*, I, 298, line 16; Nock and Festugière, eds., *Corpus hermeticum*, II, 305, line 2.

[19] *Corpus scriptorum ecclesiasticorum latinorum*, 19 (reprinted, New York: Johnson, 1965), p. 713, line 6.

[20] The discovery of the complete manuscript of Lactantius's *Epitome* was announced in *Giornale de' letterati d'Italia*, 6 (1721), 456, and *Bibliothèque ancienne et moderne*, ed. Jean LeClerc, 27 (1727), 339.

scripts of Lactantius's *Divine Institutes* containing a Latin translation of our Hermes passage.[21]

Nevertheless we are told that "Copernicus' discovery came out with the blessing of Hermes Trismegistus upon its head, with a quotation from that famous work in which Hermes describes the sun-worship of the Egyptians in their magical religion."[22] What Copernicus mistakenly believed to be a quotation is not found in the author miscalled "Trimegistus" by Copernicus, who obviously had only the slightest acquaintance with the hermetic literature, which he did not know at first hand. Yet we read that "even the impulse towards the breaking down of the old cosmology with heliocentricity may have as the emotional impulse towards the new vision of the sun the Hermetic impulse towards the world, interpreted first as magic by Ficino, emerging as science in Copernicus. . . ."[23]

Copernicus's emotional passage about the sun (*Revolutions*, I, 10) was quoted above. We are told that in it "there are perhaps echoes of Cicero's words for the sun in that famous Dream."[24] Cicero in *Scipio's Dream* (*Republic*, VI, 17) calls the sun "the universe's mind" (*mens mundi*), and Copernicus echoes *mundi . . . mentem*. When Copernicus undertook to "re-read the works of all the philosophers which I could obtain," he specifies (*Revolutions*, Dedication-Preface, I, 5) that he found a pivotal passage in Cicero. In calling the sun the universe's "ruler" (*rectorem*), Copernicus echoes *rector* in the *Natural History* (II, 12) of Pliny, from whom he took many expressions. In the cosmogonical story in the *Timaeus* (39 B) Plato's creator Craftsman kindled only one light, "which we now call the sun," in order that it might shine as far as possible throughout the entire heaven. Hence for an unswerving Platonist, as distinguished from a Neoplatonist, the sun was the universe's lantern (*lucernam mundi*),[25] the last of the five labels attached to the sun by Copernicus.

We are told that "Copernicus himself associated his discovery with Hermes Trismegistus."[26] That association, taking the form of a nonexistent quotation from a jumbled name, occurs in the company of Sophocles, Cicero, Pliny, and the Platonists.[27] In Copernicus's emotional passage

---

[21] *Corpus scriptorum ecclesiasticorum latinorum*, 19, pp. 288–289.

[22] Yates, *Giordano Bruno*, pp. 154–155.

[23] *Ibid.*, p. 156.

[24] *Ibid.*, p. 154.

[25] This description was transferred to the Virgin Mary by the tenth-century nun Hrotsvitha, *Opera* (Berlin, 1902), p. 32, line 79.

[26] Yates, *Giordano Bruno*. p. 168.

[27] Yet we have been told that Copernicus's "authorities are immediately Neopla-

about the sun, the hermetic association is a shaky one-fifth of the five associations. The three words in which it is expressed (*Trimegistus visibilem deum*) occupy less than half a line in Copernicus's manuscript of the *Revolutions*. This handwritten volume contains more than 200 folios, averaging 10 words to the line and 40 lines to the page, so that the hermetic association amounts to about 0.00002 of the *Revolutions*. Copernicus's other works and his correspondence show no hermetic association at all. Yet we are told that "Bacon's admirers have often been puzzled by his rejection of Copernican heliocentricity and of William Gilbert's work on the magnet. . . . These notions might have seemed to Bacon heavily engaged in extreme forms of the magical and animist philosophy or like the proud and erroneous opinions of a magus."[28]

4. Bruno and Copernicus

We are also told that "Copernicus might well have bought up and destroyed all copies of the *Cena* had he been alive."[29] Had Copernicus been alive in 1584, when Giordano Bruno published his *Cena de le ceneri* (*Ash Wednesday Supper*), he would have read in the *Cena's* Third Dialogue that "Copernicus didn't believe that the earth moves, because this is an incongruity and impossibility. On the contrary, he attributed the motion to it, rather than to the sphere of the stars, for convenience in computing." The spokesman for Bruno replies: "It is certain that Copernicus understood the statement as he uttered it, and proved it with all his might." This uncompromising insistence that Copernicus maintained the earth's motion to be a physical fact provokes the question why the contrary opinion is expressed "if it cannot be inferred from some statement by him." The source of this misinterpretation of Copernicus is promptly identified as "a certain preliminary Address, stuck in by an ignorant and insolent jackass."[30] Had Copernicus been alive in 1584, he might well have bought up all copies of the *Cena* in order to distribute as widely as possible its forthright denunciation of the interpolated anonymous prefatory Address which utterly falsified his geokineticism. Bruno's *Cena* first publicly exposed this fraud,[31] which nevertheless continued to fool innumerable read-

tonic." Thomas S. Kuhn, *The Copernican Revolution* (Cambridge, Mass.: Harvard University Press, reprinted 1966), p. 130.

[28] Yates, in Singleton, ed., *Art, Science, and History in the Renaissance*, p. 268.

[29] Yates, *Giordano Bruno*, p. 297.

[30] Bruno, *La cena de le ceneri*, ed. Giovanni Aquilecchia (Turin: Einaudi, 1955), p. 146, lines 4–21.

[31] This passage of the *Cena* was discussed by Frances A. Yates, "The Religious Policy

ers, including Delambre, the great nineteenth-century historian of astronomy.[32]

At Oxford University in 1583, according to a contemporary, Bruno undertook "to set on foote the opinion of Copernicus, that the earth did goe round, and the heavens did stand still; wheras in truth it was his owne head which rather did run round, & his braines did not stand stil."[33]

Fourteen years later, in 1597, upon receiving a Copernican book from Kepler, Galileo wrote to its author: "Many years ago I was converted to the theory of Copernicus. . . . I wrote out many reasons in favor of it, and rebuttals of opposing arguments. But I have not yet dared to publish them. . . . I would surely have the courage to make my thinking public if there were more people like you. But since there are not, I shall avoid such involvement."[34] Prudent Galileo was not burned at the stake like Bruno; he was merely sentenced to life imprisonment.

Although Bruno was not a professional astronomer, nobody before him understood and asserted that the sun is a star and the stars are suns.[35] This understanding was not attained by Copernicus, who was a professional astronomer amidst his other occupations. We recall having been told above that "even the impulse towards the breaking down of the old cosmology with heliocentricity may have as the emotional impulse towards the new vision of the sun the Hermetic impulse towards the world, interpreted first as magic by Ficino, emerging as science in Copernicus. . . ."[36] Further research is recommended to us: "Much more detailed 'ferreting out' of the motives behind the work of Renaissance scientists is needed before more positive statements can be made as to the influence upon them of the dominant Hermetic-Cabalist tradition."[37] No human ferret is needed to discover the motive behind the work of Copernicus, who said quite openly: "I was impelled to consider a different system of deducing the motions of the universe's spheres for no other reason than the realization

of Giordano Bruno," *Journal of the Warburg and Courtauld Institutes*, 3 (1939–40), 188, without understanding its significance for the history of science.

[32] Edward Rosen, "The Ramus-Rheticus Correspondence," *Journal of the History of Ideas*, 1 (1940), 366–367, citing I, 139–140, in Delambre's *Histoire de l'astronomie moderne*, which is being reissued by Johnson Reprint Corporation.

[33] Robert McNulty, "Bruno at Oxford," *Renaissance News*, 13 (1960), 303.

[34] *Le Opere di Galileo Galilei*, national edition, X, 68, lines 17–27; Edward Rosen, "Galileo and Kepler," *Isis*, 57 (1966), 263.

[35] Bruno, *Opera latine conscripta* (reprinted, Stuttgart-Bad Cannstatt: Frommann-Holzboog, 1961–62), vol. I, part 1, p. 212.

[36] Yates, *Giordano Bruno*, p. 156.

[37] *Ibid.*, p. 449.

that the mathematicians do not agree among themselves in their investigations of this subject" (*Revolutions*, Dedication-Preface). Copernicus's motive belongs to the internal, rather than the external, history of science. No hermetic-cabalist tradition was dominant in his mind. It was the opposition of Aristotelians and theologians that he feared.[38]

5. Modern Science and Hermetism

Mersenne's judgment of Campanella ("he will teach us nothing in the sciences"[39]) may be extended to virtually all the other persons in the hermetic-cabalist tradition. In the few borderline cases, standing with one foot in either camp, to what extent, if any, did their extrascientific beliefs affect their scientific work? Out of Renaissance magic and astrology came, not modern science, but modern magic and astrology.

[38] Edward Rosen, *Three Copernican Treatises*, 2nd ed. (New York: Dover, 1959), p. 23.
[39] Robert Lenoble, *Mersenne; ou La naissance du mécanisme* (Paris: J. Vrin, 1943), p. 41.

Philosophy of Science: A Subject with a Great Past

1. While it should be possible, in a free society, to introduce, to expound, to make propaganda for any subject, however absurd and however immoral, to publish books and articles, to give lectures on any topic, it must also be possible to *examine* what is being expounded by reference, not to the *internal* standards of the subject (which may be but the method according to which a particular madness is being pursued), but to standards which have the advantage of being simple, commonsensical, and accepted by all. Using such standards as a basis of judgment we must confess that much of contemporary philosophy of science and especially those ideas which have now replaced the older *epistemologies* are castles in the air, unreal dreams which have but the name in common with the activity they try to represent, that they have been erected in a spirit of *conformism* rather than with the intention of influencing the development of science, and that they have lost any chance of making a contribution to our knowledge of the world. (The medieval problem of the number of angels at the point of a pin had some rather interesting ramifications in optics and in psychology. The problem of "grue" has ramifications only in the theses of those unfortunate students who happen to have an engruesiast for a teacher.) This is my opinion. Let me now give some reasons for it.

2. The scientific revolution of the sixteenth and seventeenth centuries is characterized, among other things, by a close collaboration between science and philosophy. This is a direct consequence of the way in which science was debated both in antiquity and in the Middle Ages. The reaction against "medieval science," which in many cases was but a reaction against certain petrified aspects of it, leads to the development of new philosophical principles. It does not lead to a split between science and philosophy. The new philosophy that is being gradually developed is of course used to expose and to remove the hardened dogmas of the schools.

However, it has also a quite decisive role in building the new science and in defending new theories against their well-entrenched predecessors. For example, this philosophy plays a most important part in the arguments about the Copernican system, in the development of optics, and in the construction of a new and non-Aristotelian dynamics. Almost every work of Galileo is a mixture of philosophical, mathematical, and physical principles which collaborate intimately without giving the impression of incoherence. This is the *heroic time* of the scientific philosophy. The new philosophy is not content just to *mirror* a science that develops independently of it; nor is it so distant as to deal just with alternative *philosophies*. It plays an essential role in building up the new science that was to replace the earlier doctrines.[1]

3. Now it is interesting to see how this active and critical philosophy is gradually replaced by a more conservative creed, how the new creed generates technical problems of its own which are in no way related to specific scientific problems (Hume), and how there arises a special subject that codifies science without acting back on it (Kant). One can say, without too much simplification, that the change is essentially due to *Newton*. Newton invents new theories, he proposes a radically empiricist methodology, and he claims that he has obtained the former with the help of the latter. He supports his claim by a manner of presentation that seems indeed to suggest, at least at first sight, that his optics and his celestial mechanics are the perfect results of a perfect method perfectly applied. Having convinced most of his contemporaries, he creates additional support both for his science (it has been obtained in a methodologically sound way and must therefore be free from major mistakes) and for his methodology (it has led to perfect scientific results and must therefore be the correct method).[2] Of course, his presentation is quite misleading, it is full of holes, fallacies, contradictions, and he himself violates *every single rule* he proposes. Yet it was influential enough to have blinded scientists, historians (including some very recent students of the history of optics, such as Westfall), and philosophers alike.[3] 'Experience,' from now on, means

---

[1] For details concerning Galileo and his difference from Descartes and Bacon, see my essay "Bemerkungen zur Geschichte und Systematik des Empirismus," in Paul Weingartner, ed., *Grundfragen der Wissenschaften und ihre Wurzeln in der Metaphysik* (Salzburg: Pustet, 1967).

[2] The development in the quantum theory from 1927 to about 1955 was of exactly the same kind.

[3] For optics see Goethe, *Theory of Colours*, which contains a very perceptive account of the ideological development just mentioned; V. Ronchi, *Histoire de la lumière* (Paris: Colin, 1956); A. I. Sabra, *Theories of Light from Descartes to Newton* (London:

Paul K. Feyerabend

either the results of Newton's experiments as described by him (optics), or the premises of his deductions (celestial mechanics), but it *also* means, by virtue of Newton's connecting maneuver, the safe, irrevocable, and gradually expanding basis of scientific reasoning. Small wonder that thinkers who seemed to sense a flaw but who lacked either the patience or the talent to combine their critical intuition with spectacular scientific discoveries were not heard, and were increasingly isolated.[4] To survive, they changed their target from science to philosophy and thus created (or, rather, continued—for there was always a tradition that developed philosophy out of its own problems and with only the most tenuous relation to science) a self-sufficient subject, content with discussing its own problems. Science, on the other hand, being separated from philosophy, had to rely on intuitions of a different and much more narrow kind. The possibility of a fundamental criticism became more and more remote. In this respect, too, the situation was surprisingly similar to the situation that exists in certain parts of science today.[5] However, there is one difference. The nineteenth century produced one philosopher who was not prepared to accept the status quo, who was not content to criticize science from the safe distance of a special subject either, but who proceeded to suggest concrete means for its change. The nineteenth century produced Ernst Mach.

4. Ernst Mach's "philosophy"[6] contains a general criticism of the science of his time, including the house philosophy of the contemporary Newtonians, and a philosophy of science that completely abandons the idea of a foundation of knowledge. The criticism and the positive views are illustrated by his work in the history of science where factual and epistemological considerations are once more merged in perfect harmony, and they are given strength by the exhibition of shortcomings right in the center of the most advanced theories of the nineteenth century. The criticism and

Oldbourne, 1967); as well as my discussion of Newton in "Classical Empiricism," in *The Methodological Heritage of Newton*, ed. R. E. Butts and J. W. Davis (Toronto: University of Toronto Press, 1969).

[4] An exception was Faraday, but his background philosophy remained almost completely unknown.

[5] The similarity becomes even greater in view of the fact that the famous chocolate-layer-cake model of scientific knowledge that has been developed by Nagel, Hempel, and others is nothing but a more sophisticated (and less clear) repetition of Newton's views, down to the last mistake. Cf. "Classical Empiricism," footnote 8, as well as my review of Nagel in the British Journal for the Philosophy of Science, 17 (1966), pp. 237–249.

[6] "Above all, there is no Machian philosophy; there is, at the most, a methodology of science and a psychology of knowledge and like all scientific theories these two things must be regarded as preliminary and incomplete attempts." *Erkenntnis und Irrtum* (Leipzig: J. A. Barth, 1905), p. vii, footnote (against Hoenigswald).

Mach's own positive suggestions have been extremely fruitful, both in the sciences and in philosophy. In science Mach's suggestions have contributed to the development of the general theory of relativity and they play an essential role in more recent discussions, now that interest in general relativity has been revived. They had also a decisive influence, not always beneficial, on the founders of the quantum theory (even Schrödinger was often heard to say, quite emphatically: "Aber wir koennen doch nicht hinter Mach zurueckgehen!"). In philosophy it was a different story, as will be seen below.

In order to understand Mach we must distinguish most carefully (and more carefully than Mach himself did on various occasions) between his *general methodology* (s.v.v.) and the more specific *hypotheses* he used as starting points of research.

A general methodology is independent of any particular assertion about the world, however trivial, and however obvious. It is supposed to provide a point of view from which *all* such assertions can be judged and examined. It will not assume a dichotomy between an objective world and a perceiving subject who explores the world (using his mind and his senses) and gradually increases his knowledge of it. Such a dichotomy is presupposed by almost all science, it is the instinctual basis of everyday behavior (at least in Western societies), and it has been professed with almost religious fervor by thinkers who otherwise pride themselves on having made criticism a principle of science and of philosophy. Yet—is it not possible that this view is mistaken? Is it not possible that it neglects or misrepresents phenomena of an intermediate nature which show that the boundary is rather ill-defined, and perhaps altogether nonexistent? And if we *admit* this possibility, must we not ask ourselves how such a failure of realism can possibly be detected without relying on realism in the process? At this point recourse is usually made to *sensations*, and this is quite appropriate if the existence of sensations, bundles of sensations, lawful connections between sensations, is regarded as an *alternative hypothesis* rather than as an eternal measuring stick of any subject that is not explicitly about sensations. For just as the existence of the real world is a topic for critical discussion, in the very same manner the existence of sensations is a topic for critical discussion. General methodology, therefore, must refer neither to the one nor to the other (although it may provide rules for playing them off against each other) and, indeed, *no such reference is made by Mach*. According to Mach the task of science is to find simple and regular connections between *elements*. Let us analyze the various parts of this assertion.

5. All regularity, says Mach, is imposed, or constructed, yet it never fits all relevant cases. This assertion is not entirely without content. It is assumed that there exists a domain where regularities are produced and another domain which has idiosyncrasies of its own and which can never be fully comprehended, or tamed, by the imposed laws. Every rule, every law, even most precise formulation dealing with events carefully prepared is bound to have exceptions, and even a perception which at first sight seems perfectly symmetrical loses this symmetry on closer inspection. This idea of the two domains comes forth very clearly in passages such as "What is constant, the rule, a point of departure does not exist except in our thinking," which have a decidedly Kantian flavor.[7] "It is we who glue things together, not nature."[8] This seems to push us irrevocably toward a theory of sensations *but it does not*, for the elements which are related in this more or less regular manner are carefully distinguished from sensations. Speaking of sensations, Mach says quite explicitly, *already entails acceptance of a one-sided theory.*[9] Elements are not sensations. They are not perceptions either, for perceptions are rather complex entities, containing memories and attitudes as well as "natural habits"[10] of the human species. They are certainly not material objects. *They are open places, to be filled and refilled by the results of research.* All that is asserted is therefore that complexes consisting of elements which may in turn be complexes (though perhaps complexes of a simpler kind, at least at a particular stage of knowledge) are assembled into higher units whose stability is always in question and may be upset either by new theoretical procedures, or by a change of our ordering habits, or by the realization that essential things have been left out, or by a change in the elements, and so on. Now if we want to remove as many particular assumptions as possible, if we want to arrive at a truly general methodology, then we must abandon this opposition between an ordering mind and an ordered material also and must restrict ourselves to stating a *development* of prima facie simple elements which arrange and rearrange themselves, dissolve and recombine in

[7] Notebook III, February 1882, p. 82. Quoted from Hugo Dingler, *Die Grundgedanken der Machschen Philosophie* (Leipzig: J. A. Barth, 1924).

[8] Notebook I, May 1880, p. 58: "Nur wir kleben zusammen, die Natur nicht."

[9] "Da aber in diesem *Namen* ('sensations') schon eine *einseitige Theorie* liegt, so ziehen wir vor, kurzweg von *Elementen* zu sprechen . . ." *Analyse der Empfindungen* (Jena, 1900), p. 15, italics in the original.

[10] *Ibid.*, p. 137.

different patterns, a view that has great similarity with what is explained in Hegel's *Logik* (except that it does not contain the more specific hypothesis of modified preservation).[11]

Now research proceeds by first filling this general scheme with more specific content and then developing consequences, always remaining critical of the particular hypotheses used. The hypothesis to which Mach appeals rather frequently identifies the elements with *sensations*. This hypothesis plays an important part in Mach's criticism of contemporary physics and philosophy. *But it is never regarded as being above criticism.* It functions rather like, let us say, the principle of Lorentz invariance which is constantly used for the criticism of theories without being itself exempt from criticism.

That sensations cannot be an absolute basis for Mach becomes clear from the very title of his book *The Analysis of Sensations.* Sensations are to be *analyzed.* The complexity that hides behind the simple appearance is to be discovered and reduced to other and perhaps as yet unknown elements. These new elements are again in need of analysis, "they must be further examined by physiological research,"[12] and so on. The hypothetical character of sensations also becomes clear from Mach's attempt[13] to put them in their proper place and to free science from their dominance. In debates with Hugo Dingler whom he praised in the Foreword to the seventh edition of his *Mechanik* he admits to being a "non-empiricist" or a "not-only-empiricist,"[14] and he follows with interest Dingler's earlier attempts to eliminate experience from applied geometry. This partial anti-empiricism of Mach (which in truth is nothing but his universal criticism applied to the empiricist ideology) is a fascinating topic for research.

Now having *adopted* a *particular cosmological hypothesis* which seems plausible to him (elements are sensations) he *applies* it, and introduces the principle that science should contain only such concepts which can be connected with sensations. (It is useful to again compare this principle with the principle that scientific theories should be Lorentz-invariant.) But it is interesting to see that he never relies on this "empirical" criticism alone. Since the use of sensations is based on a hypothesis it is necessary to

---

[11] For details see section 3 of my essay "Against Method," in vol. IV of the *Minnesota Studies in the Philosophy of Science*, ed. M. Radner and S. Winokur (Minneapolis: University of Minnesota Press, 1970).

[12] *Analyse der Empfindungen*, p. 20.

[13] Notebooks, vol. II, p. 16 from back.

[14] H. Dingler, *Die Grundgedanken der Machschen Philosophie*, p. 61, footnote.

check it at every point by independent arguments. The objections to absolute space and to atomism are a case in point.

Absolute space is nonempirical, it cannot in any way be connected with sensations. Are there perhaps reasons which nevertheless force us to adopt it? There is Newton's bucket argument. This argument is invalid not because it makes appeal to things which are not related to sensations, but because it rests on a false assumption. Centrifugal forces, it is said, arise even if there is no material around with respect to which rotation can be asserted. This is untrue, for there are the fixed stars. If one now asserts that the fixed stars have no influence one asserts what has to be shown, viz., that centrifugal forces are not due to a particular relation to the rest of the universe. In this form the argument is circular *in addition to* yielding a metaphysical conclusion. Is it perhaps possible to decide the question by experiment? In order to advance in this direction Mach develops an alternative theory of his own in which inertial forces depend on the presence of matter and can be changed by the movement of large neighboring masses (Friedländer's experiments were entirely in Mach's spirit). This alternative theory still survives and is now in the center of discussion.

Mach's criticism of the atomic theory is another example of the way in which he combines his sense-data hypothesis with other, and more concrete, arguments. Again the metaphysical criticism is supported by difficulties of the existing atomic theories (reversibility objection; recurrence objection; stability of the atom). There is no reason to discredit Mach for his unwavering opposition to atoms.[15] For the mechanical atom of the nineteenth century has indeed disappeared from the scene.

6. To sum up: Mach develops the outlines of a knowledge without foundations. He introduces cosmological hypotheses as temporary measuring sticks of criticism. The cosmological hypothesis he appeals to most frequently assumes that all our knowledge is related to sensations. Using this hypothesis he criticizes physical theories such as the atomic theory and

---

[15] There has arisen the belief that Mach changed his mind in the last few years of his life and that he finally accepted atomism. This is based on a story told by Ernst Mayer, the head of the Radium Institute in Vienna at the time, who showed Mach a scintillation screen and reports him as saying, "Jetzt glaube ich an die Atome." There is no reason to doubt Mayer's story (Mayer was an honest man). But there is much reason to doubt the *interpretation* that is usually put on it (this has been pointed out to me by Dr. Heinz Post of Chelsea). In itself the story is rather implausible. Why should Mach be convinced by a peripheral phenomenon such as the scintillation screen? There is also the fact that the optics which was published after the incident shows him as hostile to atoms as ever. Dr. Post assumes that Mach's reaction was simply the reaction of a kind man to an enthusiastic (and rather deaf) demonstrator.

Newtonian mechanics. The cosmological criticism is strengthened by a more specific examination and by the suggestion of alternatives. In this way Mach achieves unity between science and philosophy that had been lost as the result of the dominance of Newton's physics and of his philosophy.

7. The unity is achieved *and is at once lost again* by the empiricist successors of Mach. No longer is there a combined effort to criticize and to improve physics, nor is the critical attitude retained that was such an essential part of Mach's research. It is quite impossible to narrate and to explain all the developments which led from Mach through the Vienna Circle (plus the Berlin group, plus the Scandinavian groups, and so on) to the contemporary situation. Here is a fascinating field of study for the historian of ideas. All I can do in the present short note is to make a few scattered remarks and to formulate some questions.[16]

The first and most noticeable change is the transition from a critical *philosophy* to a *sense-data dogmatism*: elements are replaced by sensations not just temporarily, and as a matter of hypothesis, but once and for all. Sensations are regarded as the solid foundation of all knowledge. It would be interesting to know how this replacement came about and how it happened to be connected with the name of Mach. Dr. Laudan has a simple suggestion and I am inclined to follow him (in the history of ideas the profound ideas are not always the best ones) : Mach was either not read at all or read with little care. As a result the important distinction between basic philosophy and special hypotheses was never discovered, let alone understood. Besides, the end of the nineteenth century teems with sensationalistic philosophies. Of course, there were also more tolerant philosophies (Neurath; Carnap), but they never got beyond the abstract statement that the selection of an observation language (sense-data language; physical-thing language) was a matter of *choice*. There was never any analysis of the concrete steps and of the concrete arguments which might favor one choice rather than another.[17] *In practice* one assumed that a certain language was given and proceeded to discuss, reform, evaluate *theories* on its basis. And it did not last long, and the abstract principle of tolerance

[16] Dr. Laudan is now studying the history of the early Vienna Circle and he has already arrived at some very interesting results. The following brief sketch is influenced by some of these results, though it is written much more carelessly than Dr. Laudan would dare.

[17] I have discussed such arguments in a particular case in sections 5–10 of "Against Method."

179

was abandoned also and was replaced by a more restrictive philosophy.[18]

Secondly, criticism of *science* is replaced by *logical reconstruction* which, in plain English, is nothing but a highly sophisticated brand of conformism. This development is partly connected with the first: if sensations are the foundation of knowledge, then the chains which connect, say, Maxwell's equations with the basis should be clearly exhibited and should be formulated as precisely as possible. Having found a reformulation of the chains one hoped to be able to rewrite the equations themselves, using the same precise language. Now there *did* exist a language which seemed to provide ready translations for mathematical formulas, viz., the language of *Principia Mathematica*. The logical reconstruction of science therefore amounted to the attempt to rewrite science in PM and to clearly exhibit all the links with the "basis."

Now an attempt of this kind may be understood in at least two different ways; *critically*: what cannot be reconstructed in this fashion must be eliminated; *conformistically*: what cannot be reconstructed in this way shows that the methods of reconstruction are faulty and must be revised. The critical version survived for some time—for had not Mach changed science by using a vague form of the verifiability principle and was not the theory of relativity the glorious result of this newly found philosophical method? But being isolated, and without help from more specific arguments (no one thought of inventing more concrete criticisms in addition to the arguments flowing from the reconstruction program), the verifiability criterion lacked strength to survive a showdown with science. In his famous debate with Planck Mach refused to revise his view of science just because it differed so much from the actual thing. Science has become a church, he said, and I have no intention to be a member of a church, scientific or otherwise. I therefore gladly renounce the title of a scientist: "Die Gedankenfreiheit ist mir lieber."[19] He argued from a strong position having both a plausible cosmological hypothesis and specific difficulties of the existing science on his side to prove his point. Not a single positivist was bold enough to wield the verifiability criterion in the same spirit. Driesch was criticized, yes, but this was a rather trivial case. There was no attempt to cut physics down to size in a similar manner. As a result the idea of a logical reconstruction became comformistic. The task was now to correctly

[18] For a sketch of this development see my essay "Explanation, Reduction, and Empiricism" in vol. III of the *Minnesota Studies in the Philosophy of Science*, ed. H. Feigl and G. Maxwell (Minneapolis: University of Minnesota Press, 1962).

[19] *Zwei Aufsaetze* (Leipzig, 1919).

present rather than to change science (it did not take long and this attitude was extended to the language of common sense also).

8. This task in turn was soon transformed into problems of a different kind, some of which were no longer related to science at all. Trying to imitate the scientific procedure (or what one thought was the scientific procedure), one started with the discussion of simple cases. Now a simple case, in this context, was not a case that looked simple when viewed from inside science. It was a case that looked simple when formulated in the language of Principia Mathematica. In addition one concentrated on the relation to the evidence, omitting all those problems and aids which arise from the fact that every single statement of science is embedded in a rich theoretical net and chosen to fit this net in one way or another (today a "simple" physical theory is a theory that is relativistically invariant). Now I do not wish to say that the properties of nets are not being discussed, for there is a large literature on precisely that point. All I wish to assert is that there exists an enterprise[20] which is taken seriously by everyone in the business where simplicity, confirmation, empirical content are discussed by considering statements of the form $(x)(Ax \to Bx)$ and their relation to statements of the form Aa, Ab, Aa&Ba, and so on and this enterprise, I assert, has nothing whatever to do with what goes on in the sciences. There is not a single discovery in this field (assuming there have been discoveries) that would enable us to attack important scientific problems in a new way or to better understand the manner in which progress was made in the past. Besides, the enterprise soon got entangled with itself (paradox of confirmation; counterfactuals; grue) so that the main issue is now its own survival and not the structure of science. That this struggle for survival is interesting to watch I am the last one to deny. What I do deny is that physics, or biology, or psychology can profit from participating in it.

9. It is much more likely that they will be retarded.

This can be shown both theoretically, by an analysis of some rather general features of the present state of the program of reconstruction, and practically, by exhibiting the sorry shape of the subjects (sociology; political science) which have made a vulgarized version of the program their chief methodological guide. I shall restrict myself to a brief discussion of two theoretical difficulties.[21]

[20] Prominent textbook: I. Scheffler's Anatomy of Inquiry (New York: Knopf, 1963).
[21] For more detailed analyses involving historical material the reader is referred to my essays "Problems of Empiricism" in Beyond the Edge of Certainty, ed. R. G. Colodny (Englewood Cliffs, N.J.: Prentice-Hall, 1965); "Problems of Empiricism, II," in Aim

Paul K. Feyerabend

The empirical model assumes a language, the "observation language," that can be specified independently of all theories and that provides content and a testing basis for every theory. It is never explained how this language can be identified nor are there any indications how it can be improved. Carnap's rule that the language should be used by a language community as a means of communication and that it should also be observational in the vague sense of containing quickly decidable atomic sentences is clearly unsatisfactory. In the sixteenth and seventeenth centuries a language of this kind would have included devils, angels, incubi, succubi, absolute motions, essences, and so on. Now this fact, taken by itself, is no objection against the scheme *provided* the scheme permits us to overcome devils in a rational manner. This is not the case. An observation language is a final standard of appeal. There are no rules which would permit us to choose between different observation languages and there is no method that would show us how an observation language can be improved.[22] We may of course try to extract the devils from their observational surroundings and to formulate a demonic *theory* that can be interpreted and tested on the basis of a different observational idiom. This procedure is *arbitrary* (why should we not invert our procedure and test Dirac's theory of the electron on the basis of an Aristotelian observation language? And it is *excluded* by the very same principles which make the model a judge of the empirical content of a notion (removing the observational embroideries of a concept changes the concept). The model is therefore incomplete at a very decisive point. And in order to remove the impression that it needs the *devil* to exhibit this incompleteness we should perhaps add that the transition from the physics of Aristotle to the physics of Galileo and Newton creates exactly the same problems. Here we see very clearly how the transition is achieved[23] and how complex are the arguments that bring it about (Galileo, for example, introduces a new observation language *in order to accommodate the Copernican theory*). Compared with this reality the double language model looks infantile indeed.

and *Structure of Scientific Theory*, ed. R. G. Colodny (Pittsburgh: University of Pittsburgh Press, 1969); "Against Method"; and my criticism of Nagel in the *British Journal for the Philosophy of Science*, 17 (1966), 237–249.

[22] In his essay "Empiricism, Semantics and Ontology," reprinted in Leonard Linsky, ed., *Semantics and the Philosophy of Language* (Urbana: University of Illinois Press, 1952), pp. 207–228, Carnap has discussed this feature with great insight. However, apart from distinguishing between external and internal problems he does not give any indication of how one should proceed.

[23] The arguments are presented in the references in "Against Method."

The second difficulty is that not a single theory ever agrees with all the known facts in its domain.[24] This being the case, what shall we make of the methodological demand that a theory must be judged by experience and must be rejected if it contradicts accepted basic statements? What attitude shall we adopt toward the various theories of confirmation and corroboration which all rest upon the assumption that theories can be made to completely agree with the known facts and which use the amount of agreement reached as a principle of evaluation? This demand, these theories, are now all quite useless. They are as useless as a medicine that heals a patient only if he is completely bacteria free. *In practice* they are never obeyed by anyone. Methodologists may point to the importance of falsifications—but they blithely use falsified theories; they may sermonize how important it is to consider *all* the relevant evidence, and never mention those big and drastic facts which show that the theories which they admire and accept, the theory of relativity, the quantum theory, are at least as badly off as are the older subjects they reject. In practice methodologists slavishly repeat the most recent pronouncements of the top dogs in physics, though in doing so they must violate some very basic rules of their trade.[25] Is it possible to proceed in a more reasonable manner?

10. The answer is of course—yes. But the remedy needed is quite radical. What we must do is to replace the beautiful but useless formal castles in the air by a detailed study of primary sources in the history of science. This is the material to be analyzed, and *this* is the material from which philosophical problems should arise. And such problems should not at once be blown up into formalistic tumors which grow incessantly by feeding on their own juices but they should be kept in close contact with the process of science even if this means lots of uncertainty and a low level of precision. There are already thinkers who proceed in this way. Examples are Kuhn, Ronchi, the late Norwood Russell Hanson, and especially Imre Lakatos who has almost turned the study of concrete cases into an artistic enterprise and whose philosophical suggestions can once more be used to transform the process of science itself. It is to be hoped that such a concrete study will return to the subject the excitement and usefulness it once possessed.

[24] For details see section 4 of "Against Method."
[25] In this, of course, they merely repeat Newton. The only difference is that the theories which Newton accepted were invented by himself.

183

Mach's Philosophical Use of the History of Science

Ernst Mach devoted a major part of his life to historicocritical investigations that are focused upon philosophical problems in science. The object of this paper is to consider how he made use of the history of science to illuminate problems in the philosophy of science that he encountered in his work as a physicist.[1]

Mach: Physicist

An examination of Mach's published papers, of which there are about two hundred, reveals that he was an ingenious experimentalist, a keen and original critic of the empirical foundations of science, and singularly alert to the value of exploring questions in those areas where physics, physiology, and psychology overlap. The importance of Mach's historical and criticoanalytical contributions has been stressed in the literature at the expense of his activity as a scientist. An appreciation of this side of his life can be achieved through a contextual analysis of his many experimental papers.

Mach's interest in physics was born of an insatiable and tender curiosity about the world of his immediate environment. Einstein wrote concerning Mach: "The unmediated pleasure of seeing and understanding, Spinoza's *amor dei intellectualis*, was so strongly predominant in him that to a ripe old age he peered into the world with the inquisitive eyes of a carefree child taking delight in the understanding of relationships." Einstein recognized that even in cases where Mach's scientific investigations were not

AUTHOR'S NOTE: It is a pleasure to acknowledge the support of the National Science Foundation for research on the life and works of Ernst Mach.
[1] A convenient source of information about Mach's life is K. D. Heller, *Ernst Mach, Wegbereiter der modernen Physik* (Vienna: Springer, 1964). For works by and about Mach see Joachim Thiele, "Ernst Mach—Bibliographie," *Centaurus*, 8 (1963), 189–237.

based upon new principles, his work invariably exhibited "extraordinary experimental talent."[2]

Mach was a physicist by training. He never wanted to be known professionally as anything but a physicist. Of course, his conception of what constitutes physics was broad from the start. In his teaching and research and in his professorial role he was not bound by conventional academic guidelines. His doctoral work at the University of Vienna was completed in 1860 with a study on electrical discharge and induction. As a privatdocent at the university he engaged in studies on the change of musical pitch and color resulting from motion, viz., the Doppler effect.[3] During the winter semester of 1861–62 he lectured on "Physics for Medical Students."[4] During the summer semester his course was entitled "The Principles of Mechanics." He was able to supplement his meager income at this time by delivering popular lectures on psychophysics, Helmholtz's theory of the sensation of tone, and optics.[5]

From 1864 to 1867 Mach was, in turn, professor of mathematics and professor of physics at the University of Graz. It was there that he developed a deep and permanent interest in psychophysiological investigations in acoustics and optics.[6] In 1865 he discovered what later was referred to as "Mach bands"—a phenomenon that relates the physiological effect of spatially distributed light stimuli to visual perception.[7]

From 1867 to 1895, the most productive period of his life, Mach occupied the chair for experimental physics at the Charles University in Prague. There he continued his experimental researches in physiological

[2] Albert Einstein, "Ernst Mach," *Physikalische Zeitschrift*, 17 (1916), 101–104.

[3] Mach's papers of 1860–62 on this subject were republished later as *Beiträge zur Doppler'schen Theorie der Ton- und Färbenänderung durch Bewegung* (Prague: Calve, 1873).

[4] These lectures formed the basis for Mach's *Compendium der Physik für Mediciner* (Vienna: Wilhelm Braumüller, 1863). Important aspects of Mach's early pro-atomistic views, as expressed in this work, are discussed in my paper "The Genesis of Mach's Early Views on Atomism," *Boston Studies in the Philosophy of Science*, vol. VI (Dordrecht: D. Reidel, 1970), pp. 79–106.

[5] "Aus Dr. Mach's Vorträgen über Psychophysik," *Oesterreichische Zeitschrift für praktische Heilkunde*, 9 (1863), 146–366 (text interrupted). *Zwei populäre Vorlesungen über musikalische Akustik* (Graz: Leuschner & Lubensky, 1865); *Einleitung in die Helmholtz'sche Musiktheorie* (Graz: Leuschner & Lubensky, 1866); *Zwei populäre Vorträge über Optik* (Graz: Leuschner & Lubensky, 1867).

[6] *Optisch-akustische Versuche. Die spectrale und stroboskopische Untersuchung tönender Körper* (Prague: Calve, 1873).

[7] See Floyd Ratliff, *Mach Bands: Quantitative Studies on Neural Networks in the Retina* (San Francisco: Holden-Day, 1965), which contains in English translation Mach's five classic papers on this subject, 1865–68, and the paper of 1906 in which he dealt with temporally varying light stimuli on visual perception.

physics and psychophysics, the most celebrated results of which were the papers of 1873–75 on the changes in kinesthetic sensation associated with physical movement, acceleration, and orientation of the human body.[8] A wide variety of optical experiments connected with refraction, interference, and polarization continued to hold his attention. He investigated the wave motion associated with mechanical, electrical, and optical phenomena. Especially influential were the novel techniques that were introduced by Mach and his students for the photographic study of wave propulsion and gas dynamics for supersonic gas jets and projectiles.[9]

Mach's active involvement in experimental physics thus covers a period of about 35 years: namely, from 1860, when he was 22, until 1895, the year in which he moved from Prague to Vienna to accept a newly created professorship at the university. As a scientist he possessed a rare independence of judgment that was not always appreciated by his colleagues. But even when scientists rejected Mach's philosophical views they respected his experimental insight and ingenuity. Wilhelm Ostwald wrote: "So clear and calculated a thinker as Ernst Mach was regarded as a visionary [*Phantast*], and it was not conceivable that a man who understood how to produce such good experimental work would want to practice nonsense [*Allotria*] which was so suspicious philosophically."[10] Arnold Sommerfeld called Mach a brilliant experimentalist but a peculiar theoretician who, in seeking to embrace the "physiological" and "psychical" in his physics, had to relegate the "physical" to a less pretentious level than physicists were accustomed to expect from a colleague.[11]

While Mach devoted substantial efforts to certain aspects of theoretical physics—notably mechanics, thermodynamics, optics, molecular spectra, wave propulsion, the theory of perception (optical and aural), and the sensation of orientation in the human body—in general he did not convince his contemporaries of the importance of his theoretical and critical deliberations. There were some exceptions. Referring to Mach's *Mechanik* of 1883 and the criticisms of Newton's views on space, time, and motion, Einstein wrote in 1916: "There you will find set forth brilliantly ideas

[8] *Grundlinien der Lehre von den Bewegungsempfindungen* (Leipzig: W. Engelmann, 1875).
[9] See Wolfgang Merzkirch, "Mach's Contribution to the Development of Gas Dynamics," *Boston Studies in the Philosophy of Science*, VI, 42–59.
[10] *Lebenslinien, Eine Selbstbiographie*, vol. 2 (Berlin: Klasing, 1927), p. 171.
[11] "Nekrolog auf Ernst Mach," *Jahrbuch der könig. bay. Akad. der Wiss.* (Munich), 1917, pp. 58–67.

which by no means as yet have become the common property of physicists."[12]

In Vienna, Graz, and Prague Mach was known as a stimulating physics teacher and popular science lecturer. Beginning in 1887 he published a series of secondary school textbooks in physics. At about the same time demonstration apparatus designed by Mach became known in Germany and the Austro-Hungarian empire. His *Popular Scientific Lectures*, which first appeared in English in 1895, was available shortly thereafter in German, Italian, and Russian. The point that deserves emphasis, in the context of this discussion, is that Mach consistently wanted to be known as a physicist. That is, he looked upon himself as a scientist who undertook experimental and theoretical investigations from the point of view of a professional physicist. About this there can be no doubt.

Mach: Historian of Science

Next to Mach's dedication to scientific investigation was his lifelong interest in the history of science. As early as 1863, at the age of 25, he was convinced that his scientific curiosity could be nurtured and promoted best if he were to become a serious student of the history of scientific ideas and practice.[13] The initial stimulus for his involvement in the history of science was a practical one intimately related to his search for sound pedagogical techniques in the teaching of science. Particularly with advanced studies he felt that students should not be expected to accept, as self-evident, propositions which had cost several thousand years' labor of thought. In 1872 in his first historical monograph, *History and Root of the Principle of the Conservation of Energy*, published in Prague, he wrote: "There is only one way to [scientific] enlightenment: historical studies!"[14]

According to Mach it was the history of science and not the logical analysis of science that would enable the scientist to consider problems without developing an aversion to them. He suggested two ways of becoming reconciled with actuality: "Either one grows accustomed to the puzzles and they trouble one no more, or one learns to understand them by the help of history and to consider them calmly from that point of view."[15]

---

[12] "Ernst Mach," p. 102.
[13] *Beiträge zur Analyse der Empfindungen* (Jena: Gustav Fischer, 1886), p. 21; English edition, *The Analysis of Sensations* (New York: Dover, 1959), p. 30.
[14] *Die Geschichte und die Wurzel des Satzes von der Erhaltung der Arbeit*, 2nd ed. (Leipzig: J. A. Barth, 1909), p. 2; English edition, *History and Root of the Principle of the Conservation of Energy* (Chicago: Open Court, 1911), p. 16.
[15] "Man gewöhnt sich an die Räthsel und sie belästigen uns nicht weiter. Oder man

He felt that the essence of a "special" classical education for the investigator of nature "consists in the knowledge of the historical development of . . . science."[16] Many years later, in 1905, in his *Erkenntnis und Irrtum* Mach wrote that, without claiming to be a philosopher, the scientist has a strong urge to satisfy an almost insatiable curiosity about the origins, structure, process, and conceptual roots of his discipline.[17]

Mach's historicocritical aims were explored first in 1872 in his *History and Root* where, with the focus clearly on history, he rejected metaphysics and sought to highlight the principle of economy of thought. It was, he remarked later, a first attempt to undertake studies epistemologically from the standpoint of the physiology of sensation (*erkenntniskritische sinnesphysiologische Studien*).[18] In this little volume he opened up ideas which were explored in one way or another for the greater part of his life: causality, the atomic theory, the criticisms of Newtonian mechanics, the meaning and purpose of scientific theories, and the criticisms of physical reductionism, metaphysics, mechanism, and materialism.

All of these matters are examined *in extenso* in Mach's historical treatises on mechanics, optics, and heat theory.[19] If we had the opportunity to follow the arguments in these works we would discover that definite recurrent themes stand out. They derive from the way in which certain unique epistemological questions are formulated: How did we acquire our present scientific concepts? Why did they take a given form among scientists and not another form that may be logically more convincing and esthetically more acceptable? What were the circumstances of analogical adaptation that led to the acceptance of a given concept?

Such questions having been posed, and clothed historically, the dominant lessons that shine through Mach's deliberations are these: Science has an incredibly rich history. Its ideas are alive. Its concepts, laws, and theories are in constant flux. They are perennially under revision and re-

---

lernt sie an der Hand der Geschichte verstehen, um sie von da an ohne Hass zu betrachten." *Geschichte,* p. 1; *History and Root,* pp. 15–16.

[16] *Geschichte,* p. 3; *History and Root,* p. 18.

[17] *Erkenntnis und Irrtum. Skizzen zur Psychologie der Forschung* (Leipzig: J. A. Barth, 1905), p. v.

[18] *Geschichte,* Vorwort, p. 3.

[19] *Die Mechanik in ihrer Entwickelung historisch-kritisch dargestellt* (Leipzig: F. A. Brockhaus, 1883); English edition, *The Science of Mechanics* (LaSalle, Ill.: Open Court, 1960). *Die Prinzipien der physikalischen Optik, historisch und erkenntnispsychologisch entwickelt* (Leipzig: J. A. Barth, 1921); English edition, *The Principles of Physical Optics* (New York: Dover; originally published 1926). *Principien der Wärmelehre, historisch-kritisch entwickelt* (Leipzig: J. A. Barth, 1896).

formulation. The conceptual products of science, always incomplete, take on a form at any particular time which reflects the historical circumstances and the focus of attention of the particular investigator—now physicist, now physiologist, now psychologist. Strictly speaking, scientific constructs therefore disclose the mode of cognitive organization of experience even when they contribute essentially nothing new to what is known. Because of the virtual infinity of the potentially relevant subject matter on the one hand, and the many-faced ways of examining it on the other, the investigator necessarily falls back upon abstraction.

Mach recognized that scientists normally place considerable stress on internal logic and consistency, maximum comprehensiveness, simplicity, and so on. Nevertheless, when examined within the historical context, scientific concepts, laws, and theories do not exhibit the tidy logical features that are supposed to represent the traditional bench marks of science. Rather, what predominantly stands out is a strong element of historical fortuitousness. With the passage of time, what is accidentally (historically) acquired is converted into the philosophically argued. For this reason it becomes mandatory for the scientist to recognize and counteract the insidious gradual process by which scientific constructs come to be regarded as philosophically necessary rather than historically contingent.

Familiarity with the internal subtleties of this movement from contingency to necessity, thinks Mach, can serve as the point of departure for developing the analytical and methodological tools to clarify and purify the conceptual components of science. This is a plea for the critical, historical, and psychological exposure of the roots of science that will reveal metaphysical obscurities and expose one-sided (e.g., mechanical) views of physics. The success of Mach's philosophical enterprise, if we may call it that, rests upon the feasibility of using the history of science to illuminate specific problems to which Occam's razor can be applied to remove metaphysical ballast and inherited anthropomorphisms and ambiguities. The over-all aim is to banish from science (and philosophy) all conceptual categories that have no analogues in experience. Scientific entities are not to be multiplied unnecessarily.

Among Mach's historical monographs the Science of Mechanics of 1883 offers a sharp-witted and provocative study of the role played by historical circumstance in the formulation of scientific concepts. The pertinence of instinctive knowledge, the function of memory, the relation between the origin of an idea and its rank in science, the psychology of discovery contra the logic of discovery, the place of induction and deduction in physics—all are explored in tangible historical terms.

The first two chapters treat the principles of statics and dynamics—in 343 pages. The over-all objective is to demonstrate that statics is prior to dynamics, historically speaking, but that logically or conceptually statics can be recognized as a limiting case of dynamics. We note that there is no attempt to cover the history of statics or dynamics in a systematic manner. Still Mach rarely misses an opportunity to probe into the furthest corner of some aspect of the history of statics or dynamics when and where it serves his purpose. The historical illumination of specific physical problems is being pursued. In the course of the analysis Mach makes evident the historical importance, for the development of mechanics, of the solution of problems in dynamics based on an appeal to static arguments and of the solution of problems in statics based on an appeal to dynamic arguments.

Here, as elsewhere, Mach is prying into what it means to say that one principle is more basic or fundamental than another. He is asking whether instinctive knowledge, which is assumed to be self-evident, is fundamentally different from clear knowledge known to have been learned by experience. To this end the analysis of the mental procedures of different investigators is instructive. For example, by what authority does one proceed (as Stevin did) from the general view to the specific case, while another proceeds (as Archimedes did) from the specific case to the general view. In reference to Daniel Bernoulli's demonstration, that the proposition of the composition of forces is a geometrical truth independent of physical experience, Mach aims to point out that Bernoulli has merely reduced what is easier to observe to what is more difficult to observe; and that is scarcely a convincing demonstration.

In other examples Mach attempts to show that a mania for proof can cover up spurious rigor and that an appeal to the higher authority of purely instinctive cognition provides, at most, fallible useful knowledge. Believing that no place should be given to proofs in science without an appeal to experience, he criticizes "proofs" that have led to a belief in necessity. Moreover, an attempt is made to indicate, historically, that prior discovery or enunciation of one principle over another is not related to primacy. Mach contends that primacy cannot be relegated to the chances of history; that greater generality of a principle does not necessarily render it more basic; and that certain principles, which are an extension in application of statics or dynamics, are equally basic. In other words, "basic" has no absolute significance when divorced from the scientific context of a problem.

The sagacious approach to scientific investigation consists in knowing when to accept instinctive knowledge as a starting point and when to embark on an abstract examination of its contents. Nature, as directly observed, leaves an imprint which is largely uncomprehended, unanalyzed, and embedded in inarticulate thought. Consequently it is risky to rest content with instinctive knowledge no matter how heuristically valuable. It is always illuminating, Mach argues, to uncover the historical conditions of the origins of knowledge.

The extended application of the principles of mechanics in relation to their deductive development is treated in the third chapter of the *Science of Mechanics*—a model of pedagogical, logical, and historical writing. Here Mach sets out to demonstrate how each of an interrelated set of principles can be applied to the solution of specific problems in the most advantageous, most economical, most logically satisfying way. He shows for a whole family of principles—exclusion of the *perpetuum mobile*, center of gravity, virtual work, *vis viva*, d'Alembert's principle, Gauss's principle of least constraint, Maupertuis's principle of least action, Hamilton's principle—that each can be derived mathematically from any other, provided that sufficient mathematical resourcefulness and industriousness are exercised.

Evidently the above-mentioned principles are related in such a way that provided one of them is accepted as true, the others can be deduced by mathematical and logical procedures alone. But what constitutes the evidence for the one which is accepted as true? Ultimately it cannot be justified except as it gives suitable answers to problems in statics and dynamics. And how was the principle discovered initially? Mach's answer, illustrated historically, is that the roots of the most primitive of such principles lie in experience—e.g., that man has not been able to construct a machine to deliver unlimited quantities of accomplishment (work) without the exertion of effort. Mach insists that none of the properties of nature can be constructed with the help of so-called self-evident suppositions alone. The suppositions must be derived from experience.

In his examination of the formal developments of mechanics Mach takes his chance to unmask the theological, animistic, and mystical elements of science, as he sees them. From a logical, economical, and practical point of view he finds no convincing reasons for the physicist to defend or rationalize the historical legacy of mechanics. True, the study of the history of science demonstrates how plausible it is for the concepts of mechanics to occupy a prestigious place among the sciences. Yet Mach believes that the applicability of mechanics to the other sciences is severely limited. He doubts, in fact, that mechanical principles will be beneficial for understanding the phenomena of heat, light, and electricity. In psychology, thinks Mach, the use of mechanics is utterly mischievous. In any case we must be wary of pressing all aspects of science into an a priori conceived mechanics mold. Or any other.

To examine critically Mach's diverse writings as a whole is to become immersed in the workings of the mind of an unbridled physicist who was intent upon clarifying certain perplexities reflected in the science of his day. In the process of pursuing this goal he not only managed to do physics and to write history, but, incidentally, to develop his own metaphysics—although he would have denied it. As a young man it seems that Mach pondered a great deal about what characterizes physics and the task of the physicist vis-à-vis new branches of science that spring up outside of physics. Through historical studies he hoped that he would learn to know how the basic physical concepts of his day had come into being and that this would provide him with the necessary directives for his own work. Mach had not the disposition to engage in the study of a scientific problem merely because it was respectable within the community of his colleagues to do so. He puzzled so deeply about the course that his own investigations should take because he was disturbed about some of the conceptual foundations of physics which were being accepted without critical examination.

It should be clear from what has already been said that Mach's physics was not meant to be the practical testing ground for his philosophical (epistemological) deliberations. Rather, the specific epistemological questions that he formulated were the logical consequence of his intense preoccupation with physics as seen from within the context of a tenacious inquisitiveness about the historical tradition inherited from the past. The exclusive function that the history of science served for Mach was the immediate result of his own scientific investigations and not the product of philosophical deliberations concerning the principles or formulas that might be prescribed for that investigation. His methodological reflections were the outgrowth of the scientific research in which he was engaged most of the time. Certainly he was concerned far less with developing a scientific philosophy than he was with the solution of specific problems that arose in connection with his attempts to keep abreast of the advances of physics and related disciplines.

If Mach eventually became involved in historicocritical investigations that had a reach far beyond his initial teaching objectives, it must be underscored that these endeavors were not undertaken with the intent of engaging in the history of science per se. The history of science was conceived rather as an indispensable critical probe for gaining an appreciation and understanding of the nature of science itself. While it can be granted that Mach was not concerned in the least with narrative, or chronicle, or reconstructing what really happened (as some historians are) it would be inaccurate to conclude that he was oblivious to the high demands of historical scholarship. If his historical information seems erroneous or superficial in places we must evaluate his work against the state of the art at the time.

It is apparent that the historical perspective of Mach's work was not primarily chronological, biographical, or even topical. He obviously made no attempt to discuss the evolution of scientific thought and practice systematically. Nor were his works conceived on the pattern of the intellectual historian who concentrates on the history of scientific ideas. Experimentation and the role of instrumentation were treated as an integral part of the critical and historical account of science that Mach sought to depict. Mach was not interested in achieving comprehensiveness of subject matter over a given area. Nor was he given to an exploration of the over-all general framework of science in relation to the organizational, institutional, or societal circumstances. He did not discuss the structure of scientific revolutions. Rather, Mach's interest in history was focused upon specific concepts, problems, instruments, and theoretical constructs the circumstances and developmental history of which, in his opinion, would help the scientist to appreciate and understand in depth the rationale of his scientific inheritance. He took up problems case by case. He managed to formulate these with unusual clarity and without overloading the issues with unessentials. He wrote no history where he saw no problem. He even was reluctant to discuss a problem when he could not offer historical commentary that would illuminate some facet of the problem.

Mach: Philosopher of Science

We have seen that Mach wanted to be regarded professionally as a physicist and that he had a vested interest in the history of science in terms of what it meant to him as a physicist. What can be said about Mach as a philosopher? His own answer to that question, without any equivocation, was that he did not want to be recognized as a philosopher. In 1886 he wrote: "I make no pretensions to the title of philosopher. I only seek to adopt in physics a point of view that need not be changed immediately on glancing over into the domain of another science; for, ultimately, all must form one whole."[20]

It is true that the specific questions Mach raised about science were essentially of a philosophical nature. Although he was never at ease with "philosophy" he was not at all averse to the examination of philosophical questions provided that such activity implied nothing more than "scientific epistemology," i.e., *erkenntnistheoretische Untersuchungen*. But we must be clear about what he understood by epistemology. In point of fact his methodological *programme*, like that of positivists in general, excluded questions that deal with the existential status of reality. This was accomplished by presupposing that experience in the form of sensations is the only category that is epistemologically meaningful.

In 1895, after having taught physics for 35 years, Mach accepted a chair in Vienna for the history and theory of the inductive sciences. We know from an examination of his correspondence that he chose this title for his professorship within the philosophical faculty in order to avoid what he disliked intensely, namely, being identified as a philosopher. When his critics referred to *die Machsche Philosophie* he wrote in 1905: "Above all there is no Machist philosophy. At most [there is] a scientific methodology and a psychology of knowledge [*Erkenntnispsychologie*]; and like all scientific theories both are provisional and imperfect efforts. I am not responsible for the philosophy which can be constructed from these with the help of extraneous ingredients. . . . The land of the transcendental is closed to me. And if I make the open confession that its inhabitants are not able at all to excite my curiosity, then you may estimate the wide abyss that exists between me and many philosophers. For this reason I already have declared explicitly that *I am by no means a philosopher, but only a scientist*. If nevertheless occasionally, and in a somewhat noisy way, I have been listed among the former then I am not responsible for this. Of course, I also do not want to be a scientist who blindly entrusts himself to the guidance of a single philosopher in the way that Molière's physician expected

[20] *Beiträge zur Analyse*, p. 21; *The Analysis of Sensations*, p. 30.

and demanded of his patients."[21] In a note of 1913, published post-humously, Mach remarked that both physicists and philosophers carried on a crusade against him at a time when he considered himself to be no more than an unprejudiced rambler with ideas of his own in various areas of knowledge.[22]

Call him what you please, Mach exerted considerable influence upon the physical thought of several generations of scientists who, as Einstein suggests, were stimulated to ask significant questions on reading his works: What characteristic objectives can be achieved by approaching physics in a certain way? To what extent are the general results of science true? What is essential, and what merely rests upon fortuitous history? Einstein says: "It is a fact that Mach's historicocritical writings . . . have exerted a great influence upon our generation of natural scientists. I even believe that those who consider themselves as adversaries of Mach scarcely know how much they, so to speak, have been born and bred to Mach's outlook."[23]

Nominally, it was not philosophy of science and it was not history of science that Mach wished to pursue; it was the history and theory of the *inductive* sciences. At least that was a title Mach felt he could live with while being engaged in what sounds to me like the history and philosophy of science. In fact, one is tempted to conclude that Mach is a submerged philosopher formulating and reformulating, or wishing he could formulate, a theory of knowledge that would appeal to scientists in general and not just to physicists—an ambitious undertaking in the days before the philosophy of science was an established discipline.

The *Erkenntnis und Irrtum* of 1905 supplies ample evidence of Mach's progressive drift from physics to philosophy of science by way of the history of science. This work is the most comprehensive analytical and systematic treatment of Mach's philosophical views—although it is seldom read and almost never referred to by historians of science. It contains little history of science. For a person intimately acquainted with Mach's historical works it is not difficult to relate his mature philosophical reflections in this work to his historical approach. Still, there are only hints, and it is best to go searching for the connections in his earlier works.

Mach was preoccupied with historicophilosophical problems for almost 50 years; and we know that he devoted himself to them without relin-

[21] *Erkenntnis und Irrtum,* pp. vii–viii.
[22] *Die Prinzipien der physikalischen Optik,* p. viii; *The Principles of Physical Optics,* p. viii.
[23] "Ernst Mach," p. 102.

quishing his enthusiasm for teaching or his concern with experimental physics. Mach's philosophy of science, if we may call it that, was dominated, as we have already seen, by the desire to explore in depth, with the aid of history, the epistemological roots of science; but he believed that this could not be accomplished meaningfully without examining the behavior of the scientist—hence Mach's insistence on the importance of the analysis of sensations, and of the fundamental relevance of psychology and physiology as a corrective to the prevailing mechanistic physicalism.

Mach was predominantly destructive of the official school philosophies of his time and he discovered that he occupied the unpopular position associated with radical and empirical skepticism, the renunciation of truth in any transcendental sense, and the support of a theory of knowledge based on biological behavior. To disclaim any intention of substituting positivism for the variegated interpretations of philosophy elaborated by Kant's idealist successors surely represented the road of least resistance for Mach. Granted that he did not want traditional philosophers to accept the task of defining the function of science or of explaining the results of science, it seems unreasonable to suppose—considering how much he wrote on methodological matters in science—that he underestimated the heuristic importance of such extrascientific activities for the intellectual life of the practicing scientist. If the strict separation of science and philosophy which Mach's thought leads to is taken for granted by most scientists and philosophers today this was not so in Mach's environment.

The ambiguous position of Mach is apparent. It is that of a physicist, engaged in psychophysiological investigations, espousing historical evidence to expound an antimetaphysical philosophy of science while disclaiming any deep interest in traditional philosophy and even rejecting positivism as a philosophical system. In view of such an outlandish congeries of attitudes it is hardly surprising to discover that Mach's contemporaries—scientists, historians, and philosophers—could not avoid taking sides while engaging in an evaluation of his writings. Positions of neutrality or indifference toward Mach were rare. They still are. In fact, it is well known that several persons radically revised their opinions about Mach's philosophy after their initial sympathetic fling with Machistic positivism.[24] For example, Planck and Einstein, in their early and productive years, were impressed favorably by Mach's sensitive perceptions. Later

---

[24] Schrödinger in his youth was attracted to Mach the philosopher, but his "guardian angel intervened . . . [and] nothing came of it." Erwin Schrödinger, *My View of the World* (Cambridge: Cambridge University Press, 1964), p. viii.

they qualified their enthusiasm for Mach's philosophy of science and eventually became quite critical as they more fully became aware of the long-range implications of so antimetaphysical and antirealistic a world view.[25]

In Mach's time physicists, at least before they turned into philosophers, exhibited a greater affinity for Mach's philosophy of science than philosophers. This is still true today. Perhaps philosophers of science can suggest what that signifies, if anything. It seems to imply that Mach was more of a physicist's philosopher than a philosopher's philosopher. That would be plausible, since Mach struggled so diligently to establish with certainty the line of demarcation between cognitive and noncognitive endeavors—between science and philosophy. Here, our principal concern is to continue to examine Mach's position as a historian-philosopher and to evaluate the pertinence of Mach's history of science for his philosophy of science.

Mach: Historian-Philosopher

A basic assumption undergirds every aspect of Mach's scientific thought: viz., that all of our knowledge concerning the world of phenomena is acquired through experience. Nature is composed of sensations as its elements. They supply the only possible immediate source of revelation for scientific information. Experience is homogeneous with respect to the elements of sensation. Whether these elements are designated as physical, physiological, or psychical, there are no antagonisms or antinomies between them.

Such is the domain of circumscription for human cognition within which Mach prefers to philosophize. He knows, of course, that the critics will expect him to furnish a rationale for rejecting the other traditional claims of knowledge. His response is always the same: that metaphysical pretensions to cognition can be ruled out of science because they are empty of meaningful content and because they merely serve to befuddle the real problems with which scientists struggle. No doubt about it—this is a circular argument. Nevertheless, Mach seems to believe that his epistemology furnishes a manageable policy of operation. History, he maintains, justifies his point of view.

The history of science, Mach believed, demonstrates that there are no sharp boundary lines separating the scientific from the nonscientific or

---

[25] For a discussion and evaluation of the differences in philosophical outlook between Mach and Planck see Erwin Hiebert, *The Conception of Thermodynamics in the Scientific Thought of Mach and Planck*, Wissenschaftlicher Bericht Nr. 5/68 (Freiburg im Breisgau: Ernst-Mach-Institut, 1968), pp. 1–106.

prescientific experience of man. Similarly, no sharp discontinuities exist between ordinary language and the highly theoretical language of science. The given world can be represented as a complex of interrelated sensations. The "given" embraces the object and the subject of the sensations: the observed and the observer, or the sensed and the senser. The dualism between subject and object disappears. The process of acquiring knowledge is structured biologically (psychophysiologically) and consists of adapting facts (*Tatsachen*) to thoughts (*Gedanken*) and of thoughts to other thoughts.[26]

According to Mach there is no such thing as Kant's *Ding-an-sich*, or at least no meaningful discourse can take place about such referents. It is an illusion to believe that a final "thing in itself" remains after all the qualities of a body have been subtracted from it. Bodies, for example, are mere mental symbols to represent the relatively permanent conjunction of elements of sensations—the conventional attributes like color and sound. The single "thing" with its many attributes is a protean pseudo-philosophical conception.

The so-called objects of our experience are thought symbols for complexes of elements (sensations). We need not know what is characteristically physical, what physiological, and what psychical among the components of experience. It suffices to discover the functional relations of the sensations to one another. To speak of something being physical means to designate only one method of cognitive organization. Everything turns on considering different ultimate variables and different relations of dependence. Thus for Mach, as it had been for Hume, the perceptions of the mind are roughly divided into two classes, being distinguished by the different degrees of their force and vivacity: first, the immediately experienced impressions (facts); and second, the ideas (thoughts), or the less lively perceptions rooted in the memory or imagination. Accordingly, the operations which constitute scientific understanding of the world are exhausted in providing relationships between matters of fact, between fact and ideas, and between ideas and other ideas.

The mind of man can organize, rearrange, reject, and synthesize what is given in experience, but deliberation about and manipulation of the empirically given cannot be relied upon to give new information about the world. This is not to say that it is pointless to predict how the world will

[26] *Erkenntnis und Irrtum*, "Anpassung der Gedanken an die Tatsachen und aneinander," pp. 164–182.

behave in the unknown, but rather that experience is the final judge about how the world in fact does behave in the unknown. With respect to our knowledge about the phenomenal world, experience is king. That king cannot be checkmated without putting an end to the game which scientists play.

Crucial for our evaluation of Mach's philosophical use of the history of science are his views on the role of the experiment in science. On such matters he presents a consistent view which correlates perfectly with his epistemology. As already intimated, it rests upon belief in the fundamental and constitutional uniformity in the method of obtaining valid knowledge for all areas of human endeavor, viz., by direct experience and by theoretical deliberation about experience. This does not mean that the sciences necessarily progress by a leveling process toward a unity which destroys the independent status of the different scientific disciplines. Quite to the contrary, Mach's scientific monism leaves no room for scientific reductionism; it is a monism of method.

What constitutes experimentation for Mach? There is no special method of investigation, discovery, or induction, in science. Rather, we recognize the substantial unity in the modes of scientific and everyday thought. Still we know that there will be enormous differences in the range of expertise and subtlety employed in the pursuit of one or another subject matter. Likewise, it is apparent that there will be striking contrasts in accomplishment for the separate branches of science, at various times, and under diverse environmental circumstances. It is these differences which must be studied and analyzed for what they reveal about the genesis of scientific constructs and ideas. It has been said often, at least since Auguste Comte, that science is a social phenomenon and that its state of organization and its composition hinge on the historical circumstances associated with its advancement. This is not what Mach had in mind. He wanted to demonstrate that the content of science as an intellectual inheritance depends upon the historical conditions under which the conceptual notions in the various sciences have developed.

The clue to fruitful experimentation rests upon a kind of operational bias strengthened by psychological and physical intuition. It represents the investigation and exploitation of functional dependencies that are suspect, and for which meaningful tests can be invented. And what is suspect? According to Mach: potentially, almost everything. Since there is no special method of scientific discovery the productive scientist will maintain an

Erwin N. Hiebert

attitude of maximum open-endedness in his search for the relevant functional dependencies. How much of nature should be placed under observation at one time? In such matters chance favors the prepared mind. In general it is wise not to draw the boundaries of a scientific investigation so broadly as to invite discouragement, but also not so narrowly as to confine the analysis to what is empty and aimless. The experimentalist should spend large fractions of his "experimenting time" in thinking about the most advantageous way to interrogate nature. The discussion of this subject comes up notably in Mach's writings under the heading of "thought experiments" (Gedankenexperimente). He asks: Who is the real experimenter? The so-called experimentalist operating with facts, or the so-called theoretician experimenting with concepts? The scientist can and must be both.[27]

What about the relationship of facts to hypotheses in the writings of Mach? Real, fixed, isolated facts exist only within our minds to be used as intellectual anchor points for building hypotheses. Reflection about these so-called facts and analysis of their conditions, parts (constituents), and consequences define the hypotheses and the conditions under which they are applicable. The hypotheses are heuristic devices that indicate the possible modes of testing our maneuvers within science; but more important, they motivate us toward other schemes of experimentation. The task which Mach set for himself in his historical treatises on mechanics, heat, and optics was to show from a critical and psychological standpoint how various ideas within these disciplines had been molded and transformed not only by the revelation of new facts but by reason of the views historically associated with them; and, further, how the general concepts of science developed from within the individual disciplines.[28]

The values that result from formulating hypotheses (as heuristic devices) are three: first, they *sometimes* reveal the correspondences that were suspected and predicted; second, they *more often* simply falsify the correspondences; third, they *most often* indicate something about a problem different from the one that was placed under investigation. Here is Mach's position on the third point: Concerning the conditions of the unknown we are not able to formulate ideas with sufficient clarity in relation to our

---

[27] Mach first discussed "Gedankenexperimente" in an article entitled "Über Gedankenexperimente," *Zeitschrift für den physikalischen und chemischen Unterricht*, 10 (1897), 1–5. The subject was explored more fully in *Erkenntnis und Irrtum*, pp. 183–200.

[28] See, for example, the Preface to *The Principles of Physical Optics*.

200

scientific objectives. So we invent temporary conditions that are conceivable or else familiar to us because of previous experience. On testing our conjectural schemes we most frequently discover something about approaching the problem from another point of view that hitherto had been unfamiliar to us. Moreover, we discover that not all of our formulated scientific problems can be solved, because they may turn out not to be problems at all. They may result from the nonsensical way in which they were posed. And thus it comes about that an experimental investigation (i.e., one in which there is experimentation with concepts or facts) leads to no new knowledge but leads the way to another mode of acquiring new knowledge.[29]

There is one other central pillar of Mach's scientific doctrine that will help us to understand how the history of science was pressed into the service of the philosophy of science. This is his principle of economy of thought, concerning which Charles S. Peirce said: "Dr. Mach, who has one of the best faults a philosopher of science can have, that of riding his horse to death, does just this with his principle of Economy in science."[30] In Mach's writings, economy of thought (die Oekonomie der Wissenschaft) shows up everywhere.[31]

The meaningful goal that science can set for itself is the simplest and most economical abstract expression of facts. It is suggested that whereas the events of nature are simply given (once), our schematic mental imitations identify "like events" in nature; that these "like events" exist with certain mutually dependent features only in our minds. In an essay on "the economical nature of physics" Mach expresses himself as follows: "Language, with its helpmate, conceptual thought, by fixing the essential and rejecting the unessential, constructs its rigid pictures of the fluid world on the plan of a mosaic, at a sacrifice of exactness and fidelity but with a saving of tools and labor."[32]

According to Mach, all communicable elements of science take advantage of economy in intellectual exertion. Scientific constructs, laws, and theories represent ordered factual information and experience. "It is the

[29] Erkenntnis und Irrtum, pp. 232–250.
[30] Essays in the Philosophy of Science, ed. V. Tomas (New York: Liberal Arts Press, 1957), p. 228.
[31] The principle of economy of thought is discussed in many places in Mach's writings but notably in History and Root of the Principle of the Conservation of Energy, The Science of Mechanics, Principien der Wärmelehre, and Popular Scientific Lectures (La Salle, Ill.: Open Court, 1943).
[32] Popular Scientific Lectures, p. 192.

object of science to replace, or save, experience, by the reproduction and anticipation of facts in thought. Memory is handier than experience, and often answers the same purpose."[33] This economical office of science not only relieves the memory; it renders superfluous the endless repetition of experiments.

The danger inherent in this biologically conditioned situation resides in the temptation to reify mental constructs like substance, force, and energy. The study of the history of science reveals that such constructs, originally provisional, in time and by successful application receive metaphysical sanctions that are virtually unassailable. Thus insulated from the ongoing scene of action of science, where criticism, revision, and rejection are the rule, these constructs acquire the status of logical necessity or receive acceptance as self-evident and intuitively known truths. For Mach there are no such invulnerable truths—at least not at a level of synthetic description that is so far removed from the phenomenological elements of pure sensation. Scientific constructs, laws, and theories never lose their provisional status.

We only begin to understand the duty that the principle of economy of thought has to perform within the historicophilosophical framework of Mach's ideas by describing, in his terms, what biological function the principle promotes. It has no ontological status. It is not a physical law. It is rather a description of the biological law that regulates the manner in which cognitive elements about the world are incorporated into man's thought. Mach claims, as Avenarius had done earlier, that the principle of economy of thought is but a description of animal behavior. The study of this behavior, i.e., the empirical study of history in the light of behavioral psychology, but liberated from metaphysical muddleheadedness, is the central task of the historian of science. The evolution of science thus becomes a special case of the larger biological process of self-preservation through mental adaptation.

In conclusion I recognize that perceptive critics will have no difficulty in pointing out that Mach does not treat the philosophy of science in a formal or methodical manner. He was not a systematic philosopher, and was not occupied with constructing scientific systems. There are in his works lacunae, as well as redundancies and inconsistencies; but Mach is never just marking time or filling in the historical details for some aspect of science unrelated to the methodological point he sets out to make. He

---

[33] *The Science of Mechanics*, p. 577.

simply claims to be exploring scientific puzzles and discovering that some of them vanish and take their places among the shadows of history.

A more serious objection to Mach's epistemological formulation of cognition than its nonsystematic character is that it violates his own philosophical position. Since he believes that empirical methods cannot be employed to establish the truth of any theory—and that includes a theory of knowledge—Mach would have to admit that his own epistemological formulation of the cognitive process rests upon philosophical presuppositions of some kind. But this is just what he denies. He claims that it is possible to analyze objectively the cognitive process in man through the empirical study of history.

It could be said that Mach should have learned that antimetaphysical protestations and all that goes with trying to establish an epistemology without overtly recognizing some philosophical presuppositions lead nowhere if not to metaphysics. Mach never conceded that he had to become a philosopher or metaphysician in order to rationalize his scientific position. Toward the end of his life he did become more tolerant about philosophy and even expressed appreciation for the fact that "a whole host of philosophers—positivists, critical empiricists, adherents of the philosophy of immanence and certain isolated scientists as well—have all, without any knowledge of one another's work, entered upon paths which, in spite of all their individual differences, converge almost towards one point."[34] Surely Mach, the positivist, who tried to restrict the data of his science to what is given in the elements of sensation, thereby showed his philosophical hand.

I have tried to explain how Mach's diverse investigations were the outcome of a search for the underlying foundations and meaning of scientific puzzlements which commanded his attention as a young physicist. The more Mach deliberated about these matters and the more he wrote in order to clarify them, the more his views crystallized and took on a dogmatic form. In doing so he had moved toward working out philosophical arguments which were no longer primarily focused on the historical explanation of scientific enigmas. Mach, the puzzled physicist who had turned historian, had been transfigured into a philosopher.

[34] *The Analysis of Sensations*, p. xli (Preface to the 4th edition of 1903). See also Mach's comments in the Preface to the 7th (1912) edition of *The Science of Mechanics*.

History of Science and Criteria of Choice

Although falsifiability, confirmation, and predictive power—all involving appeals to observational data, however conceptualized—are important criteria for the choice and acceptability of hypotheses, they are not sufficient, even if one does not go so far as to hold them not to be necessary either. The belief that more is required has a very long ancestry. Writers and scientists of many ages—explicitly and implicitly—have searched for "explanations of the phenomena" at a "deeper" level, more solidly entrenched, even if this sometimes means no more than integrating their hypotheses within a well-established explanation scheme, or making reference to general conceptual classifications or to some purely logical aspects. There is a good example of this at the start of the "modern" period of science in certain comments on the system of Copernicus made by Kepler in his *Mysterium Cosmographicum*. In a somewhat critical vein Kepler there remarks that Copernicus had really achieved no more than to "derive his system" from the phenomena of nature, from "the effects." Now, Kepler goes on to say, "if this system was correct," this only showed that Copernicus had to a certain extent been "lucky." By contrast, "what is necessary" —and this Kepler here claims as his own achievement—was to show that one's system has been "articulated in accordance with grounds of reason which precede the natural phenomena and which can be understood from its causes, from the pure concept involved in its creation."[1]

Now these "causes" to which Kepler here alludes are not, as one might perhaps have expected, a set of "higher-level" physical explanations. Instead, Kepler introduces considerations of certain privileged concepts. Much in the vein in which Descartes later was to argue, he notes that in choosing hypotheses we must start with some limitations of or constraints on our concepts; begin with the most basic and simple ideas, such as the

[1] J. Kepler, *Die Zusammenklänge der Welton* (Jena, 1918), *Mysterium Cosmographicum*, sections 1–2, pp. 141–149.

204

straight, the circular, or (among the solids) the "platonic" bodies (regular polyhedra)—all of which moreover Kepler (somewhat fancifully) relates in turn to certain esthetic, mathematical, and even theological concepts, the relation being established under the guidance of the free play of analogy. For the sake of what is to follow, let us note especially that—as in this case—the response to the search for a "reason" or "cause" of some phenomenon (basic or derivative) is not always conceived in terms of a physical mechanism, but rather relates to some general notions and principles of a more "philosophical" kind. These are introduced because they are believed to yield a more substantial foundation, with the further heuristic function of serving as signposts and guidelines, as "anticipations" of what is not only hoped to be a correct theory but also felt to provide a more penetrating or "deeper" explanation. Such approaches are usually most prominent during periods of gestation and crisis, transitions from one conceptual scheme to another, when the novel ideas are still being questioned, frequently because of clashing with received modes of thought.

Thus, to take some examples from the case study that forms the subject of this paper, Newton will worry endlessly about the "cause or reason" of gravitational attraction. But one of the many and varied solutions of this problem which he entertains, as we shall see, turns out to provide a "reason" by alleging that gravitational phenomena exhibit teleology, and that the "cause of gravity" is its having the character of a "final cause." Again, Boscovich will be found to introduce his theory of force points as an application of the principle of continuity. Maupertuis in his attempt to secure acceptance for action at a distance—that "metaphysical monster" of the philosophers (as he calls it)—argues in its favor by noting that it satisfies the requirements of uniformity and analogy, for good measure adding the observation that the "effects" of attraction seem subject to the law of least action, a principle which in his eye not only provides the most all-embracing explanation of many of the laws of the physical sciences but possesses also a special rationale since it appears to him to imply a divine teleology.[2] Or, to mention one more example, Fechner's *Atomenlehre*, after giving an account of classical Newtonian atomic theory, regarded as a physical explanation, proceeds to a "philosophical" defense of Newtonian conceptions, by presenting the idea of simple material point-atoms with intervening forces as one which lies at the intersection, so to speak,

[2] Maupertuis, *Oeuvres* (Lyon, 1756), I, 46, 169; II, 257.

Gerd Buchdahl

of a number of associated conceptions, for instance, simplicity, discontinuity, infinity, indivisibility, denumerability.[3]

All such justifications evidently involve an appeal to a supplementary set of ideas: maxims of simplicity and economy; considerations of an esthetic nature; principles of continuity or discontinuity; linkages with general metaphysical notions as for instance "the real does not change," "nothing comes from nothing," "the effect is equivalent to the cause"; or more generally, maxims like those of homogeneity, affinity (or the "analogy of nature"), teleological or alternative preferred explanation schemas, and even certain theological conceptions. Nor need we dispute in this context over the term "metaphysical"; it is perfectly possible to regard such ideas as having purely methodological or "regulative" significance, to use Kant's term for notions which (on his view) constitute the methodological equivalent of what had previously been misconstrued as possessing "metaphysical" status. Or we may even admit with Mach that such maxims have their source in an expression of "powerful instinctive yearnings" rather than a "metaphysical point of view."[4]

Let us mark the existence of such ideas as a separate vector of inductive thinking by labeling this supplementary component of choice the "architectonic component," to note that it contains supplementary rules for the construction of a scientific theory and its concepts, though we need not regard it as sharply separated from the other components that we shall distinguish.

In fact, considerations taken from the architectonic component, though usually forming part of the structure of a completed theory, frequently affect also a second component which relates more directly to the *concepts* employed in the construction of a scientific theory, and which involves what we may call an "explication" of their meaning. For the sake of brevity, we shall also say that while the architectonic component determines the "rationale" of a theory, or of a theoretical concept, explication is concerned with "intelligibility." Of course, the meanings of many terms entering into theories antedate their formulation; still, and especially in a developing science, such meanings are often found to undergo considerable change—whether in the light of considerations connected with the theoretical and experimental field forming part of the theory in question, or of more general considerations, will depend on the individual case. Cer-

[3] G. T. Fechner, *Ueber die Physikalische und Philosophische Atomenlehre* (Leipzig, 1855), p. 152.
[4] E. Mach, *The Science of Mechanics* (LaSalle, Ill.: Open Court, 1960), p. 608.

tainly, in some important cases we find that explication of meaning frequently precedes the construction of theories. Also, we should note that quite often such explications, too, have been labeled "metaphysical," i.e., "metaphysical foundations" of the science in question, though once more we shall not worry overmuch about the terminology. Thus Galileo begins his disquisitions into mechanics by addressing himself to questions concerning the proper meaning of velocity, acceleration, etc.[5] Similarly Descartes's cosmology is preceded by a study of dynamics which starts with an investigation into what may properly be meant by space and motion.[6]

It is a mark of the idea that this is not only a separate inquiry, but somehow again providing a special "basis," that the axioms arising out of such semantic investigations are frequently felt to be more firmly entrenched than purely empirical generalizations; witness for instance Einstein's reflections on the proper meaning of simultaneity, or Descartes's "a priori" elucidation of the laws of motion by reference simply to conceptual considerations concerning the notion of motion as a state, the principle of causation, maxims of simplicity and conservation, etc.[7] For many seventeenth- and eighteenth-century writers—including Locke, Newton, and Kant—such laws and properties are frequently described as constituting "the essence of matter"; witness Newton's reference to the "innate properties" of matter, including its "passive laws of motion" formed together with that matter "by God in the beginning."[8]

A considerable section of writings in the history of philosophy is in fact taken up by a critical evaluation of the very notion of an "a priori" or "metaphysical" or "conceptual" explication of the foundations of science. Thus for Kant the basic laws of matter and motion are a specially tightened form of the transcendental principles which themselves are held to be an inherent aspect of all empirical judgments.[9] Still later, for instance

[5] Galileo, *Dialogues concerning Two New Sciences* (New York: Dover, 1914), Third Day, pp. 153–169.
[6] Descartes, *Principles of Philosophy*, II, sections 1–40, in G. E. Anscombe and P. T. Geach, *Descartes: Philosophical Writings* (London: Nelson, 1964), pp. 199–218.
[7] G. Buchdahl, "The Relevance of Descartes' Philosophy for Modern Philosophy of Science," *British Journal for the History of Science*, 1 (1963), 229–249. For a classical statement and detailed exemplification of this view of the axiomatic and privileged character of conceptual axioms, see William Whewell's theory of "Ideas" as developed in his *Philosophy of the Inductive Sciences* (1840, vol. 1; *Works*, London: Cass, 1967, vol. 5).
[8] I. Newton, *Opticks* (New York: Dover, 1952), pp. 400–401; cf. *Principia*, ed. F. Cajori (Berkeley: University of California Press, 1962), pp. 398–400.
[9] I. Kant, *The Metaphysical Foundations of Natural Science*, trans. E. B. Bax (London, 1883), Preface; chapters 2 and 3.

Gerd Buchdahl

in Poincaré, they are regarded as "conventions," and in our own day they are often viewed as vehicles for constructing the terminology of the science in question.[10] We find that the emergent concepts, when closely considered, are based partly on primitive experience, while partly they lie at the intersection of a number of laws which themselves have both empirical and normative components. The empirical content allows for possibility of change; the normative aspect yields a more static approach and thus a degree of stability. A limiting case of concept formation occurs when terms are—in the parlance of logicians—"implicitly defined" by the hypotheses of the theory into which they enter. More normally, and to allow for the possibility of testing of competing theories, the concepts have a richer basis, as just indicated, and the meanings involved straddle a number of possible theoretical situations.

This brings us to the third component which determines the choice of hypotheses, the one paid most attention to by inductive logicians, involving not only empirical confirmation but the formulation of laws and their integration into a system (nomothetic and systemic articulation). We may call this the "constitutive" component, though it, again, can clearly not be sharply separated from the other two, the explicative and the architectonic components, since they are more or less presupposed in the systematic development of theories—we need only remember the importance of analogical considerations.

There are some important precedents for such a triadic methodological structure, though it has not usually been formulated quite so explicitly. Implicitly it is contained, as already hinted, in the movement of scientific history itself. But making it thus explicit may give us a more definite guide for the interpretation and clearer understanding of some of its episodes. But for the moment, let us note some precedents in the philosophy of science proper. Of primary importance here are some of Kant's ideas, where—as I interpret him—a profound appreciation of something like the present structure first appears.[11] Kant distinguishes two factors that determine the acceptance of hypotheses; in the language of this paper they

[10] H. Poincaré, *Science and Hypothesis*, part III (London: Walter Scott, 1905). B. Ellis, "The Origin and Nature of Newton's Laws of Motion," in R. G. Colodny, ed., *Beyond the Edge of Certainty* (Englewood Cliffs, N.J.: Prentice-Hall, 1965), pp. 29–68. D. Shapere, "The Philosophical Significance of Newton's Science," *Texas Quarterly*, 10 (1967), 201–215.
[11] Cf. G. Buchdahl, *Metaphysics and the Philosophy of Science. The Classical Sources: Descartes to Kant* (Oxford: Blackwell; Cambridge, Mass.: MIT Press, 1969), chapter 8, section 4 (iv).

are equivalent to the "explicative" and the "constitutive" components. (My use of the term 'constitutive' is of course rather wider than Kant's since I include in it not only the "empirical data" but also the nomothetic and systemic aspects of theorizing.)

According to Kant, then, before we can accept a hypothesis, two requirements have to be satisfied, first, that it should account for the data in a strongly predictive fashion, without the employment of auxiliary ad hoc constructions; this determines its "probability." Secondly, the *possibility* of the concepts involved in the hypothesis must have been antecedently established.[12]

Take the concept of action during impact. As we shall see in the course of this paper, this is just one of the concepts which—though basic for mechanics—is held by many of the physicists and philosophers of the seventeenth and eighteenth centuries to be unintelligible or "inconceivable." Kant sets out to demonstrate its "possibility." First it is shown that mutual interaction as a pure concept or principle is presupposed in the form of all empirical judgments concerning the coexistence of bodies in space. This transcendental principle is then applied to the empirical concept of matter, here defined as that which is capable of exerting moving force. When this is worked out in detail—involving a "construction" which employs the distinction between "relative" and "absolute" space—the result (according to Kant) turns out to be the law of action and reaction.[13]

Another instance, even more important for what follows, is the application of the category of quality to the concept of matter, this time analyzed as that which fills space. Kant purports to prove that such a filling of space can only be conceived in terms of the joint action of repulsive as well as attractive forces, both being presupposed in order to obtain the notion of a material body with finite boundaries. Note in particular that this procedure is the opposite of the Newtonian: unlike the latter, we do not first posit matter and *then* "hypothetically" add action at a distance; rather this action is a necessary condition of the conception of matter as such, and defines its "possibility in general."[14] Kant distinguishes this kind of possibility ("real" possibility), the result of a "conceptual explication," quite

[12] I. Kant, *Introduction to Logic*, trans. T. K. Abbott (New York: Philosophical Library, 1963), pp. 75–76. I. Kant, *Critique of Pure Reason*, trans. N. Kemp Smith (London: Macmillan, 1933), pp. 241, 613 (A222, A770).
[13] Kant, *Metaphysical Foundations*, chapter 3, Proposition 4, pp. 223–230.
[14] *Ibid.*, chapter 2, pp. 170–190.

Gerd Buchdahl

sharply from mere "logical" possibility, just as it is equally distinct for him from "probability," which is a function of empirical confirmation.

Finally, for Kant all this is still further distinguished from the regulative certification of a concept. Thus, to take our last example, the conception of gravity—as distinct from the demonstration of its "possibility"—is shown also to be an expression of the demand by "reason" for homogeneity and analogy, which dictates the subsumption of the various phenomena of "dynamics," described by Galileo and Kepler, under the single conception of gravity, thus exhibiting their essential affinity relative to each other.[15] As in Newton (we shall see later) so here analogy bestows a "rationale."

Of course, the three components here distinguished are not always presented in such logically tight patterns. But something like explication and regulative articulation does evidently form an intimate part of scientific activity, and at times, especially in the early stages of development, it has much the largest share of it, rather than empirical testing. This incidentally shows that the latter cannot by itself be used as a means of demarcation between science and non-science ("metaphysics"), or between good and bad science.

Some of the later historians and methodologists of science, especially those influenced by Kantian thought patterns, follow a similar classification. Whewell distinguishes explication of concepts, colligation of facts, and finally consilience of inductions together with simplification of theories. Explication of concepts is here exhibited in particular as a powerful element of scientific development, and in a number of ways, whether this concerns settling on the right sort of candidates for satisfying some general principle such as conservation, or the justifiable model for atomic constituents, or a suitable definition of species, or what have you. Whewell moreover makes it quite clear that all this is no mere dictionary activity but at all stages involves attention to observed facts, so that again one cannot too rigidly separate out these three aspects of what jointly he calls induction.[16] And this has perhaps become even clearer in the recent discussions concerning "paradigm" formation and change.[17]

To mention another precedent, Heinrich Hertz, in the introduction to

[15] Kant, *Critique of Pure Reason*, pp. 544–555 (A662).
[16] W. Whewell, *The Philosophy of the Inductive Sciences*, vol. 2, book XI, chapter 2, section ii, pp. 11–16.
[17] Cf. T. S. Kuhn, *The Structure of Scientific Revolutions* (Chicago: University of Chicago Press, 1962).

his *Mechanics*, again employs a classificatory triad. His three criteria of choice are permissibility, empirical adequacy or correctness, and suitability or appropriateness of hypotheses (or conceptions or models). "Impermissible" is what contradicts "the laws of thought." (Hertz may have had in mind here not only the "laws" of formal logic, but also the "laws" of Kantian "transcendental logic," i.e., categorial principles of the understanding.) For instance, in Newtonian mechanics the idea of "force" leads to the perplexing notion of "centrifugal force" which bodies seem to possess in addition to their inertia.[18] Such conceptual explications quite frequently lead to the rejection of one notion in favor of another. Hertz rejects "force" as a primitive concept; Faraday discovers a putative "contradiction" in the Newtonian conceptions of atoms and empty space, which (he contends) seems to involve the consequence that sometimes space acts as an insulator, and sometimes as a conductor, a contradiction which (he argues) shows that the original conception "must be false."[19]

If we grasp that all three of our components determine the choice of hypotheses at any stage of a science, we shall be even more prepared to find explicative and architectonic functions emphasized at periods of "anticipation," when hypotheses have yet to be formulated, and conceptual patterns have still to be developed. Yet even A. E. Burtt, in his seminal *Metaphysical Foundations of Modern Physical Science*, otherwise so sympathetic to our view, was still apt to censure Kepler's habit of linking his physics to a background of religious speculation—thus misunderstanding its real purpose, which was that of linking it to a set of regulative notions like those of simplicity, unity, economy.[20] And older whiggish historians of science have shown of course even less sympathy, for instance, with an Aristotle who (in *De Caelo*) could define the number of possible "natural motions" by reference to quasi-grammatical considerations. All this suggests that an awareness of the existence and significance of a richer structure of methodological components supplies us with a set of historiographical tools that may yield more sympathetic insights resulting from historical investigations. Furthermore, let us note that it is a study of the classical writings of philosophy that is now found to yield some important

[18] H. Hertz, *The Principles of Mechanics* (New York: Dover, 1956), Introduction, pp. 2–6.
[19] M. Faraday, *Experimental Researches in Electricity* (London, 1844), vol. 2, pp. 286–287.
[20] E. A. Burtt, *The Metaphysical Foundations of Modern Physical Science*, rev. ed. (London: Routledge and Kegan Paul, 1932), chapter II (D), pp. 46–52.

items for our more complex historiographical toolbox. And finally, per contra, evidently aspects of scientific methodology can throw fresh light on some of the basic interests and assumptions of classical philosophy.

Still, there is a possible objection to regarding our explicative and architectonic criteria as determinants of hypotheses. Such criteria may be useful and perhaps even necessary devices for the *selection* of hypotheses, but they can hardly—so it will be said—affect the final acceptability of hypotheses, their factual strength, or more bluntly, their "truth." I will leave the question of "truth" for later. For the moment, we will reply that it is no mean or unimportant virtue of the supplementary criteria if they narrow down the choice of hypotheses. One may of course urge that before any maxims of selection can get a grip, one needs first a general principle of finitude in order to limit the potentially unlimited number of alternatives. But this is more an objection to the general possibility of a rational reconstruction of scientific reasoning, and I propose at any rate to leave its consideration for later too. Certainly, even at a more limited level, voices are divided. Thus, it is not uninteresting to note that some contemporary logicians of science, such as W. C. Salmon, have attempted to construct a model of reasoning (in terms of a "Bayesian" approach) where these kinds of supplementary criteria are held to define at least the *notion* of creating a finite antecedent probability for a hypothesis.[21] On the other hand, in a methodology like Kant's, not only is explication of "possibility" not assumed to have any significance for posterior probability, but it is not clear whether it affects explicitly his judgment of even prior probability. And it is doubtful whether Kant thought of explication as providing a straightforward old-fashioned foundation of induction; to this question, too, I shall return later. Similar considerations hold for the architectonic component: the task of regulative principles consists in guiding our search for system and explanatory depth, and he sometimes seems to hint that our criterion of truth *is* just such "success" as answers to our need, as though "by a lucky chance,"[22] to provide systematic integration of our empirical laws,[23] and "reason" tells us how we *ought* to set about doing this by means of regulative ideas.[24] Moreover, since there is no "world-in-itself"

[21] W. C. Salmon, "The Foundations of Scientific Inference," chapter 7, in R. G. Colodny, ed., *Mind and Cosmos* (Pittsburgh: Pittsburgh University Press, 1966), pp. 257–265.

[22] I. Kant, *Critique of Judgment*, trans. J. H. Bernard (New York: Hafner, 1951),
[23] *Ibid.*, p. 16.
[24] *Ibid.*, p. 10; cf. *Critique of Pure Reason*, pp. 538–539, 544 (A651–A653, A660–A661).

outside the forms of language created by explication and beyond systematic integration due to the regulative activity of the rational scientist, such relative "success" as we have will define or be "the touchstone" of the empirical truth of our rules.[25] In a similar way, many modern logicians would want to replace the "old problem of induction" by a systematic search for conditions defining models of scientific inquiry that explicate notions of reasonableness and predictive power. It has certainly become obvious that one cannot transfer any simpleminded notion of truth to the level of scientific aims without some considerable adjustments and reinterpretation.

II

I shall return to this theme toward the end; for the moment, the best defense for any methodological structure like the one here proposed is to observe it at work; to indicate the relative weights which attach to the different components within the ebb and flow of scientific development; to see what clarification this produces for our understanding of the conflicting strands of that development. There can certainly be no doubt that the satisfaction or nonsatisfaction of such criteria has often determined acceptance or rejection of hypotheses, and this is perhaps a stronger support for their being criteria than some abstract, purely logical set of definitions of what is to count as "confirmation" or "acceptability." And unless such a timeless set, together with its "justification"—not to mention justification of the notion of "justification" itself—is produced, any charge of "subjectivism" or "psychologism" or "historicism" against such an approach must remain somewhat vacuous.

The historical case that I want to consider in more detail is the conception of gravity, partly because of its excellent documentation, and partly because one can judge it now through the method of hindsight.[26] But my remarks are not to be taken as anything like complete or even partial history. I shall only select a few logical aspects, and consider them in the light of tensions, proffered solutions and responses, on the part of a handful of scientists, philosophers, and logicians, mostly belonging to the periods that precede twentieth-century sophistication both in the field of recent science and in that of its philosophy, at a time when worries about action at

[25] Ibid., pp. 535, 556 (A647, A680).
[26] See M. B. Hesse, Forces and Fields (London: Nelson, 1961), chapters 7–8, for a critical history; also A. Koyré, Newtonian Studies (London: Chapman and Hall, 1965), chapter 3, appendixes A–C.

a distance were still a relatively live issue. And I shall consider the whole case of course through the light of my three components. Also, and incidentally, one of the tests for our triadic classification will be whether it can shed light on some of the apparent contradictions in Newton's own pronouncements on this subject.

The constitutive component can be dealt with very briefly. We shall regard it as formalized by Newton's account in the *Principia*, representing what he labels his "experimental philosophy," and whose achievement is summed up by him in the General Scholium of 1713 as having "explained the phenomena of the heavens and of our sea by the power of gravity."[27] Fundamentally, the enterprise is deductive and systematic, basing itself on certain empirical laws, e.g., those of Kepler and Galileo, using certain axioms such as the laws of motion, jointly yielding the formula of gravitation. By invoking in addition certain "Rules of Reasoning,"[28] including principles of causality, analogy, and induction, the scope of the gravitational formula is generalized so as to yield the law of universal gravitation, which is thereupon tested or confirmed by showing it to yield further positive predictions of its applicability. The crucial conception involved is that of gravitational attraction at a distance.

In the light of what follows, it is important for a moment to consider the language in terms of which gravitational force occurs in the body of the *Principia*. Book I, Proposition ii, speaks of bodies being "urged by a centripetal force directed towards" the center of rotation (page 42).[29] Page 56 (I, xi) speaks of the "law of centripetal force tending to the focus of the ellipse." Page 406 (III, iii) speaks of "the force by which the moon is retained in its orbit" as tending "to the earth." Page 407 (III, iv) says that the "moon gravitates towards the earth, and by the force of gravity is continually drawn off from a rectilinear motion." Finally, page 414 (III, vii) speaks of "a power of gravity pertaining to all bodies." This is pretty strong language; and its strength is attested by Newton's claim, as we have found him say, that the phenomena can be "explained . . . by the power of gravity" (page 546). Moreover, page 407 seems to suggest that quite often Newton thinks of this "power" as "drawing" bodies from their inertial path. It is, however, well known that Newton shied away from interpreting this "drawing" as a "pull" exerted *by the bodies* on other bodies

---

[27] *Principia*, p. 546.
[28] *Ibid.*, pp. 338–340, 406–415.
[29] References are all to *Principia*.

thus "drawn" toward them, an action, moreover, which would—as far as concerns the model implied by the *Principia*—take place across empty space.

Now the reason for Newton's abstemiousness has something to do with his own and his contemporaries' explication of the concept of matter. It emerges at its most uncompromising in the famous letter to Bentley: Action of "brute matter . . . without mutual contact," or without something "which is not material," is there declared to be "inconceivable."[30] This view, in line with that of Newton's Leibnizian critics, is echoed also with special emphasis by Locke, when he writes that gravitational attraction "cannot be conceived to be the natural consequence of that essence [namely, essence of matter; and that it cannot be explained or made] conceivable by the bare essence . . . of matter in general, without something added to that essence which we cannot conceive."[31]

Evidently these protests have something to do with contemporary explications of the concepts of matter as well as of action, force, and—more generally—causation. Before turning to this, one or two minor points. It would of course be perfectly possible to hold that explication, and, with this, interpretation of the formalism of the *Principia*, is no more than an extra, and that the deductive development of the quantitative relationships can quite well proceed without it. Newton himself sometimes sug-

[30] Third Letter to Bentley; in I. Newton, *Papers & Letters on Natural Philosophy*, ed. I. B. Cohen (Cambridge, Mass.: Harvard University Press, 1958), p. 302. Contrary to the interpretation of some recent commentators, I take it that Newton's emphasis here is on the *impossibility* of the action of matter on matter without an intervening *material* medium, and *not* the *possibility* of such action by means of an intervening *immaterial* medium. The phrase "which is not material," in *this* context, does not mean "which is immaterial." This latter interpretation has sometimes been proposed because (as we shall also see later in this paper) Newton—even in the same letter to Bentley—and more explicitly some of Newton's disciples such as Bentley and Clarke did also hold that attraction could be given a rationale by regarding it as due to a spiritual bond whose foundation is in God. But my point is that this escape route belongs to the architectonic component, and that it never affected Newton's assessment of the explication of the concept of matter. It is because of that explication that recourse is taken to God and teleology, but that recourse never canceled the explication, nor was there any inconsistency with it, and consequently Newton did not have to abandon his views expressed in the Bentley letter. The architectonic escape—by yielding a "rationale"—is provided in order to reconcile the continued use of the concept of attraction at a distance, which certainly Newton did not give up at any stage, with the explication of the concept of matter which seemed to render it "unintelligible." The teleological approach was a halfway house between total rejection of gravity and a fullfledged physicalist interpretation.

[31] Quoted in Koyré, *Newtonian Studies*, p. 155. Some further background to the relations between Newtonians and anti-Newtonians I have also given in my "Gravity and Intelligibility: Newton to Kant," in R. E. Butts *et al.*, eds., *The Methodological Heritage of Newton* (Toronto: University of Toronto Press, 1969).

215

gests this, as when in his comments on Definition VIII he says that he considers terms like "attraction" and "impulse," and any such "forces," "not physically, but mathematically."[32] This is the approach later summed up (for the case of electromagnetic theory) in Hertz's famous dictum that "Maxwell's theory is Maxwell's system of equations."[33]

However, it is difficult not to regard your symbols as pointing to a minimum of physical reality, and for the most part—as our quotations from *Principia* strongly suggest—Newton does employ the term 'force' as denoting a physical cause, as "an action exerted upon a body" (Def. IV).[34] What otherwise can he mean when he says in Query 31 of the *Opticks* that the "word" attraction is meant "here to signify only in general *any force by which bodies tend towards one another*"?[35] So the question of what interpretation ("meaning") can be given to the expression 'gravitating matter' has not been banished; the skepticism implied by "only" is quite evidently due to Newton's analysis of the concept of matter and action which still continues to play a part. The italicized words in fact look in two different directions, corresponding to two different paths.

1. Interpreting "any force by which" as "any law of force in accordance with which" gives substance and meaning to Newton's continued assertion that "gravity does really exist, and acts according to the laws which we have explained,"[36] with the possible implication that Newton wishes to restrict the concerns of his "experimental philosophy" to the lawlike kinematical phenomena involved. And this reduction of causal efficacy to rules of motion is of course the approach subsequently made familiar in the philosophy of Locke, Berkeley, Hume, and Kant, not to mention later thinkers in a similar strain.

2. The second path is Newton's ever-recurring remark that his reference to gravitational attraction still awaits an account of "how these attractions may be performed."[37] But here again the conceptual analysis casts its

[32] *Principia*, p. 5.
[33] H. Hertz, *Electric Waves*, trans. D. E. Jones (London, 1893), p. 21. Hertz was primarily concerned to assert the independence of Maxwell's equations, whether they are themselves given an interpretation or not, from any hypotheses concerning the particular structure of the ether transmitting electromagnetic radiation. But one may also understand his slogan in formalist fashion, so as to imply neutrality concerning the question of the physical nature of the "waves," and hence, of the "meaning" of the symbolism of the equations.
[34] *Principia*, p. 2.
[35] *Opticks*, p. 376.
[36] *Principia*, p. 547.
[37] *Opticks*, p. 376.

shadow (see footnote 30), for Newton usually makes it plain that the explanation must avoid an account that would "attribute forces, in a true and physical sense, to certain centres."[38] How is he going to avoid this? Our methodological trichotomy suggests that the tension to which Newton is subject may be resolved in more than one way.

a. The first type of solution belongs to the "constitutive" component, involving physical or mechanical hypotheses. To be sure, Newton seeks jealously to preserve the formal purity of his axiomatic account in the *Principia*. Unlike the work of his continental competitors, for instance Leibniz and Huygens, he does not permit "contrary hypotheses" (putatively forced upon us in virtue of the conceptual impossibility of action at a distance) to impugn the "truth" of "propositions inferred by general induction from phenomena," as he formulates this in the fourth rule of reasoning.[39] On the other hand, physical hypotheses concerning "the manner of action" are permissible, provided that for the time being they do not get mixed up with the official systemic account. Such a hypothesis was for instance introduced by Newton into the 1717 edition of *Opticks*, expressly stated to "show that" he did "not take gravity for an essential property of bodies."[40] This was the hypothesis of an ether, itself possessing a particulate structure. But its introduction involves fresh difficulties. Although the hypothesis avoids attractive forces, it does operate with "forces" "by which those particles . . . endeavour to recede from one another"[41]—a curious makeshift. Our perplexity (unless we charge Newton with sheer muddle-headedness and contradictory attitudes) is perhaps somewhat lessened if we take him to offer a mechanism which reduces attraction between large bodies to repulsive forces acting between small particles at short distances, possibly in order to provide a unitary (and hence, *simpler!*) mechanism, incidentally thereby also perhaps seeming to answer his Leibnizian critics.[42]

b. Still, it suggests itself that there must be some further considerations for resolving the puzzle of how Newton could—in the light of the Bentley letters—continue to operate with the conception of distance forces as such, if we ignore his occasional formalist escape route. Such an alternative is

[38] *Principia*, p. 6.
[39] *Ibid.*, p. 400.
[40] *Opticks*, Advertisement II.
[41] *Ibid.*, p. 352.
[42] As suggested, for instance, in Joan L. Hawes, "Newton's Revival of the Aether Hypothesis and the Explanation of Gravitational Attraction," *Notes and Records of the Royal Society*, 23 (1968), 210.

provided by criteria belonging to the architectonic component, analogy, harmony, teleology being here the relevant ideas. One of the arguments which belongs to this category is stated almost explicitly in Query 31, which harks back to Newton's prolonged studies in the phenomena of electric attraction. He asks whether it is not true that "small particles of bodies" have "forces, by which they act at a distance," and goes on to say: "For it is well known, that bodies act one upon another by attractions of gravity, magnetism, and electricity; and these instances shew *the tenor and course of nature,* and make it *not improbable* but that there may be more attractive powers than these. *For nature is very consonant and conformable to herself.*"[43] Evidently, then, the existence of analogous cases adds its own strength.

c. This reference to nature being conformable to herself occurs in many places, e.g., in the comments on Rule III, with its reference to "the analogy of nature, which is wont to be simple, and always consonant to itself."[44] The force of this architectonic component seems then to supply us with an answer to the riddle how Newton could escape between the horns of the dilemma of his talk about forces between small particles—or, for that matter, large bodies—and the impossibility of action at a distance. What he does is to fall back on "final causes";[45] and nowhere is this clearer than in the 1706 edition of *Opticks,* in Query 28, written at a period when he was temporarily rejecting the hypothesis of an ether altogether. To the question How are we to explain attraction if we reject an ethereal medium, he replies with a counterquestion: But surely, "nature does nothing in vain"? And the subsequent context of that Query refers us to evidence of design in both animate and inanimate nature, ending with the comment that it most certainly "appear[s] from the phenomena" that there is a divine being, the source of the harmony of nature.[46] Now as Kant held later, this reference to God as the origin of design may be interpreted as a pseudo-constitutive object, which we place in the "idea" of design, to simulate what is in fact a *regulative principle* whereby nature is regarded *as if* it were designed in accordance with the systematic activity of scientific reason.[47]

This emphasis on support from harmony is also to be found in some of

[43] *Opticks,* p. 376.
[44] *Principia,* pp. 398–399.
[45] *Ibid.,* p. 546.
[46] *Opticks,* pp. 369–370.
[47] Kant, *Critique of Pure Reason,* pp. 556, 559, 561 (A681, A686, A688).

Newton's unpublished notes for some sections of the *Principia*, as has been emphasized recently by McGuire and Rattansi, where we find Newton relating gravitation to the general harmony of nature, appealing to the authority of the Pythagoreans' view of divinity, "inspiring this world with harmonic ratios like a musical instrument,"[48] an attitude which represents a considerable current of Neoplatonic ideas in seventeenth-century thought.

Providing a "rationale" for gravitation is, however, not the same thing as pronouncing it to possess physical reality—contrary to what some commentators have recently maintained.[49] To use a form of speech implied in Clarke and Bentley, to "spiritualize" gravitational force is not to physicalize it outright.[50] While "God in the beginning"—to use the language of Query 31—created the atoms and endowed them with certain innate qualities such as solidity, hardness, impenetrability and inertia together with the relevant "passive laws of motion"—making up, as we shall discuss presently, the "essence" of matter—"active principles" such as gravity, fermentation, and cohesion have a more ambivalent status. Sometimes Newton employs the phenomenalist route, when he says that he regards these principles only as "general laws of nature," whose "causes" are "not yet discovered."[51] But two pages previously, he has already referred to the "active principle" of gravity as "*the cause of gravity*,"[52] indicative of his ambivalent approach. What is really involved we can guess from the fact that no sooner has he said that "the causes of those principles are not yet discovered" than he again abruptly shifts his text to a discussion of "the counsel of an intelligent agent"—clearly a reference to analogy and teleology once more.[53]

To sum up Newton's procedure, we find that the architectonic component acts as a resolvent of the tension set up between the denial of attraction consequent to his conceptual analysis and the urge to provide a physicalist interpretation for the constitutive component (the axiomatic portion of *Principia*). And as Newton says there: to "discourse" of God "from

---

[48] J. E. McGuire and P. M. Rattansi, "Newton and the 'Pipes of Pan,'" *Notes and Records of the Royal Society*, 21 (1966), 108–143.

[49] Cf. A. R. Hall and M. Boas Hall, ed., *Unpublished Scientific Papers of Isaac Newton* (Cambridge: Cambridge University Press, 1962), p. 197. See also footnote 30, above.

[50] For Clarke, cf. *Rohault's System of Natural Philosophy* (London, 1723), p. 54n (Clarke's note). For Bentley, cf. *Eight Sermons* (Oxford, 1809), Sermon VII, pp. 227, 234, 238.

[51] *Opticks*, pp. 401–402.

[52] *Ibid.*, p. 399.   [53] *Ibid.*, p. 403.

the appearances of things," or as we might say, to provide such a teleological rationale, "does certainly belong to natural philosophy," as distinct from "experimental philosophy," i.e., our constitutive component.[54]

Here again, the history of subsequent philosophy can provide more explicit texts, though perhaps no timeless defense, for the attempt of the historian of science to interpret Newton's ideas. Newton's problem of providing a gravitational bond, without endowing bodies with a corresponding innate property of attraction, or without physicalizing a rational entity like attractive force, is profoundly generalized by Berkeley whose basic metaphysics of itself already leads to the denial of all "necessary connections" and "active forces" between physical things, the latter being regarded by him as no more than so many "passive ideas." Here, too, the missing link is provided by God's causal activity. But more: Berkeley is aware of a further problem we alluded to before and generalizes it: why interpolate a hidden mechanism (such as Newton's ether) if all change is really due to the direct action of a spiritual type? Berkeley's answer is classical: such a mechanism makes possible a rational and harmonious explanation of the phenomena in accordance with the laws of nature—i.e., explanation needs continuity; secondly (he adds), the production of the phenomena in accordance with the "rules of mechanics" is an expression of the "wise ends established" by the deity: a teleological principle.[55] So a preferred explanation form (here teleology) once again provides a rationale, where the nexus of *physical* continuity has been snapped, partly because it had come to be interpreted (in Berkeley, and after him, Hume) as a "metaphysical bond." Similarly, Maclaurin, the historian of Newton's "philosophy," who was profoundly influenced by Berkeley, *both* introduces an ether to account for gravity *and* operates with the general principle that all power and efficacy derives from God. And the problem of the apparent superfluousness of the mechanistic medium is explained in similar fashion by arguing that God does not act through immediate volition, but employs "subordinate instruments and agents" to "perform the purposes for which he intended them."[56]

III

Our reference to a "metaphysical bond" brings us back to the point whose discussion we deferred, when meeting the protest that attraction of

[54] *Principia*, p. 546.

[55] G. Berkeley, *Principles of Human Knowledge*, sections 62, 72.

[56] C. MacLaurin, *An Account of Sir Isaac Newton's Philosophical Discoveries* (London, 1775), pp. 408–409.

matter for matter *in distans* was inconceivable to Newton and Locke, and not of course just to them: there is a stream of writers reaching right down to the end of the nineteenth century who repeat the complaint, as for instance still expressed uncompromisingly in Stewart and Tait, who insist as late as the 1880's that it is "impossible . . . for a moment to admit the possibility of such action."[57] And we know from Faraday's writings that the idea, to quote Tyndall, "perplexed and bewildered him," and that he "loved to quote Newton upon this point," "over and over again" coming back to and quoting the Bentley passages already mentioned.[58]

As I showed before, all these scientists and philosophers are assuming here some sort of interpretation of the mathematical formalism of the theory. Further, they clearly assume that there is *some causal relation* holding between bodies subject to the law of gravitation. But up to now we have only considered those attempts which, when faced by the supposed difficulties involved in the concept of attraction, imagine either that we must look for an intervening mechanism or that we must limit ourselves to a phenomenological account; and the latter, again, either by confining our attention to the requisite laws of the system, with their "causes or reasons" unexplained, or by providing such a "reason" through a teleological model—all three evidently being solutions espoused by Newton and some of his successors. What is still outstanding is an investigation of the explicative component as such. For clearly it would be possible to resolve Newton's difficulties by readjusting the explication of the concept of matter, and—as soon became evident—by providing some clarification of the concept of causal action involved. This, as we have already seen, was Kant's approach, almost coincident in time with a similar attempt by Boscovich, though the conceptual ramifications involved do not emerge there as clearly.[59] Similar hints of such an approach we find later also in Faraday, where he distinguishes between "experimental philosophy" and "speculations" whose purpose is that of "rendering the vague idea more clear," one of his instances being the concept of action at a distance.[60] Indeed, his answer to the question "what are gravitation and solidity" is remarkably similar to Kant's, although there were of course many other contemporary influenc-

[57] [B. Stewart and P. G. Tait], *The Unseen Universe* (London, 1885), p. 146.
[58] J. Tyndall, *Faraday as a Discoverer* (London, 1877), p. 82.
[59] Cf. Roger J. Boscovich, *A Theory of Natural Philosophy* (Cambridge, Mass.: MIT Press, 1966), especially part I, sections 16–31.
[60] M. Faraday, *Experimental Researches*, vol. 3, p. 408.

es: "Certainly not the weight and contact of the abstract [atomic] nuclei. [Rather] . . . one is the consequence of an *attractive* force, which can act at distances . . . and the other is the consequence of a *repulsive* force. . . ."[61]

Evidently such conceptual investigations are not purely a priori. They borrow heavily from the context of empirical and theoretical researches. Thus, Kant, seeking to repel the suggestion that bodies can only act when they touch one another, counters this with the question "What do we mean by touch?"; he answers this by noting that the required analysis (which—as he stresses—is empirical in nature) does not involve matters of pure geometry, but rather the fact that when we touch a body we experience a force of repulsion, which suggested that it was force that is the primary concept here involved.[62] As a consequence he espouses the solution—anticipated by some Leibnizians—of packing force (or action) into the very conception of matter, just as is subsequently done by all those who make field action the core of the whole phenomenon. But to do this, one had to break through a whole host of preconceptions concerning the concept of matter and of action, i.e., causation. Let us therefore survey some of the controversies that stemmed from the Newtonian approach. This will give us an opportunity to observe (at least in one instance) what is involved in conceptual explication.

For Newton, as we saw, matter is characterized by "passive laws" only; moreover, he imagines a closed set of "innate" properties, which he thinks of as its "essence," e.g., impenetrability, extension, mobility, inertia—but not of course gravity, which would endow matter with an active principle,[63] something which this generation was trying to avoid, perhaps because it seemed to reintroduce hylozoic conceptions. Its rejection is expressed in the strongest possible terms, as we have seen, by claiming it to be *inconceivable* that matter should possess attractive properties.

[61] *Ibid.*, p. 449.

[62] Kant, *Enquiry concerning the Clarity of Principles*, in G. B. Kerferd and D. E. Walford, trans., *Kant Selected Pre-Critical Writings* (Manchester: Manchester University Press, 1968), p. 20. Needless to say, this analysis was not accidental. The choice of the empirical element of force was suggested precisely by the importance which that concept had come to acquire in the Newtonian corpus. We see here clearly how empirical and theoretical developments come to affect the "meanings" of scientific terms.

[63] *Principia*, pp. 399–400; *Opticks*, Advertisement II. So instead of being "essential," gravity is said to be "universal"; only universal, because although it is a lawlike property, its magnitude varies, depending on the distance, and so is not a constant property (cf. p. 400), but still something, i.e., a universal property, because sanctioned by the constitutive and the architectonic components.

But in what sense "inconceivable"? Locke and Clarke occasionally suggest logical inconceivability, when they say that action at a distance involves a "contradiction,"[64] but we should not make too much of this, since the notion of "logical contradiction" is not at that time too sharply distinguished from other kinds of contradiction. People quite generally use "inconceivable" to refer not only to what is self-contradictory, but also to what clashes with possibilities to which not only traditional evidence but currently received modes of thought seem overwhelmingly opposed, a position which Kant's "metaphysical foundations" expresses admirably, if in a somewhat formal way. In fact, it is just because the situation is logically so fluid that we find such an extreme variety of positions vis-à-vis our concept. Altogether, these are too numerous and complex to deal with here in more than bare outline.

How then do the various schools of thought seek to avoid the putative difficulty of the conception of attraction at a distance, if we now discount the appeal to intervening media, as well as phenomenalist or teleological escapes?

There is first of all the school which includes such diverse figures as Locke, Maupertuis, and Whewell. All three agree in the belief in an "essence" of matter, which includes properties held to belong to matter in some sense "necessarily" or "primordially."[65] Locke and Whewell also seem to share the view that this necessity is in some way "transparent" to our intellect. And both Locke and Maupertuis contrast this with the case of gravity, which they regard as a property whose admitted possession by matter they hold to be "inconceivable." Surprisingly, however, none of them conclude that the possession is impossible. The pressure from the constitutive component, the "inductive success" of the theory, prevents this conclusion. Rather, they interpret "inconceivable" to mean "nonintuitive," not transparent to the intellect.[66] In another terminology, the gravity of matter is now viewed as a synthetic a posteriori truth.

[64] For Clarke, see H. G. Alexander, ed., *The Leibniz-Clarke Correspondence* (Manchester: Manchester University Press, 1956), Clarke's Fourth Reply to Leibniz, section 45, p. 53. Clarke says that attraction "without any intermediate means, is . . . a contradiction: for 'tis supposing something to act where it is not."

[65] Whewell, *Philosophy of the Inductive Sciences*, vol. 1, book I, chapter 4; book III, chapter 9, especially p. 258. Maupertuis, *Oeuvres*, I, 95, 101, speaks of "primordial properties," which are "regarded" as "essential properties of matter."

[66] For Locke, see Koyré, *Newtonian Studies*, p. 155, quotation from a letter to Stillingfleet; for Maupertuis, see I, 97–98: "The manner in which properties reside in a subject is always inconceivable to us" and "it is experience to which we owe our knowledge" of both the primordial and all other properties.

Gerd Buchdahl

This conclusion will frequently appear under a skeptical guise, nowhere more than in Locke, who generalizes Newton's problem of how a body can act at a distance into the question of how it can act—or rather: how it can be seen, or be conceived, to act at all, by impact or otherwise, since "all matter is essentially passive."[67] And a similar escape is suggested by Maupertuis when he writes that "the manner in which the properties reside in a subject is always inconceivable for us. . . . Attraction is simply a question of fact. . . . It is in the system of the world that one must look whether a principle has an effective place in nature. . . ."[68]

Here the complaint has become metaphysically generalized. In the case of attraction, it might perhaps have been contended that the phenomenon itself cannot be "directly perceived" to take place; that it is only a theoretical interpretation; but not so for impact—unless you take a starkly phenomenalist stance, foreign to Locke's thinking. And so he writes that although "by daily experience" we observe the production of motion by impulse, the idea is totally "obscure and inconceivable": "the manner how hardly comes within our comprehension," so that we have only a very "obscure idea of an active power."[69] Here it becomes obvious that this is not just a difficulty engendered by the construal of matter as "passive," but an incipient critique of the notion of transient causality.

In the sequel there were two possible responses to all this. A prominent member of the first type of response was Mill. He accepts—but without Locke's skeptical undertones—the synthetic a posteriori status of the phenomenon of attraction, as he does of all empirical relations, and (commenting on Newton's letter to Bentley) indignantly rejects the charge of "inconceivability," which he simply puts down to the results of "past history and habits of our own minds,"[70] indeed labeling it a "fallacy of simple inspection." (In the same vein he also castigates as "fundamental errors of the scientific enquirers" most of the regulative ideas and maxims that we have mentioned before, such as principles of conservation, economy, simplicity.[71]) On the other hand, the price paid is the inordinate weight that has to be placed on induction pure and simple (part of the constitutive

[67] Cf. Locke, Essay, ii.21.4 and 23.28.
[68] Maupertuis, Oeuvres, I, 98, 103.
[69] Locke, Essay, ii.23.28.
[70] J. S. Mill, System of Logic (London, 1879), vol. 1, book II, chapter 5, paragraph 6, p. 275.
[71] Ibid., vol. 2, book V, chapter 3, paragraph 3, pp. 319–320.

component), as a logical operation for which Mill has to claim some kind of absolute "validity."[72]

Not all the empiricists go that far. Hume for instance clamps certain constraints on causal action, such as the requirement of spatiotemporal contiguity, which in particular he applies to the case of action at a distance, where he insists that we should always "presume" that a "chain of causes" exists which links "distant objects" in contiguous fashion;[73] and he bestows considerable praise on Newton's "hypothesis" of "an ethereal fluid" as one such method of explanation.[74] Perhaps this "presumption" was also motivated by a conceptual point, for Hume held that "a vacuum and extension without matter" are strictly speaking "impossible to conceive."[75]

It is thus always the case that the weight of scientific reasoning will be found distributed in different directions. This brings us to the second response to the charge of inconceivability. This involves a growing realization that one may always modify a troubling concept and thereby remove the contradiction. Behind this there lies, however, a more general point, namely, that in the metaphysics of science (and metaphysics in general, for that matter) one cannot start with rigid definitions as in mathematics, but instead one needs to employ a more empirical and tentative method when approaching the basic concepts of a science.

A very interesting early instance of such a method can be found in the Newtonian John Keill's *Introduction to Natural Philosophy*. Though we need to start with definitions, he declares, a proper method construes such definitions not of an "intimate essence," in terms of "genus and difference," but as "descriptions" obtained from "acquaintance" with "bodies or their actions," or "deduced" from "some of their properties . . . discovered by" the "senses."[76] Thus, "solidity" must be defined, not in the old Aristotelian way of "impenetrability," or the "inability of two bodies to be in the same place at once," but as "that property of a body whereby it resists all other bodies" that press on it "on every side."[77]

Keill was among those influencing Kant, who took, as already noted, a

[72] *Ibid.*, book III, chapter 21, paragraph 2, p. 101.

[73] D. Hume, *A Treatise of Human Nature*, ed. L. A. Selby-Bigge (Oxford: Clarendon Press, 1946), book I, part III, section II, p. 75.

[74] D. Hume, *An Enquiry concerning Human Understanding*, ed. L. A. Selby-Bigge (Oxford: Clarendon Press, 1902), section VII, part I, p. 73n.

[75] Hume, *Treatise of Human Nature*, book I, part II, section IV, p. 40.

[76] John Keill, *An Introduction to Natural Philosophy* (London, 1720), Lecture I, pp. 7–8.

[77] *Ibid.*, Lecture II, p. 12.

precisely similar approach.[78] Displaying the "possibility" of a concept turns out to be a process whereby this is related to other basic concepts, both empirical and categorial; but the difference between this and the approach of empiricists like Mill consists in the sharp distinction between constitutive and explicative inquiry, the latter providing a special kind of "foundation," a bit like insisting that one's theoretical terms need to have an independent meaning before they can play their part in the construction of a scientific theory. In the case of action at a distance, the change of meaning might be summarized—as was done later by Stallo—not in the slogan that a body acts where it is—the old view—but rather, it is where it acts. For the "inconceivability" claimed for the concept of the action of matter, says Stallo, was an "incompatibility of new facts or views with our intellectual prepossessions"; not—as he argues against Mill—just a fallacious habit. And he adds that "a reconstruction of our familiar concepts of material presence" would of course lead to a new conception of matter, different from the Newtonian one of hard atomic particles separated from one another by empty space.[79]

The lack of interest shown by writers of this group in the "problem of induction" is testimony that the center of gravity of the weights determining the choice or acceptability of hypotheses has shifted toward the region of conceptual explication. In this way, the developments and disputes in the history of science illuminate the intentions of philosophers and methodologists, just as our understanding of *their* doctrines provides a new key for our grasp of some of the scientific disputes.

A summary of the structure of my argument so far may be helpful. I have distinguished three components that appear to determine the acceptability of scientific hypotheses, where each component evidently itself has a particulate structure:

1. Conceptual explication:
   intra-theoretical
   pre-theoretical
2. Constitutive articulation:
   formal presentation of system
   informal interpretation
3. Architectonic determination:
   regulative ideas and maxims

[78] Cf. note 62.
[79] J. B. Stallo, *The Concepts and Theories of Modern Physics* (London, 1885), chapter 9, p. 145.

226

preferred explanation types
consilience

Applying the schema above to the evaluation of the Newtonian conflict situation, I have distinguished between the original situation as it first presents itself to Newton and the subsequent attempts at mediation in order to provide a resolution of the conflict.

I. Original situation.
1. Constitutive component: its informal interpretation yields an explication (intra-theoretical context) which suggests action at a distance.
2. Conceptual explication: derived from pre-theoretical contexts and suggests "matter cannot act on matter at a distance."
II. Proposals for mediation.
1. Manipulations of the constitutive component:
   i. formalist response: "we are concerned only with the calculus, the mathematical aspects."
   ii. phenomenalist approach: "we are concerned only with the kinematical and lawlike aspects of the phenomena."
2. Recourse to the architectonic component:
   i. Alternative explanation types:
      a. hypothesis of an inter-phenomenal ether.
      b. conception of a supra-phenomenal or teleological setup: God as the "cause or reason" of gravity.
   ii. regulative ideas: the power of analogy.
3. Reconstruction of conceptual component:
   i. reinterpretation of significance of "inconceivability" as "contingent" (synthetic a posteriori).
   ii. reinterpretation of concept of matter (borrowing from intra-theoretical context) as reducible to a center of repulsive and attractive forces (from "matter acts where it is" to "matter is where it acts").

IV

There is still left a question outstanding which I have briefly broached once or twice before. Suitable or "satisfying" conceptual explications, or the adoption of preferred explanation schemes, so it will be said, may lead one to believe in having got hold of the right theory, but they can hardly provide support in fact; that can only come from the constitutive factor, e.g., empirical confirmation. Such objections, however, clearly overlook a number of difficulties and facts. Facts: because it simply is the case that scientific theorizing has a lot to do with aiming to "satisfy" our quest for "understanding," and this evidently stands in need of our two other components. Difficulties: because the objection assumes both a simple view of scientific truth and an optimistic assessment of the nature of scientific inference, and of some transcendent "inductive foundation."

At this point it will be best to take the view that we cannot expect a solution of our problem if it is looked at through the spectacles of the "old problem of induction."[80] If instead we take such a position as espoused in Goodman's *Fact, Fiction, and Forecast* (to choose just one among many contemporary variants), our task can be viewed as that of providing satisfactory models for the explication of the conception of admissible "projections."[81] The provision of such models is to some extent a *descriptive* exercise, subject to certain conditions, e.g., that projections should be in accord with our normal "inductive intuitions," and that no paradoxes should arise. According to Goodman, the main condition that must be satisfied is that projected predicates should exhibit a certain degree of "entrenchment" with respect to normal inductive processes.

For Goodman, "entrenchment" was supposed to be a guarantee that such admissible predicates would yield lawlike projections, lawlikeness being thus linked in an essential way with entrenchment.[82] Goodman never offered a very satisfactory account of what might constitute "entrenchment," but the reference to "lawlikeness" provides us with a useful bridge to what has preceded in this essay. As we have noted, for Kant what I have called "conceptual explication" is equivalent to providing "metaphysical foundations." Now his view on the ultimate function of, as well as limitations on, "metaphysical foundations" is well expressed in one of his *Reflexionen* where he writes that the understanding can provide no more than the "ground of lawlikeness" of empirical laws, and not the "ground of particular laws," i.e., their empirical foundation. And he adds: "All metaphysical principles of nature are no more than grounds of lawlikeness."[83] This gives us a hint for a formal bridge between the metaphysics of nature (in the sense of explication) and entrenchment. Our conceptual and architectonic criteria could then be viewed as providing those

---

[80] Some recent writings have laid a similar stress on supplementary components; e.g., G. Schlesinger, *Method in the Physical Sciences* (London: Routledge and Kegan Paul, 1963); R. Harré, *Matter and Method* (London: Macmillan, 1964); *The Anticipation of Nature* (London: Hutchinson, 1965); Gerald Holton, "Presupposition in the Construction of Theories," in H. Woolf, ed., *Science as a Cultural Force* (Baltimore: Johns Hopkins Press, 1964).

[81] Nelson Goodman, *Fact, Fiction, and Forecast* (London: University of London, 1954), pp. 52ff.

[82] *Ibid.*, pp. 74, 75, 77, 95.

[83] *Kant's Schriften* (Ak. ed.), XVIII, 176 (No. 5414).

constraints which would assure a degree of "entrenchment" for the terms of a theory, it being understood that these criteria are only necessary but not sufficient conditions; for quite evidently we always require the check from the constitutive criterion.

Here a new objection will arise. It will be said—as I have indeed noted a moment ago—that these criteria are the result only of an empirical, and hence descriptive, survey of methodological activities arising out of the context of historical episodes. And it would then seem that such an account provides insufficient grounds for ascribing the character of "rationality" to the scientific procedure defined thereby. Now insofar as this expresses once again a complaint by an "old-fashioned" inductivist against the "new view of induction," I will not discuss that any further, since it would lead us here too far afield. But one might perhaps suggest that insofar as Kantian explications link up (however weakly) with categorial principles, such a procedure offers us at least a *simile* of rationality, with the further assumption that such a foundation has a raison d'être because it offers us "a way round the world." And the same remark would go for the architectonic principles, where they are viewed as heuristic methods without which no science would ever get off the ground. (Not for nothing did Kant regard them as "demands of reason"!)

But here a more substantial objection may be brought forward: such transcendental structures may perhaps enable us to construct a possible world, but they are no guarantee that they catch the sense of the "actual world." Once again, I will overlook the concealed old-fashioned inductivist note in this objection. Rather, what it suggests is that there are certain criteria, for instance, falsification, which seem more directly designed to facilitate an approach to the actual (observational and inferential) structure of nature. Now, partly, there lies behind this the consideration that to the falsification schema there corresponds a logically valid formula of reasoning (*modus tollens*), whereas the same cannot be said of, say, the Peircean schema of abduction, to which corresponds in fact only an invalid formula (the so-called "fallacy of affirming the consequent").

Now the correct comment on this seems to me that such contentions involve a confusion between a methodological schema and a logical formula. As has become abundantly clear in recent years, falsification as such (however much surrounded by additional provisos) is no guarantee of arriving at any "true" conclusions, including instantaneous rejections. What is correct is that the existence of apparently falsifying circumstances forces us to reconsider our ideas on a subject; but it is at best only one among many considerations. Certainly, we should not be tempted to slide from the compulsiveness of the logical formula into a belief in any conclusiveness of its application in a scientific (i.e., empirical) field.

Corresponding remarks apply to some of our other criteria. As my case study shows, conceptual explication does not move in a logical vacuum. For instance, we noted that the Kantian explication of matter borrows heavily from the suggestiveness of the Newtonian theory, i.e., the constitutive component.

It must be admitted, however, that there are obvious differences in degree of "depth" with which our different criteria attach to the final result, the construction and adoption of a theory. I purposely included "preferred explanation-type" among these criteria (with the prominent example of teleological explanation) because such a criterion seems particularly apt to provoke the objection of arbitrariness and historicity. But if the reader will reflect carefully over the whole complex situation, he will realize that the extent to which the different criteria express rational schemata varies only by degrees. And if it is thought that the existence of a formal correspondence (as in the case of falsification) should bestow the ceremonial title of rationality on a procedure, then the considerations here advanced suggest that—with the proviso of differences of degree in depth—the same privilege may be accorded to the other criteria as well. And it is for this reason that in this essay I have as a first step attempted a somewhat more systematic enumeration of, and comparative assessment of the different tensions between, such criteria when "at work" in an actual historical context.

COMMENT BY LAURENS LAUDAN

Gerd Buchdahl's paper is an extremely interesting one; more to the point, it is probably an extremely *important* one. With considerable skill and subtlety, he shows how inadequate—and how irrelevant to actual science—have been those philosophies of science which argue that the major (if not the only) determinant of scientific choice is empirical success. He says explicitly what many of us have felt but never expressed so clearly; namely, that the evaluation and assessment of a theory occur on several levels simultaneously, and that empirical adequacy (what Buchdahl calls the "constitutive" vector) constitutes only one of the many levels on which a theory can be debated and criticized. Thereby, he provides a rationale for the oft-observed, but generally unexplained, fact that although theories are (as Feyerabend, Kuhn, and Lakatos argue) "born refuted," they are nonetheless "rationally" accepted and compared. Buchdahl points to the kind of strategies and ploys which scientists can and do resort to when confronted with anomalies, criticisms, and exceptions. He shows that scientists can have reasons, apart from Kuhnian collective commitment, for retaining a theory in the face of contrary empirical evidence. He also indicates why a theory which is perfectly sound empirically may nonetheless be felt to be unsatisfactory and even unacceptable, which empiricists have never been able to explain. The fact that theories are challenged and defended at these different levels—the architectonic, the explicative, and the constitutive—reveals the question of scientific choice to be a much more complicated and intricate affair than is commonly thought. As Buchdahl points out vis-à-vis Newton, attacks and weaknesses on one level can often be compensated for, or at least "neutralized," by strengths at another level (and, of course, vice versa). Thus, even if Newton could not show that action at a distance was intelligible,[1] he could at least show that it suited the "analogy of nature," and that it was a fact about the world: that, in Buchdahl's language, it had an architectonic and constitutive foundation.

I do not have space here to consider the very perplexing question whether nonempirical arguments at the architectonic and explicative level are methodologically justified or whether (as some intransigent empiricists would maintain) they should have no place in the appraisal of scientific theories. What I do want to explore here is the suitability of these Buchdahlian vectors for explaining the kind of appraisals which scientists have *in fact used*, whether justifiably or not. Putting it another way, I want to consider briefly the usefulness of Buchdahl's schema to the historian of science who wants to understand the nature of scientific debate.

Buchdahl has taken a major step in this direction himself, with his treatment of the debate (between 1687 and 1850) concerning action at a distance and force. His case study already testifies to the explanatory power of Buchdahl's categories of analysis. But I think it also points to an ambiguity in Buchdahl's classification of components which I want to discuss here.

As Buchdahl makes clear, the strongest thing Newton's theory had going for it was its enviable empirical success, its confirmed explanatory

---

[1] I shall qualify this remark below.

Gerd Buchdahl

power. On the "constitutive" level, therefore, the theory of gravitation had no serious rivals. Where the major problems arose, as everyone knows, was in the interpretation of action at a distance. Was it the case that bodies could really act where they were not, or was gravity a direct result of the will and action of God, or was gravitation rather an appearance to be explained in terms of kinematic collisions between ether particles and macroscopic bodies? These options, as well as others, were open to Newton, and to his critics and followers, and most of the debate about action at a distance from Newton to Faraday was an attempt to decide between such alternatives.

Most of the criticisms to come from Newton's contemporaries were of the following kind: Action at a distance is occult, unintelligible, mysterious, and unmechanical. (All more or less synonymous![2]) If Newton was to make his theory acceptable, claimed his critics, he must show how gravity could be reduced to, and explained in terms of, impacts, collisions, or other species of *contact action*.

Before we turn to look at Newton's attempts to deal with this criticism, I think we should consider what Buchdahl makes of the debate at this preliminary stage. On his view, the Leibnizian and Cartesian demand for a mechanical explanation of gravity is an argument exemplifying the *constitutive* component. (See his page 217.) I disagree fundamentally with Buchdahl here, and I think our differences are more than verbal ones. Let me try to explain why. Suppose that someone had argued (as Euler would forty years later) that Newton's gravitational theory gave inaccurate predictions for the motions of the moon, or for the perturbations of Venus. Such arguments clearly challenged the empirical adequacy and explanatory power of the theory of gravitation and thus belong to the constitutive component. Equally clearly, such arguments were fundamentally different *in kind* from the mechanist's argument against Newtonian theory. The latter did not draw attention to empirical difficulties, but to conceptual ones. It did not charge that Newtonian theory is empirically inadequate but that it is intellectually inadequate. It is too obvious to require elaboration that these cases are very different and that lumping them together

[2] That these terms were seen as more or less identical in meaning is confirmed by that famous passage in the *Leibniz-Clarke Correspondence* where Leibniz chides Clarke (and, by implication, Newton) as follows: "That means of communication [of attraction] (says he [Clarke]) is invisible, intangible, not mechanical. He might as well have added, inexplicable, unintelligible, precarious, groundless and unexampled. . . . 'Tis a chimerical thing, a scholastic occult quality." H. G. Alexander, ed., *The Leibniz-Clarke Correspondence* (Manchester: Manchester University Press, 1956), p. 94.

under the "constitutive component" artificially coalesces two very different kinds of criteria of choice under a single head. But what is the alternative?

One might at first think that the mechanists were criticizing Newton's theory at the architectonic level, because Buchdahl has defined the architectonic level to include everything from "theological conceptions" to "maxims of simplicity and economy." When the mechanical philosophers charged that action at a distance was occult and unintelligible, their claim was (I believe) that Newton had not shown *how* actions at a distance could be reduced to the ontological categories of the mechanical or corpuscular philosophy. These categories were thought to specify exhaustively the kinds of things which made up the world. Unless Newton could show that action at a distance was deducible from (or reducible to) the kinds of entities postulated by the mechanical philosophy, then his theory was prima facie unacceptable. This is clearly a kind of criticism very different from what might be drawn out of considerations of simplicity or teleology. Indeed, it is sufficiently different in character to make it positively misleading to lump them both together under the same head. But it might still be said that Buchdahl's nomenclature is sufficient, and that the mechanists' critique falls under the explicative component. I am doubtful of this too; for, with the notable exception of Locke, none of the mechanistic opponents of action at a distance were confronted with a problem of meaning which required explication. On the contrary, they knew *precisely* what action at a distance meant, but wanted no part of it. When Leibniz wrote to Clarke that "a body is never moved naturally, except by another body which touches it and pushes it . . . any other kind of operation on bodies is either miraculous or imaginary,"[3] he was not saying that he did not know the meaning of action at a distance. He was claiming rather that he knew what it meant but believed it did not happen *in rerum natura* because "all the phenomena of bodies must be explained mechanically."[4]

Thus, the charge that gravity was occult is no more regulative or explicative than it is constitutive. I think the solution is to be found in a *quartum quid*. There is, in short, a fourth important criterion of choice which, for want of a better term, I shall call the "metaphysical" or "reductive" component. The reductive component generally specifies a set of entities, qualities, and modes of interaction which are regarded as both real and

[3] *Ibid.*, p. 66.
[4] Letter to Arnauld, 1686, *Die Philosophischen Schriften von G. W. Leibniz*, ed. C. I. Gerhardt (Berlin, 1875–90), II, 78.

basic. These privileged entities are viewed as "unexplained explainers" (to use Sellars's language) and a scientific theory or hypothesis is adequate (at the reductive level) if and only if it can be expressed or explained in terms of the privileged ontological categories. Thus, when the mechanical philosophers criticized the *Principia* because it did not show how gravity could be explained by impacts or collisions, their criticism involved not the explicative or constitutive but the reductive component.

We can now return to consider the question posed earlier; namely, how did Newton deal with the charge that his theory was unintelligible, because not reducible to the categories of the mechanical philosophers? As we have seen, Newton's critics realized that his theory had its major strength at the constitutive level, and its greatest weakness at the reductive level. Assuming (for the moment) that our four vectors are exhaustive of the kinds of criteria for theory acceptance, we can see that Newton had a number of alternative stratagems open to him, most of which he utilized at one time or another:

1. He could proclaim himself a proto-positivist, dismissing the importance of architectonic, explicative, and reductive components, and say it is a sufficient test of his theory that it is empirically well corroborated.

2. He could admit that the reductive argument against gravitation was a compelling one, and seek to eliminate actions at a distance by explaining them in terms of a mechanical ether.

3. He could argue that action at a distance was intelligible and that, in terms of a new nonmechanist ontology, distance forces are primary. This would be a counterargument (as opposed to a concession) at the level of the reductive component.[5]

4. Conceding that he has some minor difficulties at the reductive level, Newton could produce arguments for his theory taken from the architectonic and explicative levels.

Shrewd strategist that he was, Newton on occasion falls back on each of those ploys to defend himself. Thus, his famous disclaimer against making hypotheses, especially mechanical ones, has Newton adopting pose 1.[6] Haughtily dismissing the mechanistic reservations of his critics, he tells

---

[5] Still another example of a counterargument at the reductive level would consist in showing that contact action was unintelligible and incoherent. Although Newton himself does not take this option in his public writings, he does in unpublished manuscripts, as do several of his followers including Boscovich.

[6] Recall the classic passage from the penultimate paragraph to the General Scholium: "hypotheses, whether metaphysical or physical, whether of occult qualities or mechanical, have no place in *Experimental Philosophy*."

234

them that gravitation is a fact, whatever its cause might be,[7] and intimates that this is all his *Principia* requires or presupposes. At other times, however, he will opt for strategy 4, and it is one of the strengths of Buchdahl's analysis that it gives a coherent reconstruction and rationale for Newton's references to teleology, the analogy of nature, "nature doing nothing in vain," and so forth. We can now see that these are attempts to answer the charge of unintelligibility by falling back on an architectonic justification.

But what about options 2 and 3? These are both courses of action at what I have called the reductive level. Moreover, it is difficult to imagine Newton's having seriously pursued both options, for there is an obvious tension between them. In case 2, Newton would have to concede that the mechanists were right, and modify his system so that all apparent actions at a distance were viewed as the result of contact forces. In case 3, however, Newton would undercut the mechanists' argument by offering a new ontology of forces, where action at a distance is a "respectable" concept.

There are very great exegetical problems here, and it would be silly to act as if Newton unequivocally went one way or the other. I have no doubt that Buchdahl would agree with me on this point. Even granting the textual difficulties, however, I think that there is an interpretation of what evidence we do have which is rather different from Buchdahl's. With certain qualifications, Buchdahl maintains that Newton took what I have called option 2. Being himself more or less a mechanical philosopher, Newton—on Buchdahl's view—shared the general doctrine that *actio in distans* was inconceivable; accordingly, he offered a number of etherial hypotheses in hopes of explaining away action at a distance. Thereby, he sought both to silence his critics and to ease his own conscience.

Buchdahl's account, shared to a greater or lesser degree by Newtonian scholars such as A. R. and M. B. Hall, J. E. McGuire, Joan Hawes, and I. B. Cohen, seems to me to be a dubious reconstruction. I am inclined to suggest that Newton generally preferred option 3, thereby conceding little or nothing to the mechanical philosophers. The basic question at issue here, put in its most general terms, is this: Why, from 1675 until the last set of Queries to the *Opticks* in 1717, did Newton persistently return to the theme of an etherial explanation for gravity? Buchdahl, in consort with Cohen, McGuire, and the Halls, answers: because Newton was convinced that action at a distance was unintelligible and saw in a mechanical

---

[7] In Query 31 of the *Opticks*, for instance, Newton insists that attractive forces such as gravity are "manifest Qualities, their truth appearing to us by Phenomena, though their Causes be not yet discovered."

ether a device for rendering such actions "intelligible." But the supporting evidence for this answer is ambiguous at best. As Buchdahl himself points out, and as I shall argue in more detail elsewhere, almost all of Newton's many etherial hypotheses depend *essentially* on actions at a distance. In 1717, for instance, *the particles of Newton's ether act at a distance, not only on one another, but also on larger bodies.*[8] This is scarcely the kind of ether calculated to quiet the doubts of a mechanist who can only "conceive" contact action! Given the structure of Newton's various ethers, almost all of them depending on particles acting at a distance, it simply cannot be maintained that Newton's etherial speculations were designed to eliminate distance forces. But there is still the famous Bentley letter, and Buchdahl goes to some length (see especially his note 30) to argue that that document, if no other, establishes that Newton was fundamentally opposed to action at a distance. And in order for me to establish that Newton preferred option 3, I must discuss that letter briefly here.

The relevant portion of the letter goes as follows: "It is inconceivable, that inanimate brute Matter should, without the Mediation of something else, which is not material, operate on, and effect other Matter without mutual Contact . . . that one Body may act upon another at a Distance thro' a Vacuum, without the Mediation of anything else, by and through which there Action and Force may be conveyed from one to another, is to me so great an Absurdity, that I believe no Man who has in philosophical Matters a competent Faculty of Thinking, can ever fall into it."[9] As Buchdahl allows, there are two very different interpretations possible here. One reading (which Buchdahl supports) has Newton arguing that it is impossible for one bit of matter to act on another without acting through an intervening *material* medium. On this view, the letter to Bentley is a forceful expression of the inconceivability of action at a distance. But there is (at least) one other interpretation, viz., that bodies *can* act at a distance on one another but only if there is an intervening nonmaterial medium. The clarification of this issue would require much more space than I have here and would include a discussion of the technical term 'brute matter' as well as what Newton might mean by a nonmaterial medium. It is suffi-

[8] See especially Queries 21, 28, and 31 to the *Opticks*. The Halls seem to have completely missed the significance of this point. They could not be more incorrect when they state that "any hypothetical aether that Newton was prepared to entertain in order to facilitate the explanation of natural forces was itself a mechanical fluid." *Unpublished Scientific Papers of Sir Isaac Newton* (Cambridge: Cambridge University Press, 1962), p. 207.

[9] *Four Letters from Sir Isaac Newton to Doctor Bentley* (London, 1756), pp. 25–26.

cient, for my case, to point out that the overwhelming objection to Buchdahl's interpretation is that it does not fit with Newton's behavior when dealing with distance forces. As far as short-range forces are concerned, Newton never offers etherial or mechanical explanations, being perfectly content to deal with distance forces as basic.[10] Moreover, as we have seen, in dealing with long-range forces where Newton does postulate an ether, it is almost always an ether governed by distance forces. Whatever Newton might have written to Bentley in 1692, his scientific works (and their apologetics) show no symptoms of a man especially uneasy about action at a distance. Nor is this merely to be explained by a lack of imagination on Newton's part. From the 1690's onwards, there were several well-known and genuinely mechanistic ethers for explaining gravity (e.g., Fatio's[11]) which Newton could have adopted if he had really wanted a mechanistic hypothesis. The fact that he did not endorse these, and that his own hypotheses were invariably nonmechanical, is reasonably convincing evidence that he took option 3.

But Newton was not one to close any option prematurely, and he always allows for the possibility that someone may find a mechanistic explanation for gravity. Thus, in his Queries to the Opticks, he frequently reiterates that he is not ruling out of court the possibility that gravity is due to the action of a genuinely mechanical ether. But this is no more than a possibility as far as Newton is concerned, and it is manifestly not one which Newton thinks worth pursuing himself. While preferring to work exclusively with forces acting at a distance, and insisting that doing so is coherent and intelligible, Newton is not so confident an opponent of the mechanical philosophy (in the Cartesian-Boylean sense) as to deny altogether the possibility of a mechanistic analysis of these distance forces.

But one crucial question still remains. If Newton's various etherial hypotheses were not devices to reduce action at a distance to a mechanical model, then why did Newton persistently return to etherial theories of one kind or another? If he felt actions at a distance were genuinely acceptable, then why did he regularly try to explain gravitation in terms of the action of an ether? The answers to these questions are many-sided. Perhaps it is sufficient to point out here that Newton found the ether to be a concept

[10] This point is clearly established by Joan Hawes in her as yet unpublished work on Newton.
[11] A detailed discussion of the "mechanical" ether of Fatio de Duillier may be found in B. Gagnebin, "De la cause de la pesanteur. Mémoire de Nicholas Fatio de Duillier présenté à la Royal Society le 26 Février 1690," Notes and Records of the Royal Society of London, 6 (1949), 106–160.

of high explanatory value, and relied on it increasingly as he found more and more kinds of phenomena which it seemed to explain. As he makes clear in the Queries to the *Opticks* and in his many unpublished manuscripts, the ether could explain phenomena as diverse as (1) the transmission of heat through a chamber void of air (Query 18); (2) the damping of a motion of a pendulum in a "vacuum" (Query 28); (3) various properties of light, particularly, the "fits of easy reflection and transmission" (Query 19); (4) the transmission from the sense organs to the brain (Query 23); and several other phenomena. It is not surprising that Newton, having already filled space with an (unmechanical) ether for purposes of optics, thermometry, etc., should speculate whether this ether might be also responsible for gravitational attraction.

If I am right, the account Buchdahl has given of Newton's defense of his system needs to be modified in certain respects. For instance, he sees Newton as unable to answer the action-at-a-distance objection head-on and, consequently, "escaping" (the term is Buchdahl's) to the architectonic level to give a rationale for his theory. While I agree that the architectonic component *is* there and is important, I believe that Newton responded to the reductive argument of the mechanists in its own terms. Action at a distance, on his view, is intelligible and defensible per se, without falling back on arguments about design, the harmony and tenor of nature, and the like. The latter were of course used to strengthen the case for action at a distance, but they were not the only weapons Newton had at his disposal.

Needless to say, none of my remarks undermines Buchdahl's general theses about the different kinds of criteria governing the choice of scientific theories. On the contrary, by modifying his criteria slightly I hope I have strengthened those theses. There is much else in Buchdahl's paper which deserves serious discussion and elaboration. I am confident that my reply will be only the first of a steady stream of reactions to his extremely suggestive essay.

COMMENT BY HENRY SMALL

The message of Gerd Buchdahl's paper for the historian of science is clear: do not overlook the role of philosophical criteria in directing the course of scientific change. This is certainly sound advice, since historians of science have often stressed the empirical criteria at the expense of the philosophical, which are usually less visible. But having identified such criteria in historical cases, where do we then proceed?

It appears that the fundamental question which has not been answered is how do the criteria derive their strength; why is the force to conform greater for some than for others? That such a hierarchy exists is evident from studies of how scientists deal with anomalies. In modifying theory there are some assumptions which they are less willing to compromise than others.

But the relative strengths of criteria cannot be understood if they are considered in isolation from the idea system in which they play a part. The problem is to determine how this strength depends on the position of a criterion within the system. The solution will, I think, reveal the deeply-lying psychological basis of science: not as irrational commitment to criteria, but as commitment based on the perception, rather than on the logical analysis, of a complex structure of ideas.

REPLY BY GERD BUCHDAHL

In order not to run into lengthy disquisitions, I will take Dr. Laudan's very perceptive comments more or less in the order in which they appear. His contentions will enable me to clarify certain things and to elaborate on others which obviously needed further elucidation.

Dr. Laudan imputes to me the contention that "the Leibnizian and Cartesian demand for a mechanical explanation of gravity is an argument exemplifying the *constitutive* component." Here we need to distinguish between the *considerations that lead to* the formulation of a theory employing such and such explanatory principles, and the *form of* that theory itself. What I say on the page referred to is that the forms of Newton's theory and those of his continental rivals differ, but it is obvious that the considerations determining for instance the Leibnizian formulation (involving propositions concerning the ether) arise out of conceptual (or, as Laudan would have it, "metaphysical") explications that differ from Newton's. What in fact I meant to emphasize was that Newton's conception of matter (hostile, as I take it to be to action at a distance) did not at that stage infiltrate the *form* of the theory, whereas Leibniz's conception did dominate the latter's formal presentation. The formulation of the conceptions belongs to the explicative component but that of the theories to the constitutive.

239

This brings me to the question of the appropriateness and sufficiency of the notion of "conceptual explication" (see Laudan, page 233). I think it will be clear from the context of my paper that this notion is intended at *least* to stand for Laudan's "metaphysical" or "reductive" component, for it will be fairly evident that "conceptual explication" in my use of that term is the offspring of a group of mental operations and argumentations of which Descartes's and Kant's are given as examples, and which by Kant, for instance, were admittedly labeled "metaphysical foundations." As I use this notion, it covers a very broad spectrum. It yields for instance a position about what are the "essential properties" of matter, what is the essential nature of motion, and so on. I might easily have chosen the *term* 'metaphysical explication' but I deliberately avoided this for a number of reasons. "Metaphysics" is a hard-worked concept. That God exists, man is free, and his soul immortal were typical metaphysical expressions according to the Kantian corpus. But Kant's "metaphysical" construal of the concept of matter as capable of exerting repulsive force which can be represented through vectorial construction is clearly a very different piece of "metaphysics," just as are those inquiries into the "deep structure" of language represented by Kant's transcendental principles (in the *Critique of Pure Reason*), called by him sometimes "the transcendental part of the metaphysics of nature" (cf. *Metaphysical Foundations*, page 140), or by some of the writings of Wittgenstein (*Tractatus*), Strawson, and Chomsky in our time. Still more appropriately, one might say that a notion like Locke's "simple idea," or indeed the whole theory of "ideas" as interposed between man and the world of "real things," is a more typical piece of "constructive metaphysics."

Against this, I needed a term that was rather more general and that would denote metaphysical considerations and constructions perhaps, to be sure, but that would also remind us in the end of the far looser and more tentative nature of the intellectual processes here in question, and which eventually lead to decisions about the scientific concepts involved. "Metaphysical" is a notion at once both too "constructive" and "existentialist" and too much reeking of putative achievement and conclusiveness.

Not to stray too far from the context of my paper, consider for instance Kant's construal of metaphysics in *Enquiry concerning the Clarity of Principles*, referred to on page 222 (see also footnote 62). There Kant attempts a contrast between the procedures of "mathematics" and of "metaphysics" (defined as "the first principles of our knowledge"). In mathematics we begin with "explanations" or "definitions," because the meaning of the signs employed is already certain. In metaphysics we can never begin in this way. The reason given is that in mathematics "I have no concept at all of my object until it is given by the definition." By contrast, and this is the really crucial point, "in metaphysics I have a concept given to me already, although it is a confused one. My duty is to search for the clear, detailed and determinate formulation of this confused concept" (*Enquiry concerning the Clarity of Principles*, pages 14–15; and cf. my reference to the influence on Kant by Keill, page 225 of my paper). In the example of the method which Kant here recommends (and to which I allude briefly in the paper), he notes that there is no need to "dispute the correctness" of the Newtonian law of gravitation. What is necessary still, partly in order to secure this law against false imputations on the part of metaphysicians —an allusion to the denial of action at a distance on "conceptual" grounds —is to inquire into the meaning of notions such as "being at a distance," "touch," "resistance," "impenetrability," "force" (*Enquiry concerning the Clarity of Principles*, page 20). These kinds of "investigations into meaning" (as distinct from "truth") have become a commonplace in our day through the writings of Moore and Wittgenstein. To be sure, initially there may be a sense in which we "know precisely what action at a distance meant" (Dr. Laudan), but as the debates of several centuries show, in *another sense*, namely, the sense intended by Kant when he says that "in metaphysics I have a concept given to me already, although it is a confused one," further "clarification" may still be called for, of the kind which for instance William Whewell likewise included under the notion of "explication of conceptions" and "clarification of ideas." It was in this sense that I meant to urge that a conceptual investigation was required into the notions of "matter," and of "action" and "causation," in order to placate those who continued to insist that the action of matter on matter was "inconceivable."

Questions into the "nature" of causation and the "causal tie," for instance, are frequently framed in terms of alternative expressions, as either "what do we mean by causal action?" or "what is that entity?" or "what is the ontological significance of the claim involved in causal statements?" Already in Hume we get this dual approach. Hume begins by assuming that we know, in one sense, what is meant by causation, i.e., spatiotemporal contiguity, etc., as well as the existence of a necessary connection. But then he begins to ask questions about the genesis of the idea of necessary connection and to discuss this question by looking for the quasi-meta-physical entity ("impression") corresponding to the "idea" of necessary connection, an investigation at the same time not infrequently couched in the language of "what do we *really* mean when we claim the existence of a causal connection?" So my "conceptual explication" involves drawing attention both to the existence of the particular "reductive components" or "unexplained explainers" of Dr. Laudan's, and (where the occasion is appropriate) to the philosophical analysis of these reductive elements, on the lines indicated.

I agree with Laudan that there are "great exegetical problems" in trying to reconcile Newton's invocation of the ether hypothesis of 1717 with his teleologically based construal of gravitation as "basic," i.e., of reconciling the "inter-phenomenal" with the "supra-phenomenal" escape routes. One answer to the problem is that Newton felt tentative toward all these solutions and that therefore they may all be regarded as "queries" in the Newtonian sense of that term; and the notion of "query" is indeed intended to remove the need for ultimate consistency. To this we may add the consideration that Newton offered the ether hypothesis of 1717 as a didactic device, its professed aim being to stop anyone imputing to him a belief in the physical reality of action by matter on matter. Such a suggestion makes sense of course only if we grant my contention that providing an architectonic (teleological) rationale was not meant to yield a "realist" interpretation of the concept of attractive forces. And as I note, it was certainly not impossible for *some* of the writers during this period (e.g., MacLaurin) to reconcile a teleological approach with a supposed need to supplement Newton's theory with an ether hypothesis, a hypothesis possessing—as Laudan notes—"high explanatory value" in regions where we were dealing with an unfamiliar phenomenon, i.e., action at *great* distances.

Now I grant (and I note in my paper) that Newton does speak of attractive forces, and even employs this idea in his ether hypothesis of 1717. What we cannot do, pace Laudan, is to ignore—in the face of this—the doctrine of the Bentley letters and the explicit invocation against gravity as an essential property of matter (in the *Opticks* of 1717, already alluded to). To fall back on the conception of a "nonmaterial medium" or of gravity as "a direct result of the will and action of God" only increases the mental fog which undoubtedly prevailed in Newton's time—my Kantian interpretation of this "teleology" as a regulative device was meant to clear the fog a little (cf. page 218 of my paper)—and it is even true that the very notion of an "ether" is not always so clearly mechanistic in character as to make it impossible to believe that it may not have served at times to close the gap between apparent inconsistencies. My teleological interpretation of Newton's intentions has at least the merit of making the best possible case for an admittedly confused and intermediate stage between straight-forward acceptance of a phenomenological theory, on the one hand, and a realist interpretation of attraction, on the other.

I do note that Newton couples his views about short-distance forces with a reference to nature's being "very consonant and conformable to herself." Now this may just be a way of linking the acceptability of magnetic and electric forces with that of gravitational force. But this leaves the question open as before, since the nature of these forces in respect of being ultimate distance forces in the realist sense of that term was still an unsolved problem too. Moreover, as I meant to imply with my citation of the passage just referred to, the allusion to the "consonance of nature" evidently links the whole thing to Newton's teleological approach to large-distance gravity once again. Their existence, one and all, testifies to the harmony of nature, a conception through which Newton seeks, as I have tried to show, to escape from the suggestion of the informal interpretation of the constitutive component that gravity is basic in a realist sense. Indeed, it is just because Newton's interpretation focuses on the harmony of nature that we should expect him to fill in the *details* by recourse to ether explanations in the variety of fields Laudan notes on page 237, a fact which would provide a final and decisive proof that Newton's teleological approach implied to the end a nonrealist interpretation of forces in general.

So I am unrepentant. I can see no sense in regarding Newton as accepting action at a distance as "intelligible" (Laudan), in the face of his avowed denials, unless we hold that for the most part as a practicing scientist he held this view but on occasion felt scruples which he then equally temporally sought to assuage by his provision of the architectonic escape route. But perhaps ultimately Newton did fail to be as consistent as logicians with their tidy minds may try to make out. It is not unusual to find people holding inconsistent views when these are sufficiently "unclear" (in Kant's sense) and in process of change, battling against received tradition and subject to stress from new developments. It is precisely then that alternative explications come to the surface. And I would not therefore wish to give the impression that I hold my own views to be "right," and Dr. Laudan's mistaken. Such simple juxtapositions are not possible in a field where historical evaluation and logical appraisal do mutual battle with one another.

I now turn briefly to the question posed by Mr. Small: where does the exposition of such a finer structure of appraisal of criteria eventually get us? I am slightly uneasy about the tenor of this question, for it suggests that criteria like the ones here put forward are being viewed as means for the discovery of hypotheses, or at least as judges guiding instant decision-making processes with respect to acceptability. I would not want to claim exhaustiveness for my list. Moreover, since when applied it may lead to contradictory advice, and hence conclusions, it clearly cannot operate mechanically. My very case study shows that elements of structure may display mutual oppositions and tensions, perhaps indicative of the existence of warring "paradigms," though one may also interpret this as an indication that there never are any self-contained paradigms, and that the tensions involved in the struggle to get a new theory accepted are deeper and frequently contain more general aspects including low-lying philosophical assumptions which may cut across centuries of debate.

One of my general intentions had in fact been to add further considerations in favor of our growing awareness that existing purely "inductivist" discussions of the foundations of science are insufficient as grounds for the explication of the "logic of science." And I have purposely cast my net of criteria widely and have intentionally included criteria that may seem at first sight excessively historicist, subjective, almost arbitrary, in order to highlight the problem of "rationality" of scientific "inference."

This has the initial result of making any desired solution even more seemingly hopeless than our experience with purely inductivist theories of scientific inference might suggest. But, on the other hand, I have chosen my criteria (or components) so as to make it possible to indicate some relevant connections with purely philosophical discussions which have usually not been thought to relate any longer to modern inductivist problems—witness the threads of connection drawn between "entrenchment" and "lawlikeness" and "metaphysical foundations" (conceptual explication) toward the end of my paper.

Beyond this I would be the first to admit that a more systematic presentation of criteria is desirable. For instance (as Dr. Laudan also hints) the list of criteria lumped together under the "architectonic component" is too all-embracing. On the other hand, the interconnections between the different components are not always clear; thus a glance at my summary at the end of section III of my paper shows the constitutive component as "suggesting" action at a distance on the basis of an "informal interpretation" of the formalism of the theory (I.1); a sort of "conceptual explication" matched against the ex cathedra explication I.2, displaying the contrary conclusion. I do not think, however, that excessive tidiness in one's classificatory schemes is necessarily a virtue. Such schemes can never be regarded as more than mnemonic devices, reminding us of the underlying critical debates which necessarily must contain more "ragged edges" and more ambiguous crosscurrents than any such schemes can explicitly display.

Non-Einsteinian Interpretations of the Photoelectric Effect

Few, if any, ideas in the history of recent physics were greeted with skepticism as universal and prolonged as was Einstein's light quantum hypothesis.[1] The period of transition between rejection and acceptance spanned almost two decades and would undoubtedly have been longer if Arthur Compton's experiments had been less striking. Why was Einstein's hypothesis not accepted much earlier? Was there no experimental evidence that suggested, even demanded, Einstein's hypothesis? If so, why was it met with such profound skepticism?

Questions such as these, which relate to the processes by which scientific theories gain acceptance, are less popular with historians of science than they ought to be. A good deal of insight into the nature of science, especially into the interplay between theory and experiment, may be obtained from a study of these transition periods. This point may be illustrated immediately by briefly sketching Einstein's arguments for his light quantum hypothesis, its relationship to the photoelectric effect, and the experimental evidence on the photoelectric effect available to physicists in the early years of this century.

Einstein gave two powerful arguments for light quanta, one negative, the other positive. His negative argument was that classical radiation theory was fundamentally inadequate; it could not account for the ob-

AUTHOR'S NOTE: I am especially grateful to Professor Martin J. Klein for his numerous and illuminating comments, and I should also like to thank Professor Howard Stein for his helpful suggestions.
[1] "Über einen die Erzeugung und Verwandlung des Lichtes betreffenden heuristischen Gesichtspunkt," *Annalen der Physik*, 17 (1905), 132–148; translated in H. A. Boorse and L. Motz, eds., *The World of the Atom*, vol. 1 (New York: Basic Books, 1966), pp. 545–557. For a detailed historical discussion, see Martin J. Klein, "Einstein's First Paper on Quanta," *The Natural Philosopher* (New York), 2 (1963), 57–86; "Thermodynamics in Einstein's Thought," *Science*, 157 (1967), 509–516. My brief discussion of Einstein's paper draws heavily on Klein's work.

served spectral distribution for black-body radiation, and it led inexorably to the blatantly false conclusion that the total radiant energy in a cavity is infinite. (This "ultra-violet catastrophe," as Ehrenfest later termed it, was pointed out independently and virtually simultaneously, though not as emphatically, by Lord Rayleigh.[2])

Einstein developed his positive argument in two stages. First, he derived an expression for the entropy of black-body radiation in the Wien's law spectral region (the high-frequency, low-temperature region) and used this expression to evaluate the difference in entropy associated with a change in volume $v - v_0$ of the radiation, at constant energy. Secondly, he considered the case of n particles freely and independently moving around inside a given volume v_0 and determined the change in entropy of the system if all n particles accidentally found themselves in a given subvolume v at a randomly chosen instant of time. To evaluate this change in entropy, Einstein used the statistical version of the second law of thermodynamics in conjunction with his own strongly physical interpretation of the probability. He found that the expression for this change in entropy is formally identical to the one he had derived earlier for black-body radiation, and from this fact he drew what to him was the unavoidable conclusion: monochromatic black-body radiation in the Wien's law spectral region behaves with respect to thermal phenomena as if it consists of independently moving particles or quanta of radiant energy. It was also clear from his arguments that the energy of each quantum is proportional to its frequency.

Einstein's light quantum hypothesis (his "heuristic point of view") was therefore a necessary consequence of very fundamental assumptions: in no sense did he propose it in an ad hoc fashion to "explain" certain experiments. If, however, under certain circumstances light indeed exhibited a quantum structure, this ought to be experimentally verifiable. Einstein gave a detailed discussion of three experimental applications, one of which was the photoelectric effect. The famous equation he derived showed that the maximum kinetic energy T of the photoelectrons depends linearly on the frequency v of the incident radiation. In modern notation, $T = hv - w_0$, where h is Planck's constant and w_0 is the work function of the photosensitive surface. (It is worth noting that Einstein himself used a combination of other constants instead of the single constant h, which merely emphasizes the fact that Einstein did not, as many believe, build

[2] "The Dynamical Theory of Gases and of Radiation," *Nature*, 72 (1905), 55.

on Planck's earlier work.) Einstein also pointed out that the intensity of the incident radiation should determine only the number and not the energy of the ejected photoelectrons, predictions which were consistent with Lenard's "trailblazing" 1902 experiments.[3]

But what about Einstein's extremely bold prediction that the maximum photoelectron energy depends linearly on the frequency of the incident radiation? Here the experimental situation was highly uncertain. Thus, Lenard had found only a general increase in electron energy with incident frequency; and several years later in 1907 Ladenburg concluded[4] that the energy varied quadratically with the frequency. While Joffé was quick to point out[5] that Ladenburg's experiments were also consistent with Einstein's linear relationship, it developed that they were also consistent with a frequency to the ⅔ power variation proposed in 1911 by F. A. Lindemann.[6] A year later in 1912 O. W. Richardson and K. T. Compton,[7] as well as A. L. Hughes,[8] reasserted the validity of the linear relationship. But their experiments were immediately challenged by Pohl and Pringsheim,[9] who maintained that the best fit to Richardson and Compton's data could be obtained with a logarithmic relationship. It was not until 1915–16 that R. A. Millikan settled the issue.[10]

Now the remarkable fact is that these disparate experimental results, as we shall see, did not lead those who opposed Einstein's interpretation of the photoelectric effect to seriously question the validity of Einstein's linear relationship. Their opposition stemmed from a different quarter. They failed to see that Einstein's light quantum hypothesis was a necessary consequence of Einstein's assumptions, and they tended to take ac-

[3] P. Lenard, "Ueber die lichtelektrische Wirkung," Annalen der Physik, 8 (1902), 149–198.

[4] E. Ladenburg, "Über Aufangsgeschwindigkeit und Menge der photoelektrischen Elektronen in ihrem Zusammenhange mit der Wellenlänge der Auslösenden Lichtes," Berichte der deutschen physikalischen Gesellschaft, 9 (1907), 504–515; Physikalische Zeitschrift, 8 (1907), 590–594.

[5] A. Joffé, "Eine Bemerkung zu der Arbeit von E. Ladenburg," Bayrische Akademie der Wissenschaften, München, Sitzungberichte, mathematisch-physikalische Klasse, 37 (1907), 279–280; Annalen der Physik, 24 (1907), 939–940.

[6] "Über die Berechnung der Eigenfrequenzen der Elektronen im selektiven Photoeffect," Berichte der deutschen physikalischen Gesellschaft, 13 (1911), 482–488.

[7] "The Photoelectric Effect," Philosophical Magazine, 24 (1912), 575–594.

[8] "On the Long-Wave Limits of the Normal Photoelectric Effect," Philosophical Magazine, 27 (1914), 473–475.

[9] "On the Long-Wave Limits of the Normal Photoelectric Effect," Philosophical Magazine, 26 (1913), 1017–24.

[10] "A Direct Photoelectric Determination of Planck's 'h,'" Physical Review, 7 (1916), 355–388.

ceptance of quanta to imply rejection of classical electrodynamics—this in spite of Einstein's 1909 proof that a necessary consequence of Planck's law is the coexistence of quanta and waves in black-body radiation.[11] (Einstein's proof rested on his analysis of energy and momentum fluctuations in black-body radiation, a proof his contemporaries were unprepared to accept or even understand.) Maxwell's theory had been repeatedly confirmed, by Hertz's electromagnetic wave experiments, Lebedev's (and Nichols and Hull's) radiation pressure experiments, and all interference and diffraction experiments.[12] The heuristic value of Maxwell's theory was beyond question. To believe that it would not eventually encompass photoelectric phenomena appeared to most of Einstein's contemporaries —the exception was Johannes Stark[13]—to be most unreasonable.

Given the belief (to a large degree unjustified by the experimental evidence, as we have seen) in the validity of the linear relationship between energy and frequency, as well as the disbelief in Einstein's interpretation of it, a natural question arose: Were there perhaps other, less radical, interpretations of the photoelectric effect? A historical examination of the answers proposed to this question will form the main thrust of my paper, but before I begin I should like to make two general remarks. First, we shall evidently be discussing theories formulated not primarily in response to a troublesome experimental observation,[14] but rather in response to an unpalatable interpretation. Second, while revolutionary new theories tend to be generated by young minds, and while it is therefore natural to attribute any resistance to them to the "old guard," we do not seem to have such a situation here. We do not seem to have an example of Planck's contention that "a new scientific truth does not triumph by convincing its opponents and making them see the light, but rather because its opponents eventually die, and a new generation grows up that is familiar with

[11] "Zum gegenwärtigen Stand des Strahlungsproblems," Physikalische Zeitschrift, 10 (1909), 185–193; "Über die Entwicklung unserer Anschauungen über das Wesen und die Konstitution der Strahlung," Physikalische Zeitschrift, 10 (1909), 817–826. See also Martin J. Klein, "Einstein and the Wave-Particle Duality," The Natural Philosopher, 3 (1963), 3–49.
[12] For discussion and references, see E. T. Whittaker, A History of the Theories of Aether and Electricity, vol. 1 (New York: Harper, 1960), pp. 275, 319ff.
[13] "Zur experimentellen Entscheidung Zwischen Ätherwellen und Lichtquantenhypothese," Physikalische Zeitschrift, 10 (1909), 902–913. See also 11 (1910), 24–25, 179–187.
[14] For such examples, see T. S. Kuhn, The Structure of Scientific Revolutions (Chicago: University of Chicago Press, 1962), pp. 52–90.

Roger H. Stuewer

it."[15] We do not seem to have this situation because the opponents in this case were H. A. Lorentz, J. J. Thomson, Arnold Sommerfeld, and O. W. Richardson, ages 57, 54, 42, and 31, respectively, in 1910. Needless to say, apart from the great spread in ages, the most noteworthy characteristic of these physicists is their eminence.

II

Correspondence deposited in the Einstein Archives in Princeton shows that between 1909 and 1911 Einstein used H. A. Lorentz as a sounding board for his developing ideas on the nature of radiation. Thus by the time Lorentz was invited by the Wolfskehl Commission to deliver six lectures at Göttingen in October 1910, he was completely familiar with Einstein's work. In his fifth lecture, Lorentz treated Einstein's interpretation of the photoelectric effect in detail, and, notwithstanding our earlier remarks, maintained that the predicted linear relationship between energy and frequency was "confirmed by experiment." Lorentz then told his audience that "in spite of [this evidence] the speaker holds the light quantum hypothesis to be impossible, if the quanta are regarded as completely incoherent, an assumption which is the most natural one to make."[16] The interference and diffraction difficulties appeared insurmountable to Lorentz, the physicist who (to quote Ehrenfest) had brought clarity to the "kind of intellectual jungle"[17] that was Maxwell's Treatise.

Fortunately, Lorentz noted, there was another possible explanation of the photoelectric effect, namely, Haas's. According to Lorentz, Haas imagined a light wave to be incident on an electron in a Thomson atom, thereby setting the electron into oscillation. As long as the electron remained inside the positive sphere, the incident wave would only be scattered or dispersed; if, however, the electron received more than some threshold energy, it would be ejected from the positive sphere and a discrete amount of energy would be simultaneously abstracted from the incident wave. Planck's constant entered into Haas's theory by assigning the quantum energy $h\nu$ to the electron at the boundary of the positive sphere, and Lorentz noted with satisfaction that application of Haas's theory to a gas, Argon, yielded order of magnitude agreement with Planck's own value.

---

[15] Scientific Autobiography, trans. Frank Gaynor (New York: Philosophical Library, 1949), pp. 33–34.

[16] "Alte und neue Fragen der Physik," Physikalische Zeitschrift, 11 (1910), 1249.

[17] Quoted in G. L. de Haas-Lorentz, ed., H. A. Lorentz: Impressions of His Life and Work (Amsterdam: North-Holland, 1957), p. 55.

Lorentz concluded: "In Haas's hypothesis, the riddle of the energy elements is combined with the question of the nature and action of positive electricity, and it may be that these different questions will, for the first time, together find their complete solution."[18]

The crux of Lorentz's argument was that it was far more reasonable to assume that the photoelectric effect would become understandable on the basis of atomic structure considerations than to believe it entailed abandoning Maxwell's theory. I should like to urge the reasonableness and naturalness of Lorentz's point of view. It was reasonable because in 1910 little of a definite nature was known about the dynamics of the atom; it was natural because it was not new—already in 1902 Lenard had proposed that the incident electromagnetic radiation stimulated the emission of electrons by "triggering" atomic disruptions.[19] To a certain degree, therefore, Einstein's interpretation must have appeared radical on two counts: not only did he explicitly reject a Maxwellian approach; he also implicitly rejected Lenard's approach. Since not only Lorentz but J. J. Thomson also embraced Lenard's approach in 1910, Einstein's interpretation might be viewed as a temporary detour from an established route.

III

J. J. Thomson was one of the great atom builders of all time, the traditional "Thomson Atom" being only one of a rather formidable collection of models. As his biographer Lord Rayleigh remarked: "J. J. was not inclined to be dogmatic about his atomic theories, and indeed he was quite prepared to change them, sometimes without making it altogether clear that he had wiped the slate clean."[20] Believing that "the chief value of any theory was as a basis for further experiments,"[21] Thomson concentrated on what a theory would explain and not on what it would not. This attitude led him to proliferate atomic models—J. G. Crowther termed them "semi-manufactured scientific goods"[22]—and it is only by understanding Thomson's attitude toward theory construction that we can understand how he could propose two completely different theories of the photoelec-

[18] "Alte und neue Fragen," p. 1253.
[19] "Ueber die lichtelektrische Wirkung," pp. 169–170.
[20] The Life of Sir J. J. Thomson, O.M. (Cambridge: Cambridge University Press, 1943), p. 141.
[21] Ibid.; see also p. 136.
[22] British Scientists of the Twentieth Century (London: Routledge and Kegan Paul, 1952), p. 12.

tric effect between 1910 and 1913 based on two completely different atomic models.

Thomson published his first theory[23] a few months before Lorentz delivered his Wolfskehl Lectures. Thomson postulated that in the atom there are electric doublets, each with an electron ("corpuscle") circling around it below its positive pole. Proceeding directly from the electron's equations of motion, Thomson proved that the electron's kinetic energy is proportional to its frequency of revolution and that it rotates on the surface of a cone of half-angle 55° ($\tan^{-1}\sqrt{2}$) with apex at the center of the doublet. Since, however, the electron's distance from the apex is not fixed, and since its frequency of revolution depends on this distance, there will be electrons of many different kinetic energies present in a metal plate containing many atoms. That was all Thomson required, for now an incident wave of given frequency would certainly find an electron rotating at the same frequency, so that resonance would occur. According to Thomson, this would twist the electric doublet, and if it were twisted enough, it would cast off its electron. Since at the time of ejection, the electron's kinetic energy would be proportional to its frequency of rotation, it would by the principle of resonance also be proportional to the frequency of the incident radiation. From an argument involving Wien's displacement law, Thomson was able to infer a value for the electric moment of the doublet, which in turn enabled him to calculate the constant of proportionality—which turned out to be of the same order of magnitude as Planck's constant h! To Thomson, the implication was perfectly clear: he wrote that "we cannot regard Ladenburg's experiments as proof of the unitary structure of light. . . . [My] theory enables us to explain the electrical effects produced by light, without assuming that light is made up of unalterable units, each containing a definite and, on Planck's hypothesis, a comparatively large amount of energy, a view which it is exceedingly difficult to reconcile with well-known optical phenomena."[24]

J. H. Jeans criticized[25] Thomson's doublet-electrons on the grounds that they are dynamically unstable, but Thomson answered[26] that their long-term stability was unnecessary for the success of his theory. Perhaps, however, Thomson did not entirely forget Jeans's criticism, because three

[23] "On the Theory of Radiation," *Philosophical Magazine*, 20 (1910), 238–247.
[24] *Ibid.*, p. 246.
[25] "On the Motion of a Particle about a Doublet," *Philosophical Magazine*, 20 (1910), 380–382.
[26] Letter to Editors of *Philosophical Magazine*, 20 (1910), 544.

years later in 1913 Thomson proposed a second theory[27] based on an entirely new atomic model. In his new model, Thomson postulated the coexistence of two forces: a radial inverse cube repulsive force "diffused throughout the whole of the atom," and a radial inverse square attractive force "*confined to a limited number of radial tubes in the atom.*"[28] Thus, inside such a radial tube both forces would be present, and by setting up the equation of motion of an electron in it, Thomson readily demonstrated that the electron could oscillate about an equilibrium position with a frequency depending on the force constant of the repulsive force. Once again, that was all Thomson required, for now an incident wave would certainly find an electron with which it could resonate, and if, after being set into oscillation, some "casual magnetic force" moved it laterally out of the tube, it would come under the "uncontrolled action" of the repulsive force and be expelled from the atom. Thomson proved that in leaving the atom the electron (charge e, mass m) would pick up energy $T = \pi \cdot \sqrt{Cem}\, \nu$. The quantity C is the repulsive force constant, which Thomson, in a completely ad hoc manner, fixed at the value $h^2/me\pi^2$, so that by substitution $T = h\nu$. "Thus," Thomson wrote, "we see that the kinetic energy with which the corpuscle is expelled is proportional to the frequency of the light and is equal to the frequency multiplied by Planck's constant." "This," he concluded, "is the well-known law of Photo-Electricity."[29]

While Thomson's theories attracted some attention and stimulated some experimental work, particularly at the Cavendish Laboratory, the ad hoc character of many of Thomson's basic assumptions must have been apparent to many of his contemporaries. Indeed, W. H. Bragg at the University of Leeds regarded certain features of Thomson's first theory as "fantastic";[30] in general, Bragg considered attempts like Thomson's to squeeze quantum manifestations out of classical theories to be retrogressive. Thomson's second theory actually took on an even more contrived

---

[27] "On the Structure of the Atom," *Philosophical Magazine*, 26 (1913), 792–799.
[28] *Ibid.*, p. 793.
[29] *Ibid.*, p. 795.
[30] Letter to A. Sommerfeld dated February 7, 1911, on deposit in the Archive for History of Quantum Physics, American Philosophical Society Library, Philadelphia; depositories also in Copenhagen and Berkeley. For more on Thomson's work, see Russell McCormmach, "J. J. Thomson and the Structure of Light," *British Journal for the History of Science*, 3 (1967), 362–387. John L. Heilbron and Thomas S. Kuhn discuss the influence of Thomson's first theory on Niels Bohr in "The Genesis of the Bohr Atom," in *Historical Studies in the Physical Sciences*, vol. 1, ed. Russell McCormmach (Philadelphia: University of Pennsylvania Press, 1969), pp. 226–229.

cast a few weeks after he proposed it when he found it necessary to replace the radial attractive cones with attractive cylinders![31] But while Thomson's theories (if indeed, this term is fully appropriate, considering their ad hoc features) rested on insecure grounds, a basic point Thomson was making should not be lost. Like Lorentz, Thomson showed it was possible to envision atomic models upon which a quantitative theory of the photoelectric effect could be based. This very same point was made in 1911 in a more sophisticated way by Arnold Sommerfeld.

IV

Sommerfeld's motivation and method form an interesting contrast to Thomson's. Whereas Thomson was basically unsympathetic toward the new quantum theory, Sommerfeld was convinced (as he wrote in 1913) that "Planck's discovery of the universal quantum of action has been called to heal the momentary sufferings of theoretical physics."[32] Thomson, in his second theory, had introduced Planck's constant at the end of his calculation in a strictly ad hoc manner; Sommerfeld introduced it at the outset by postulation. He contended that "an electromagnetic or mechanical 'explanation' of [Planck's constant] h seems to me to be just as unnecessary and unpromising as a mechanical 'explanation' of Maxwell's equations. It would be much more useful to pursue the h-hypothesis in its various consequences and trace other phenomena back to it."[33] One of these phenomena was the photoelectric effect, concerning which Sommerfeld wrote Einstein was "at present [1911] not able to maintain his completely audacious point of view."[34]

Sommerfeld postulated that in "every purely molecular process a definite and universal amount of action"[35] is taken up or given up. He introduced Planck's constant by fixing the exact amount by the condition

$$\int_0^\tau (T - V)dt = \frac{h}{2\pi}$$

where T and V are the relevant kinetic and potential energies, and τ is the time during which the process takes place. Sommerfeld applied this con-

---

[31] Letter to Editors of *Philosophical Magazine*, 26 (1913), 1044.
[32] "Theorie des lichtelektrischen Effektes vom Standpunkt des Wirkungsquantums," *Annalen der Physik*, 41 (1913), 873.
[33] "Das Plancksche Wirkungsquantum und seine allegemeine Bedeutung für die Molekularphysik," *Berichte der Deutschen Physikalischen Gesellschaft*, 9 (1911), 1092.
[34] *Ibid.*, p. 1074.　　　　[35] *Ibid.*, p. 108.

dition to the photoelectric effect as follows. Assume that the incident radiation—electromagnetic waves—sets a bound electron into oscillation by resonance, and that after a certain time τ the electron's energy becomes large enough for it to be released from the atom. (Sommerfeld later[36] estimated τ to be on the order of 10^{-5} second, a small, and in 1911, unobservable, delay, but nonetheless one that is inconsistent with Einstein's interpretation.) Substituting the electron's kinetic and potential energies into the above condition, transforming the result by partial integration, and introducing the electron's equation of motion, Sommerfeld proved that $T = h\nu_0$, where ν_0 is the natural frequency of the electron in the atom. At resonance this frequency is equal to the frequency of the radiation, ν, so that $T = h\nu$; and this, concluded Sommerfeld, is "Einstein's Law."

(Neither Thomson nor Sommerfeld was ever particularly concerned that neither of their photoelectric effect equations, in contrast to Einstein's, contained an additive constant, namely, the work function. Sommerfeld expressed the reason for his lack of concern as follows: "This [the work function] is obviously foreign to the pure molecular process and would only appear if we were to follow the photoelectron [after its ejection from the atom] along on its subsequent path through the surface of the metal."[37] Thomson no doubt wholeheartedly endorsed this statement.)

Sommerfeld drew attention to both a similarity and a difference between his theory and Einstein's. The theories were similar in that both predicted that as the electron was ejected from the atom, a discrete amount of energy would be abstracted from the incident radiation. The theories were different in that Sommerfeld, unlike Einstein, but like Thomson, envisioned a resonance phenomenon. This difference had observable consequences. Whereas Einstein's equation predicted that the electron's energy should be entirely independent of any atomic frequencies, Sommerfeld's theory predicted that a plot of energy versus frequency "should possess, for each natural frequency of the atom, a maximum." (Had Thomson not concentrated only on the case of complete resonance, he would have been driven to the same conclusion.) Which theory fit the existing data better, Sommerfeld's or Einstein's? Sommerfeld had the answer. J. R. Wright in Millikan's laboratory at the University of Chicago had recently shone ultraviolet light on aluminum and had demonstrated "with certainty . . . that the maximum photoelectron energy does not

---

[36] "Theorie des lichtelektrischen Effektes," p. 885.
[37] Ibid.

vary approximately linearly with the frequency." Furthermore, Wright had found evidence that the photoelectric effect depends on the plane of polarization of the incident radiation. "With respect to both points," Sommerfeld concluded, "our theory is in better accord with Wright's measurements than Einstein's light quantum theory."[38]

Sommerfeld, who first proposed his theory in mid-1911, soon had a chance to discuss it—actually, a refined version of it—personally with Einstein at the Solvay Conference in Brussels. From Einstein's comments, which are recorded in the proceedings of the conference,[39] it appears that he was at best lukewarm toward Sommerfeld's theory. More positive reactions, however, came from other conferees, notably Planck, Lorentz, and Brillouin.[40] At any rate, Sommerfeld was in general apparently encouraged by the reception accorded his theory—Max Born later termed it his "wild adventure"[41]—because, aided by his former student Peter Debye, he continued to refine it, eventually publishing a 58-page paper on it[42] in the Annalen der Physik of 1913.

V

The theories of Lorentz, Thomson, and Sommerfeld shared one salient feature: each rested on a definite hypothesis on the structure of the atom. Consequently, the acceptance of Bohr's model in the years following 1913 forced their abandonment, which, incidentally, illustrates rather nicely how an advance in one area of science generally has considerable ramifications in a related area. No one today would be tempted to advocate the theories of the photoelectric effect we have examined so far.

Of a very different character is the theory O. W. Richardson began developing at the end of 1911.[43] Since I believe Richardson's theory merits study even today, the question arises why only historians have heard of it. The answer is, I think, that Richardson's approach, which he felt possessed a great advantage, strikes us now as possessing a great disadvantage. For Richardson adopted not a microscopic, but a macroscopic, approach. As he wrote: "For the present, I wish to avoid discussion of the vexed ques-

[38] "Das Plancksche Wirkungsquantum," pp. 1090–91.
[39] P. Langevin and M. de Broglie, eds., La Théorie de Rayonnement et les Quanta (Paris, 1912), pp. 373–392, especially pp. 373, 381–382, 384, 390.
[40] Ibid., pp. 390–392, 451.
[41] "Arnold Johannes Wilhelm Sommerfeld," Biographical Memoirs of Fellows of the Royal Society, 8 (1952), 283.
[42] "Theorie des lichtelektrischen Effektes," pp. 873–930.
[43] "Some Applications of the Electron Theory of Matter," Philosophical Magazine, 23 (1912), 594–627.

tion of the nature of the interaction between the material parts of the system and the aethereal radiation and to confine my remarks to the conclusions which may be drawn from the existence of a statistically steady condition of the aethereal and electronic radiations."[44] Thus, Richardson would adopt the methods of thermodynamics, which explicitly avoids hypotheses on the micro-world.

In the photoelectric effect, radiation incident on a metal plate causes the emission of electrons from it, and ordinarily one does not worry about the possibility of any electrons returning to the plate and being absorbed by it. Richardson, however, did worry about these returning electrons: he set as his goal the determination of the equation describing equilibrium between the rates of electron emission and absorption. To analyze this problem, Richardson carried out a thought experiment using the most common of all thermodynamic thought apparatus—a piston in a cylinder. He assumed that the only photosensitive surface in the system was the bottom of the cylinder; further, he assumed the piston to be both completely impervious to electrons and completely transparent to radiation. If, now, the cylinder is filled with radiation and the piston, initially in contact with the bottom of the cylinder, is slowly raised, the radiation will pass through the piston from above and eject electrons from the bottom of the cylinder, thereby causing the space below the piston to fill with electrons. By slowly moving the piston up and down, the electrons could be pumped in and out of the bottom of the cylinder in a reversible manner. It was this process that Richardson analyzed in detail thermodynamically—a later treatment[45] involved the direct application of the Clausius-Clapeyron equation to it, considering the electron emission to be analogous to the evaporation of a monatomic gas.

Richardson of course had to specify the spectral distribution function of the radiation in the cylinder, and in his first treatment he used Wien's distribution. Later, in a more refined version of his theory,[46] he used Planck's distribution, but this change affected only the precise form of the electron energy distribution and not any of his basic conclusions. Assuming Wien's distribution, Richardson derived an integral equilibrium equation for N, the rate of electron emission as a function of frequency. By

[44] *Ibid.*, p. 617.
[45] A. L. Hughes and L. A. Dubridge, *Photoelectric Phenomena* (New York: Mc-Graw-Hill, 1932), pp. 196–199.
[46] "The Theory of Photoelectric and Photochemical Action," *Philosophical Magazine*, 27 (1914), 476–488.

Roger H. Stuewer

direct substitution, Richardson easily proved that the equation was satisfied by

$$N = \begin{cases} 0 & 0 < h\nu \le w_0 \\ \dfrac{B}{\nu^3}\left(1 - \dfrac{w_0}{h\nu}\right) & w_0 < h\nu < \infty \end{cases}$$

where B is a constant and w_0, in Richardson's interpretation, is the latent heat of evaporation per electron. We see that for radiant energies $h\nu \le w_0$, no electrons should be emitted from the photosensitive surface, while for radiant energies $h\nu > w_0$, a continuous emission should occur. Furthermore, from his equilibrium equation for N, Richardson immediately derived a second equation for T, the kinetic energy of the emitted electrons. This equation had the solution:

$$T = \begin{cases} \text{meaningless} & h\nu \le w_0 \\ h\nu - w_0 & w_0 < h\nu < \infty \end{cases}$$

It is obvious that these results are formally identical to Einstein's, and Richardson was fully aware of this significance of his theory. The equations above, he wrote, have "been derived without making use of the hypothesis that free radiant energy exists in the form of 'Licht-quanten,' unless this hypothesis implicitly underlies the assumptions:—(A) that Planck's radiation formula is true; (B) that, ceteris paribus, the number of electrons emitted is proportional to the intensity of monachromatic radiation. Planck . . . has recently shown that the unitary view of the structure of light is not necessary to account for (A) and it has not yet been shown to be necessary to account for (B). It appears therefore that the confirmation of the above equation . . . by experiment would not necessarily involve the acceptance of the unitary theory of light."[47]

Einstein, in view of his 1909 insights into the wave-particle duality which were mentioned earlier, would of course have contended that his light quantum hypothesis was, indeed, a necessary consequence of the validity of Planck's law. Furthermore, he would also have pointed out that his light quantum hypothesis simply and naturally accounted for the fact that the number of electrons emitted is proportional to the intensity of the incident radiation. That Planck's law could be derived from other, weaker, assumptions in no way affects the validity of the necessary consequences of that law.

[47] "The Theory of Photoelectric Action," *Philosophical Magazine*, 24 (1912), 574.

Richardson, however, did not appreciate the significance of this point and concluded that his derivation undercut Einstein's light quantum hypothesis. To Richardson this was an extremely important conclusion, because concurrently with his theoretical work, he and his former student Karl T. Compton were carrying out experiments on the photoelectric effect.[48] These experiments were unquestionably the most accurate and refined to date, and as I noted at the beginning of my paper, Richardson and Compton concluded in favor of the linear relationship between energy and frequency. Notwithstanding his theoretical work, this came as a surprise to Richardson: "when these experiments were started," he later wrote, "I thought it improbable that the equation would turn out to be correct, on account of the very grave objections to the form of quantum theory on which it had up to that time been based by Einstein." Fortunately, he continued, this equation "evidently has a wider basis than the restricted and doubtful hypothesis used by Einstein."[49] Indeed, his and Compton's experiments, he concluded, could be taken to "confirm the theory of the photoelectric action which was recently developed by one of the writers."[50]

It is no doubt unfair to generalize at this point and conclude that whenever a scientist has a choice between a conservative and a radical theory he will choose the conservative one (especially if he himself has developed it!) but that was clearly the case here. Actually, of course, the choice was not so much between a conservative and a radical theory, as it was between a desire to avoid hypotheses on the micro-world and a lack of that desire. But just because Richardson had succeeded in deriving Einstein's equation without explicitly invoking Einstein's light quantum hypothesis, the validity of that equation was not automatically established. The experimental issue was by no means settled. As a matter of fact, it was in response to Richardson and Compton's *experimental* work that Pohl and Pringsheim proposed the logarithmic relationship between energy and frequency that I mentioned at the beginning of my paper. Richardson and Compton's work was open to two major criticisms: first, their experiments had been carried out over a very restricted range of frequencies, so that a number of different functional relationships fit their data reasonably well; sec-

---

[48] "The Photoelectric Effect," *Philosophical Magazine*, 24 (1912), 575–594.
[49] "The Photo-electric Action of X-Rays," *Proceedings of the Royal Society*, 94 (1918), 269.
[50] Karl T. Compton and O. W. Richardson, "The Photoelectric Effect—II," *Philosophical Magazine*, 26 (1913), 550.

ond, there was, in Richardson's words, an experimental error of an "insiduous [sic] nature"[51] in their work, an error that led Richardson and Compton to put much more faith in the *average* electron energies they measured than in the *maximum* electron energies. It is, of course, the *maximum* energies that are the important ones theoretically.

All experimental difficulties were not overcome until R. A. Millikan completed his painstaking work in 1915–16. In the course of years of research, he eventually constructed what he called a "machine shop in vacuo"[52] to make his photoelectric measurements. The accuracy with which Millikan's data points fell on the predicted straight line was truly remarkable. Einstein's light quantum hypothesis was therefore finally vindicated. Or was it? Concerning the theoretical interpretation of his experiments, Millikan suggested that in the atom there were electrons "in all stages of energy loading up to the value $h\nu$," so that the incident radiation merely triggered an "explosive emission."[53] And in 1917, in an entirely fascinating passage, Millikan carefully distinguished between Einstein's *equation* and Einstein's *theory.* He wrote:

Despite . . . the apparently complete success of the Einstein equation, the physical theory of which it was designed to be the symbolic expression is found so untenable that Einstein himself, I believe, no longer holds to it, and we are in the position of having built a very perfect structure and then knocked out entirely the underpinning without causing the building to fall. It stands complete and apparently well tested, but without any visible means of support. These supports must obviously exist, and the most fascinating problem of modern physics is to find them. Experiment has outrun theory, or, better, *guided by erroneous theory* [my italics], it has discovered relationships which seem to be of the greatest interest and importance, but the reasons for them are as yet not at all understood.[54]

When the long-awaited experimental confirmation came, it was not accepted as such! The fact that Lorentz, Thomson, Sommerfeld, and Richardson had been in varying degrees successful in developing alternate interpretations of the photoelectric effect goes a long way in accounting for this remarkable situation. There were of course other factors, perhaps the most significant being von Laue's very striking 1912 discovery of the crystal diffraction of x-rays,[55] which, after many years, seemed to conclu-

---

[51] *Ibid.,* p. 567.
[52] "A Direct Photoelectric Determination of Planck's 'h,' " p. 361.
[53] *Ibid.,* pp. 387–388.
[54] *The Electron* (Chicago: University of Chicago Press, 1917), p. 230.
[55] "Interferenzerscheinungen bei Röntgenstahlen," *Bayrische Akademie der Wissen-*

sively demonstrate their wave nature. Curiously enough, the discovery of the Bohr Atom in 1913,[56] with its transitions between discrete energy levels, seems to have done little to alleviate the skepticism toward Einstein's light quantum hypothesis, though this may be due in large measure to the fact that Bohr himself was so strongly opposed to light quanta. At any rate, the skepticism prevailed until roughly 1924, until after Arthur Compton's famous interpretation of his beautiful x-ray scattering experiments[57] had been accepted. Compton's discovery possessed huge shock value and finally forced physicists to cope seriously with Einstein's hypothesis. Eventually, the photon (to use G. N. Lewis's lasting term[58]) became accepted into physics as a particle of light possessing both energy and momentum.

VI

I should like to conclude with a postscript. Most of what I have just related I was familiar with when, in the fall of 1968, I met Professor Peter Franken at the Sommerfeld Conference in Munich and learned from him that it is possible, even today, to account for the main features of the photoelectric effect along non-Einsteinian lines, that is, without assuming light quanta or photons to be incident on the atom. In brief, Franken has constructed his theory[59] from a well-established quantum mechanical base, time-dependent perturbation theory. An atom in its ground state, described by Schrödinger's equation, is assumed to be subjected to a *classical electromagnetic wave* of frequency v, which perturbs the state of the atom. Assuming that the incident wave has enough energy to bring about a transition of an electron to the continuum, time-dependent perturbation theory shows that only a definite level in the continuum will be excited. This level is of energy hv above the ground state, so that if w_0 is the ionization potential of the atom, the level corresponds to a free electron of kinetic energy $T = hv - w_0$, which is Einstein's equation. Two things are crucial to observe: first, Planck's constant h is introduced not by assumption but by Schrödinger's equation as applied to the atom; second, the frequency v is the frequency of the perturbing *electromagnetic wave*.

schaften, München, Sitzungberichte, mathematisch-physikalische Klasse, 42 (1912), 303 (with W. Friedrich and P. Knipping).

[56] "On the Constitution of Atoms and Molecules," *Philosophical Magazine*, 26 (1913), 1–25, 476–502, 857–875.

[57] "A Quantum Theory of the Scattering of X Rays by Light Elements," *Physical Review*, 21 (1923), 207, 483–502.

[58] "The Conservation of Photons," *Nature*, 118 (1926), 874–875.

[59] Private communication.

Franken also points out that in addition to the linear relationship, two other consequences follow immediately. First, "Fermi's Golden Rule" shows that the *rate* of electron emission is proportional to the square of the perturbation matrix, and since the perturbation matrix is proportional to the amplitude of the perturbing wave, we have immediately that the rate of electron emission is proportional to the *intensity* of the incident radiation. Second, since the rate is established as soon as the perturbation is "turned on," there should be no systematic time delay between the time of electron emission and the time when the radiation is incident. Thus, all known aspects of the photoelectric effect may be explained *without assuming photons to be incident on the atom.* I should point out that after Franken called my attention to this rather striking conclusion, I found similar derivations in other sources, for example, in Eugen Merzbacher's *Quantum Mechanics*[60] and in J. R. Oppenheimer's 1939 Lecture Notes on Quantum Mechanics.[61] Neither Merzbacher nor Oppenheimer, however, draws attention to the point emphasized by Franken.

Franken fully recognizes, it should be noted, that photons may be *produced,* as for example in atomic transitions. His claim is only that it is possible to interpret the photoelectric effect (and other quantum electrodynamical phenomena) without introducing the photon concept at the outset—without assuming that photons are *incident* on the atom. Thus, he would regard as highly misleading, or even erroneous, statements like that of S. Tolansky, who recently wrote that there "is absolutely no possible way of accounting for photoelectric effects . . . except by adopting the idea of the photon as a sort of particle carrying its full energy and travelling with the velocity of light."[62] Perhaps one might ask the following question: If the photoelectric effect were the only relevant experiment available to us, would we be justified in concluding from it that radiation consists of quanta? In view of our earlier remarks, it would seem that we would have to give a negative answer to this question.

But this in no way diminishes the force of Einstein's original arguments for light quanta: Planck's law entails, as a necessary consequence, the light quantum hypothesis. That one experiment—the photoelectric effect—may be explained on weaker or different assumptions in no way affects Einstein's basic arguments. And certainly nothing is more artificial than con-

[60] New York: Wiley, 1961, pp. 459–460.

[61] Notes taken by B. Peters. I should like to express my gratitude to Professor Morton Hamermesh for giving me a copy of these notes.

[62] *Revolution in Optics* (Harmondsworth: Penguin, 1968), p. 40.

centrating on one experiment to the exclusion of all other phenomena: the photon concept, like all fruitful concepts in physics, derives its validity from an interlocking theoretical and experimental matrix. The main purpose of this paper has been to display some of the difficulties that were involved in recognizing and understanding only one of the elements in this matrix.*

* EDITOR'S NOTE: Dr. Buchdahl's comments on this paper are combined with his comments on Professor Stein's paper, following the latter.

On the Notion of Field in Newton, Maxwell, and Beyond

In a strict and philosophical sense, it seems to me, there is no such subject as "field theory."[1] There is not, on the one hand, an approximately
well-defined physical subject matter (i.e., class of phenomena) that can
define such a theory—in the way light, for instance, or electromagnetism,
or heat, defines optical or electromagnetic theory or thermodynamics. (It
can be argued that today, in the quantum theory of fields, such a subject
matter does exist. This is a point of some interest, and I shall comment
upon it later. But presumably the scope of the assigned topic was not intended to be restricted to quantum field theory.) On the other hand, I do
not think there is a clearly delimited fundamental kind, or form, or structure, which distinguishes field from non-field theories. More precisely: the
historically natural use of the term "field" allows its application within
many theories, and these are characterized by several features of fundamental interest; but not all the theories share all the features; and I see no
convincing reason to designate any special set of features as necessary and

AUTHOR'S NOTE: This work has been supported by the National Science Foundation.
This paper was written on the assumption (which proved erroneous) that the procedure of the conference at which it was presented would call for a talk of one hour's
duration; this assumption determined its length, and accounts for the occasional references to the space available. When it was decided that the proceedings of the conference should be published, the schedule of publication deadlines did not allow extensive
revision; hence it was impossible to expand the somewhat cryptic and abbreviated later
sections of the paper. A more rounded exposition of the matters there touched upon is
reserved for a future occasion.
    [1] This paper when presented was introduced by (substantially) the following remark,
having particular reference to the paper delivered at a previous session by Mr. Arnold
Thackray: "One word of orientation: the principal aim of this paper is clarification—
or to promote understanding—of the content and methods of the science of physics. I
believe that science to have existed (with a coherent subject matter), and to have had
a coherent development, from (say) Galileo's day to our own. Whether work of this
sort ought to be countenanced, or damned as 'irrelevant' and out of tune with the
times, is a question I shall not discuss."

sufficient to confer upon a theory the dignity of the appellation "a field theory." I intend therefore to take as my theme, not field theory as such, but the role, in several investigations, of concepts and principles loosely associated with the notion of field. My aim is simultaneously historical and philosophical: for I believe that understanding of the issues controlling particular investigations is a necessary condition for understanding inquiry (data for the philosophy of science can come only from the history of science); and I believe that a clear appreciation of the content of the science we possess—itself, in my view, a primary object of the philosophy of science—is facilitated by, in practice even perhaps impossible without, some attention to the routes by which that science has been acquired. Toward the end of my paper, I shall suggest certain general bearings of the historical part, or of principles illustrated by it, for issues of recent methodological concern. But I must confess that, for me, the philosophical interest of particular scientific investigations does not exclusively consist in what general methodological conclusions may be drawn from them; indeed, the contrary connection is to me more compelling: that one philosophical test of methodological or epistemological principles is their ability to illuminate particular investigations.

I

Historically—and by a great margin—the first physical field to be made the subject of a coherent theory was the gravitational field (which, ironically, is today the least well understood, so far as concerns its incorporation within the set of principles that we have come to regard as fundamental in physics). It is, I think, not usual to associate Newton's name with field theory. But I should like to point out several respects in which Newton's thought has made significant use of notions which, for us, do belong to the cluster surrounding the word "field." Such uses occur at the level of heuristic detail, or "research"; at the level of major conceptual organization, or "fundamental theory"; and—in a way that has become apparent comparatively recently (thanks to the publication by the Halls of a remarkable Newton manuscript)—at a still more general level, for which I can really find no term more apt than "metaphysical."

1. At the beginning of the *Principia* Newton gives us definitions of three quantities, which he calls three different "measures" of a "centripetal force": the absolute quantity, the accelerative quantity, and the motive quantity. These definitions have certain peculiarities of detail which afford

interesting matter for textual explication; but their essential drift is altogether plain, from the discussion that follows them. By a "centripetal force," in this context, Newton means what we should call a central force field. The motive quantity measures the action of the field upon a body, and is exactly what we call the force on the body. The absolute quantity is intended as a measure of the entire field associated with a particular center. Newton does not specify how this measure is determined,[2] and such a specification would require some more restrictive condition than Newton imposes upon the general concept of a centripetal force; but in the cases that arise, there is no real ambiguity: for force fields governed by what is now called a "Newtonian potential," the absolute quantity attached to any center is the source strength at that center; for gravitational force in particular, it is gravitational mass. As for the accelerative quantity, what Newton truly means by it is the field intensity. This is not what he says in the definition; he says that the accelerative quantity measures the centripetal force by the acceleration it produces. But in explaining his intention, he tells us that the motive quantity may be ascribed to the body acted upon (i.e., as measuring that action); the absolute quantity to the center (as measuring the strength of the entire tendency toward that center); and the accelerative quantity to the *place* at which it is measured— as expressing "a certain efficacy, spread from the center to the individual places around it, to move any bodies that are in them." The conception is unmistakably that of field intensity: a function defined on a region of space, whose value at a point measures the tendency, or "disposition," for bodies at that point to be acted upon. In general, it makes no sense to require the values of this field function to be accelerations; but for the force Newton intends to deal with, acceleration is the proper measure of intensity; and (at the cost of some illogicality) he states the definition as if such were the general case.

---

[2] In this respect, the definition of the absolute quantity of a centripetal force is comparable to the much-discussed definition of the quantity of matter, or mass. Newton treats "definition" as an expository rather than a logical category; in our stronger sense, the definitions he intends are always to be gathered from his supplementary explanations and from his practice in applying the term in question. In the present case, there seems to be an implicit acknowledgment by Newton of the deficiency of his definition. Each of the other "quantities" is defined as "the measure . . . proportional to" a certain well-defined magnitude; but for the absolute quantity there occurs what seems a deliberately vaguer expression: "The absolute quantity of a centripetal force is the measure thereof that is greater or less according to the efficacy of the cause propagating it from the center through the surrounding regions." (Motte's translation here is slightly inaccurate.)

This so far is mere conceptual trappings; the question for us is, what role does this array of concepts play in Newton's investigation of the solar system and of gravitation? Of course we know that the results of the investigation can be expressed in terms of field intensity, source strength, and force. What I want to suggest is that the investigation itself makes essential use of these concepts in their interrelationship—and crucial use, in particular, of the concept of field intensity.

Indeed, the basic inductive conclusion upon which Newton's solution of the problem of the solar system and of gravitation rests is the following: There is a centripetal force, of which each major body in the solar system is a center, having the properties that (1) the field intensity is the acceleration, and (2) this intensity, for each center considered singly, is inversely proportional to the square of the distance from that center. Consider, for example, the sun and the planets. For the Copernican-Keplerian system of reference we have Kepler's laws. Newton solves these for the accelerations, and deduces the inverse-square law of acceleration, (a) for each planet separately, as its distance from the sun varies by the relatively small amount determined by its eccentricity, and also (b) from one planet to another, over the vast range of distances (but the rather small sample) provided by the actual series of planets. The induction is very convincing. The fact that the acceleration is the field intensity is critical, for the evidence comes entirely from six bodies, each exploring the field in a fixed and severely restricted range; the inductive basis would therefore be rather weak if we were not, by good luck, able to relate directly to one another purely *kinematical*—and, thus, ascertainable—parameters of the several bodies' motions. This lucky fact is not the work of Newton's definitions, but of nature. Newton's merit was to know how to use what he was lucky enough to find. And my point is that his analysis of the situation makes an essential use of the notion of a field. To regard the planets as test bodies, exploring a field, may seem a vivid and suggestive way of thinking, of heuristic value in searching for a theory—and I think that is so; but it is not the claim I am now concerned to make. My claim is both simpler and more far-reaching. That the accelerations of the planets, severally and collectively, are inversely as the squares of their distances from the sun is not the conclusion of Newton's induction; that is his deductive inference from the laws established by Kepler. Newton's inductive conclusion is that the accelerations toward the sun are everywhere—i.e., even where there are no planets—determined by the position relative to the sun; namely, directed

267

toward that body, and in magnitude inversely proportional to the square of the distance from it. And although the inductive argument is very straightforward—certainly not dependent upon tortuous constructs—that argument cannot be made, because its conclusion cannot even be sensibly formulated, without the notion of a field. From a mathematical point of view, the idea of an acceleration attached to each point in space is the idea of a function on space, hence a field; from the physical and methodological point of view, the idea of an *acceleration* characterizing *a point where there happens to be no body* makes no sense at all, unless one accepts the notion of a disposition, or tendency; subject to probing, but not necessarily probed.

This conclusion is reinforced by a careful analysis of the later phases of Newton's argument.[8] First, he arrives at a similar result for the acceleration fields explored by the satellite systems of Jupiter and of Saturn (considered separately). In these cases, the spatial reference frames are of course different; the central body is always taken as at rest. The evidence here is more slender: not only because there are fewer test bodies, but especially because the accuracy of the data allows Newton only a degenerate analogue of Kepler's first two laws; namely, that the orbits are roughly circular, traversed with constant speed (so that no evidence is present for the inverse-square law within the motion of a single body). Next he finds the same for the moon, in its course round the earth; but here the reverse restriction obtains, since there is only one test body: only over the small range of distances from the earth explored by the moon do we have direct evidence for the inverse-square law. The relation is, moreover, within this range, noticeably inaccurate—a fact which Newton (of course correctly) attributes to the influence of the sun's field. This leaves a very weak direct basis for the law as applied to the moon over such a range of distances as from its actual orbit down to the very surface of the earth. But just this application is required by Newton, in order, by comparing the moon's acceleration with the acceleration of falling bodies at the earth's surface, to infer the identity of the centripetal force of astronomy with the terrestrial force known as weight or gravity. It is plain that what justifies this procedure is the fact that the relationships established for the several systems, taken collectively, form a *single* body of evidence for the proposition I have formulated earlier (namely, that each major body is the center of an

---

[8] "Later" in the order I am sketching here; in Newton's exposition, the proposition about the satellite systems of Jupiter and Saturn is stated first. This is of no moment.

inverse-square acceleration field); so that the data from the planets and the other satellite systems support the conclusion about the moon.[4] Indeed, terrestrial evidence enters obliquely to provide more support; for while no direct astronomical evidence supports the proposition that the region about the earth carries an "acceleration field," Galileo's law of falling bodies tells us precisely that.

I shall dwell no further upon details of this investigation, but only remark that it involved the assumption that the reaction to the centripetal motive force of gravitation is exercised upon the central body—an assumption which, as Huygens pointed out and as Newton seems to have implicitly acknowledged, enters the argument as a pure hypothesis; and that it also involved the fitting together of the several fields, with their several spatial reference frames, into a single coherent spatial system—which is then demonstrably unique up to a uniform translational velocity. I have discussed these aspects of the matter in another place.[5] What I shall turn to next is what I have referred to as the level of "fundamental theory." (Of course, the theory of gravitation itself, in the definitive formulation that Newton found, has its place at this level.)

2. The general conceptual framework within which Newton placed both his own work and his program for natural philosophy is expounded by him in several places with great clarity. He takes for granted the notions of space and time, with all of Euclidean geometry and the classical kinematics of absolute motion. He assumes the notion of bodies, located in space but not necessarily—and indeed not actually—filling it; and he pictures these bodies as constituted of minute (but not punctiform) indivisible particles. (So far as Newton's positive work is concerned, the hy-

---

[4] I do not mean here to take a position on whether induction may involve general propositions in an essential way or is always from particulars to particulars. Even if a satisfactory technical analysis of the logic of induction—which, despite much very good work, we certainly do not yet have—should rest upon the latter alternative, the points I have made above would stand: (1) The particular conclusion about the moon cannot be formulated without a dispositional term or a contrary-to-fact conditional. (2) If the inference can be made at all, it must support the general law of the field in the sense of Carnap's "instance-confirmation": there is clearly no basis for a particular statement about a body transported between the earth and the moon that does not equally support an analogous statement about analogous transport within the field. (3) The conclusion must rest upon the diverse sources of data I have described; therefore the logical analysis, however its details are organized and whatever the special language in which it is couched, must allow essentially that assimilation of "kinds" which is expressed here by saying that (a) all the satellites of one body explore "the same field," and (b) the fields of all the centers involved are of the same sort.

[5] "Newtonian Space-Time," *Texas Quarterly*, 10 (Autumn 1967), 174–200.

pothesis of indivisible particles really plays no very important role.) Finally, he assumes a general notion for which he uses the word *force*, or *vis*. In Newton's most basic usage, a force is a principle of motion: to know a particular force is to know an associated law of motion. One such force is both universal and intrinsic to all bodies: the *passive* principle of motion, or force of inactivity, *vis inertiae*, whose expression in the form of law is the three famous laws of motion.[6] Another universal principle is *impenetrability* (although I do not know of a place where Newton calls this a "force"): a principle by virtue of which no two bodies can be simultaneously present in the same point of space. But the object of greatest attention in Newton's natural philosophy is the class, largely unknown (or known very imperfectly), of *active principles*. These are the forces that can produce, and more generally can change, motion; and Newton assumes—presumably as involved in the third law (namely as asserting that the force of inertia qua "resisting" force must always have an object to exercise resistance upon)—that they are principles of *interaction* of bodies. When such a force acts, or is "impressed," upon a body, Newton speaks of "an impressed force"; the latter "consists in the action only; and does not remain in the body after the action." Impressed force is, therefore, the particular and fluctuating action of "force" in its fundamental sense; the two categories are distinct, and when Newton says that impressed forces are of various origins, one of which is centripetal force, he is not committing a blunder in saying that a force comes from a force.

The chief problem of natural philosophy, then, according to Newton, is the discovery of the active principles, or forces of nature—and then the proof, i.e., test, of the discovered principles, by their application to natural phenomena. The discovery itself is to be accomplished by the study of phenomena, and especially phenomena of motion. This proposal is, of course, a direct generalization of Newton's actual performance in discovering the principle or law of universal gravitation. Newton gives many, and convincing, reasons for believing that the principal phenomena of matter

---

[6] This is not speculative interpretation of Newton's doctrine, but what he states explicitly and with precision; so, in all pedantry, the "law of inertia" for Newton is not just the first law of motion, but all three together. Cf. *Principia*, Definition III and the paragraph that follows it; but especially *Opticks*, book III, Question 31, toward the end (Dover ed., p. 401): ". . . a *Vis inertiae*, accompanied with such passive Laws of Motion as naturally result from that Force . . ."; and with no possibility of misconstruction (*ibid.*, p. 397): "The *Vis inertiae* is a passive Principle by which Bodies persist in their Motion or Rest, receive Motion in proportion to the Force impressing it, and resist as much as they are resisted."

depend upon fields of force, both attractive and repulsive, among the particles of bodies; and he devoted great effort to the attempt to gain information about these fields. The most striking example is the interrupted inquiry reported at the beginning of the third book of the Opticks. In the preceding book, Newton had made an attempt to gain from optical phenomena an estimate of the sizes of the particles of bodies,[7] and also an estimate of the power—which must be regarded as an "absolute quantity" of centripetal force—of various particles to act upon light. He had obtained some results,[8] which he clearly regarded as plausible and useful, but as insufficient, both in content and in degree of support, to serve as the foundation of a theory. In view of its place in the structure of the Opticks, and of the character of the Queries that Newton supplied as substitute for the investigations he was unable to carry out, it seems to me beyond reasonable doubt that the study of the phenomena of "inflection," or diffraction, was undertaken by Newton with the chief purpose of determining the basic law of interaction of particles of bodies and light—in other words, the law of the "optical field"; and in the hope that the success of this undertaking not only would represent a fundamental advance in optics, but would afford the means for progress in the study of the structure of matter. The method of investigation is clear: The test bodies for exploring the field in question are the rays of light. Within transparent bodies, these rays are in equilibrium, and move uniformly; at a reflecting or refracting surface, the behavior of the rays is accounted for by very general assump-

---

[7] Not of the "ultimate" particles, but of those parts of bodies, separated by interstices, "on which their Colours or Transparency depend" (Opticks, book II, part I, near the beginning; Dover ed., pp. 193–194). These are not ultimate particles, in Newton's opinion; he believes them to be transparent (ibid., part III, Proposition II; Dover ed., p. 248), hence penetrable by the particles of light, hence not the ultimate impenetrable particles. (Although in the proposition just cited Newton speaks of the "least parts" of bodies, the argument he gives applies to "very small" parts; one must take him to use the word "least" colloquially in this sense. That Newton considers these parts themselves to have an internal structure is clear from his statement—ibid., end of Proposition VII; p. 262—that the sense of vision, aided by microscopes, may eventually reach these parts; but, he fears, no further: "For it seems impossible to see the more secret and noble Works of Nature within the Corpuscles by reason of their transparency." Cf. also the discussion of Proposition VIII, where Newton alleges the probability that any rays of light that actually do strike any one of "the solid parts of Bodies" are absorbed, or as he says "stifled and lost in the Bodies"; and concludes that "Bodies are much more rare and porous than is commonly believed"—ibid., pp. 266, 267.)

[8] For instance, on the second question, he formulates the tentative conclusion that it is either exclusively or preponderantly the "sulphureous Parts" that interact with light to bend its path; hence that the power of a body to reflect and refract light is proportional to its sulphureous content—which in turn he thinks is for the most part nearly proportional to the density of the body (ibid., Proposition X; Dover ed., p. 275).

tions about the character of the field; but the complexity of the phenomena of diffraction suggested that a careful determination of the light paths very close to the edge of an opaque body would reveal a remarkable structure in the field, involving intensities alternately in opposite directions (i.e., forces which at certain distances are attractions, at others repulsions). As Newton was very well aware, forces of this type are essential for an account, in the terms of his program, of the behavior of solid bodies.

I have so far argued, first, that in Newton's investigation of gravitation the notion of a field plays a logically uneliminable role in the inductive evaluation of the evidence; and second, that that notion has a central place in fundamental theory for Newton—not only in the expression of particular fundamental laws (like that of gravitation), but far beyond this, in what (following Peirce) we might call Newton's "abductive" scheme for reducing all the phenomena of nature to order. Perhaps it will be objected that the second claim appears to give away the issue of action at a distance: if fields, or centripetal forces, constitute a fundamental category for scientific explanation, then we seem to have excluded the possibility of accounting for all phenomena by interactions through contact alone. This is not what I mean to imply. Discussion of the point will afford the transition to what I have referred to as the metaphysical level of Newton's thought.

3. Newton tells us emphatically and repeatedly that our only source of knowledge of corporeal nature is experience. This precludes dogmas, whether positive or negative, about the interactions of bodies. It also precludes the claim to direct insight into the nature of any interaction, of any sort: that is, for Newton (at least if he is consistent), the claim that we know how bodies can interact by contact, or more specifically how they can communicate motion in impact, in any other sense than that we have observed interactions of this type to occur, must be delusive. I believe that Newton is consistent, and that this is his view. In the light of this view, centripetal forces when gathered from experience—as gravity was, through Newton's beautiful analysis; and as at least the existence of other such forces, both attractive and repulsive, can be, through reasoning he presents at length—have as good a right as any other principles to legitimate status in philosophy: they are then, he says, "manifest Qualities." They are *verae causae*, and the evidence for their existence and manifold role in the processes of nature is just what confers upon them fundamental importance. But none of this excludes a search for the causes of these principles them-

selves: that is, we know they occur; but we do not know that they—or any particular ones among them—are irreducible. Nor do we know them to be reducible. The question, in each case, is therefore a proper and necessary one for natural philosophy: whether or not a particular centripetal force, such as gravitation or electricity or magnetism, whose own laws have been found out, is the effect of deeper-lying structures; and if so, what those structures and their laws are.

But what of action at a distance, and what of ultimate explanations? We know that Newton was troubled about the cause of gravity: despite his principles, as I have explained them, he was far less content to rest upon gravitation than upon impenetrability as an irreducible property of bodies; and he goes so far as to say, in his well-known letter to Bentley, that it is *inconceivable* that matter should affect other matter without mutual contact, except by the mediation of something else, or the action of some agent, material or immaterial. The last qualification, "material or immaterial," is disconcerting; and when we read in a letter by Newton's protégé Fatio de Duillier, dated thirteen months after the letter in question to Bentley, that Newton is undecided between two opinions about the cause of weight—(1) that it is caused by the impacts of streaming cosmic particles, Fatio's own hypothesis (later taken up by Le Sage); (2) that it is caused by "an immediate Law of the Creator of the Universe"—I think we are apt to be not only puzzled but annoyed, and to feel that Newton is quibbling. In point of fact, I cannot altogether acquit him of this: it is the case, as I read Newton, that bodies attracting by an immediate law of God means for him exactly the same thing as direct action at a distance. I think Newton knew this equivalence clearly, and disguised it in his public utterance to avoid unwelcome embroilments. But although his statement is (on my reading) evasive, it is characteristically precise and accurate; and if it involves a quibble, there is something interesting behind it.

What sets this whole matter in a light that seems to me to clear up all obscurities is the fragment *De gravitatione et aequipondio fluidorum*, first published (with a translation) by the Halls in 1962, in their selection from the Portsmouth Collection of Newton manuscripts.[9] The largest part of this unfinished paper is a rather extended philosophical discussion of space and motion, and of the nature of body. I have had occasion elsewhere to

[9] A. Rupert Hall and Marie Boas Hall, *Unpublished Scientific Papers of Isaac Newton* (Cambridge: Cambridge University Press, 1962).

273

refer to this document in connection with the former subjects;[10] what is relevant now is Newton's treatment of the nature of body (which he takes up in order to clarify and defend his divergence from Descartes).

Newton's exposition takes the form of an account of how God can create matter. He does not claim to establish the true method of creation, or the true and essential nature of bodies; but only a possible method, and (correlatively) a possible representation of that essential nature: for, he says, "I have no clear and distinct perception of this," and cannot therefore certainly know whether God in fact might have created beings in all ways like bodies with respect to phenomena, and yet different from them "in essential and metaphysical constitution"; although it seems scarcely credible that this should be so. (There can be no doubt that this is partly ironic; the parody of Descartes is obvious, and a little labored—a characteristic of the style of this paper, in marked contrast to Newton's usual lean and vigorous prose.[11]) To explain, thus hypothetically, the creation of a body, Newton postulates the potency of God's intellect and will; he does not pretend to make the power of God's will itself intelligible, but merely points out that we have a power of moving our own bodies, and that this power is no more intelligible to us than the power he will assume for God. This does not (as one might suppose) altogether trivialize his task: what he has supposed, in effect, is that God can realize any intelligible objective; what remains is to make the objective, "creation of a body," intelligible.

[10] See note 5.

[11] The Halls (*Unpublished Scientific Papers of Isaac Newton*, p. 89) indicate considerable uncertainty about the date of this work; but suggest, on the basis of several considerations, including this stylistic one, that it is quite early. This conclusion seems plausible enough; but I should like to suggest two reasons for considering the question further—if clearer evidence can be obtained. In the first place, the Halls suggest evidence of immaturity in the thought of this essay, as well as in its style. In my opinion, that view is unjustified. In the paper cited earlier (note 5 above) I have explained why I consider Newton's treatment of the problem of space and motion, in this essay, to be trenchant and deep; and this in explicit comparison with both Descartes and Leibniz on the same questions. In the present discussion, I make a similar assessment of the quality of Newton's treatment of the nature of body. Thus I think that this work, for all its prolixity and lack of stylistic balance, is in its content a profound piece of philosophy. As to the style, it has to be considered that the work is only an interrupted draft —Newton's drafts in general seem to be prolix and rambling, compared to his finished writings; and that it is on an unaccustomed subject, or at least in an unaccustomed mode, for Newton—a circumstance that might account for the awkwardness of the exposition.

In the second place, there is oblique historical testimony that points to a rather late date for the work. The analysis of God's creation of matter given here by Newton is the same as that which, according to the French translator Pierre Coste, is hinted at in an

For clarity, Newton assumes the world to exist already, and makes the problem the creation of one new body. What God must—or, rather, may —then do, he says, is this: (1) Make some region or other of space impervious to the existing bodies—i.e., simply choose not to allow the motions of those bodies to penetrate that region. (2) Having established a region of impenetrability, allow it, in the course of time, to migrate from one place to another; or more precisely, confer the property of impenetrability, not on a fixed spatial region, but on different regions at different times—and in such a way that the mutations of the distribution of impenetrability constitute a smooth motion in space. In doing this, moreover, ensure that the motion of the new region of impenetrability and the motions of the already existing bodies together satisfy certain laws. (3) Confer upon the mobile impenetrable region the property, or power, of affecting our minds (in sensation), and being affected by them (in volition), whenever it comes to occupy the place of some particle in (say) a brain, now possessed of that power. Newton remarks that the first two steps, the meaning of which seems to involve no obscurity whatever, already suffice to make something indistinguishable in almost every respect

obscure passage of Locke's *Essay concerning Human Understanding* (book IV, chapter x, article 18; see the edition edited by Alexander Campbell Fraser, vol. II, p. 321, n. 2, for Coste's statement). Coste received his information from Newton, who, he says, "told me, smiling, that he himself had suggested to Mr. Locke this way of explaining the creation of matter; and that the thought had struck him one day, when this question chanced to turn up in a conversation between himself, Mr. Locke, and the late Earl of Pembroke." Now, Coste may easily have misunderstood Newton: it is possible that Newton said no more than that he had suggested this thought that day, and that Coste, by immediate misconstruction or later elaboration, fancied him to mean that he had then thought of it for the first time. It is also possible that Newton exaggerated the spontaneity of his thought on that pleasant occasion with Locke and Pembroke. But the possibility that both Coste and Newton are accurate in their testimony ought not, I think, to be merely dismissed.

In any event, there can be no doubt that Locke did receive this argument from Newton, and that this is responsible for the change he made in the passage in question in the second edition of his *Essay*; therefore, Newton communicated the argument to Locke some time between 1690 and 1694; and therefore, however early its conception, he had not by then rejected it. Similarly, the agreement of the views on space, time, and motion, in this fragment, with the views stated in the *Principia* (the former being a more extended elaboration of the latter) shows that on this subject too our fragment contains the mature opinions of Newton. The interest, therefore, in the date of composition of the fragment does not have to do with assessing its relevance to Newton's mature thought; this appears to be beyond doubt; but only with the comparatively minor question of obtaining evidence for the periods at which Newton had arrived at his mature views.

I should like to add that the existence of this fragment and its connection with Locke's *Essay* were brought to my attention by a reference in Alexandre Koyré, *Newtonian Studies* (London: Chapman and Hall, 1965), pp. 91–93.

Howard Stein

from a new body: something that would interact with all matter like any particle of matter, and that would constitute a sensible particle (since it would act like any other particle upon a sense organ). The third point—direct interaction with a mind—does indeed involve obscurity; but this resides in the deficiency of our understanding of mind; and Newton reminds us that the power, however obscure, of interaction between mind and body was the starting point of his analysis. He does insist upon the essential need for the third step, on the grounds that the operations, both active and passive, of our minds seem to involve the brain, and there seems to be a continual exchange of matter between the brain and the rest of the world; so, he says, it is manifest that the faculty of union with mind is in all bodies.

In summarizing the chief points of this analysis of the nature of matter, Newton puts first the fact that he has altogether dispensed with the unintelligible notion of substance. Or rather, since in the second place he puts the point that his beings "will not be less real than bodies, nor less able to be called substances," one should say that without *positing* an *obscure* notion of substance, he has been able to *construct* a *clear* one. Newton says that "the preconception"—namely, "of bodies having, as it were, a complete, absolute, and independent reality in themselves"—"must be laid aside, and substantial reality is rather to be ascribed to these kinds of attributes which are real and intelligible things in themselves, and do not need to inhere in a subject, than to some subject which we cannot conceive as dependent—nay, more, cannot form any Idea of." He hints that a similar analysis might be possible even of the nature of God; but adds at once that "while we are unable to form an Idea of [God's power], nor even of our own power by which we move our bodies, it would be rash to say what may be the substantial basis of mind."

I want to come back to this piece of Newtonian metaphysics later, and to say why I consider it deeper philosophy than, say, the obvious analogues in Berkeley and Hume. But at this point I think it is time to make the connection with field theory.

It is not a metaphor, but a literal truth, that Newton's metaphysics of body reduces the notion of matter to the notion of field. A body, Newton tells us, is a region of space endowed with certain properties. The clarity, or intelligibility, of the properties Newton specifies consists in the fact that they are conceived ideally as testable: they are dispositional characteristics of the spatial region, like the field quantities of gravity or electromagnetism. In particular, Newton takes as basic what it is quite natural to call the "impenetrability field"—a two-valued function on space (or rather space-time), since impenetrability either is there or is not. If, as Newton assumes for simplicity, we take for granted the existence of observable test bodies, this field is ideally testable per se. But to reduce the notion of body altogether and ultimately, Newton says, we also need the notion of the faculty of interacting with minds; and we need this notion in any case to represent our experience of nature. Newton postulates, therefore, what he thinks that experience suggests, namely that the distribution of this faculty of interaction coincides with that of impenetrability. Further, according to the Newtonian abductive scheme, there must be (1) a field of inertia, whose value at a point gives the mass density there (and whose "support"—i.e., the set of all points at which it is non-zero—coincides with the set on which impenetrability exists), and (2) such other fields—or, more generally, laws of interaction—as investigation of nature discovers. In any case, what I have called "ideal testability" is, for impenetrability as for other fields, testability only in a Pickwickian sense: I have already referred to the strenuous efforts which Newton devoted to the search for information, not even about the *ultimate* particles of matter, but about what we might call the molecules of bodies. And of course the very existence of the impenetrability field is a hypothesis. Newton tells us that we know of it only through experience, and reviews the inductive evidence for it;[12] but that evidence is less convincing than Newton thought—or it seems so, at least, with hindsight, in the light of evidence obtained later. Newton knew that induction is fallible; and I suggest that it would have surprised him, but not disconcerted him, to learn that the ultimate fields of impenetrability had been replaced by conceptions of another sort. I suggest, more specifically, that the program of deriving the properties of matter from a pure field theory in which the field variables are continuous (rather than two-valued)—a program so influential in the early decades of this century—constitutes, with respect to Newton, a deep revision of what I have called his fundamental theory, but demands no essential change in what I have called his metaphysics. The received idea that the Newtonian system involved a fundamental dualism of matter and force, or of substance and field,[13] is true of the system of physics that Newton was led to de-

[12] E.g., *Principia*, book III, discussion of the third rule of philosophizing.
[13] See, for example, Hermann Weyl, *Philosophy of Mathematics and Natural Science* (Princeton, N.J.: Princeton University Press, 1949), pp. 165–177.

velop; but when applied to Newton's own most basic conceptions, that idea is wrong.

But I have promised to relate all this to Newton's curious statements about action at a distance. The point seems to me this: In one sense—at the level of metaphysics and of theology—there is absolutely no difference in status, given Newton's analysis, between action through contact and action at a distance. Neither is intelligible from the mere conception of body as extended substance (whether filling space, as Descartes would have it, or occupying parts of space, in accordance with the atomists); both are intelligible from the conception of bodies as fields of impenetrability and inertia, accompanied by interaction fields. The arbitrariness in the specification of interaction fields—represented in Newton's account by the dependence upon God's fiat—is no greater than the arbitrariness in establishing impenetrability, inertia, and laws of interaction by contact. So when Newton tells Bentley, "It is inconceivable, that inanimate brute Matter should, without the Mediation of something else, which is not material, operate upon, and affect other Matter without mutual Contact," what he says truly expresses his views; and would still do so if he left out the words "without mutual Contact." On the other hand, at the level of fundamental physics, the situation is not quite so parallel between the two modes of interaction. For Newton did consider impenetrability to be *the first basic property of bodies*; and this means that interaction by contact (should contact ever indeed occur) is a *necessity*—a direct consequence of the fact of impenetrability; whereas interaction at a distance would represent, so to speak, a *further arbitrary* decision of God. It is in this sense, I think, that Newton's repeated denials that he holds gravity to be essential to bodies have to be understood. And this consideration certainly influenced Newton to consider seriously the possibilities of such a force as gravity being caused by a material medium. But yet again on the other hand, some of the considerations advanced in the *Opticks* tend rather persuasively to the conclusion that there is no reasonable hope of reducing all the forces of nature to effects of impacts of particles. Therefore, I think, Newton's views on this deep question in physics were in a state of considerable tension. Moreover, setting aside the issue of ultimate and total reduction, the possibility (in any given case) would remain that some force of nature already discovered, some field made manifest by the analysis of phenomena, had underlying causes that were still occult; this possibility would always call for further investigation. Newton's first remark to Bentley on this subject was: "The Cause of Gravity is what I do not pretend to know, and therefore would take more Time to consider of it." When, later, he declared the inconceivability of "inanimate brute Matter" acting at a distance without immaterial mediation, he added: "And this is one Reason why I desired you would not ascribe innate Gravity to me." I have tried to suggest what his other reasons were. I think he was altogether sincere in saying that he did not know the cause of gravity, and wished to consider it further; but I think he feared that to say plainly all he thought about the question would make great trouble for him.

II

I had intended, originally, to devote as much space to the notion of field in the work of Maxwell, and as much again to developments from Maxwell to Einstein, as to Newton; but this intention has clearly been defeated. I must confine myself to a rapid series of remarks on these subjects, and a brief sketch of some general conclusions.

1. Maxwell discovered that the centripetal forces of electricity and magnetism are effects of a deeper-lying structure. To be more accurate, he discovered that all the known laws of electrical, magnetic, and electromagnetic phenomena are explicable as effects of such a structure; that some (although not all) previously obscure points in the subject are put into satisfactory condition in this new theory; that very definite new phenomena (not, however, easy to realize experimentally) are predicted; and that the same structure which has among its effects, according to the theory, the phenomena of electricity and magnetism will also account for the behavior of light.

What is the character of the underlying structure found by Maxwell? His statement is a model of lucidity: "The theory I propose may . . . be called a theory of the *Electromagnetic Field*, because it has to do with the space in the neighbourhood of the electric or magnetic bodies, and it may be called a *Dynamical* Theory, because it assumes that in that space there is matter in motion, by which the observed electromagnetic phenomena are produced."[14] It is well known that Maxwell's first account of his theory[15] was based upon a quite detailed model of the arrangement, connections, and motions of this postulated matter. Clearly, then, we have a

[14] "A Dynamical Theory of the Electromagnetic Field," *The Scientific Papers of James Clerk Maxwell* (Cambridge: Cambridge University Press), vol. I, p. 527.
[15] "On Physical Lines of Force," *ibid.*, pp. 450–513; see especially part II, pp. 467ff.

279

theory that fits entirely within the Newtonian abductive scheme. But this theory had two rather serious defects: The postulated details were far *too* detailed, in the sense that things were supposed for which there was no real basis, even of a merely suggestive kind, in the evidence (this is exactly the kind of "hypothesis" that Newton so strongly deprecated in natural philosophy: detailed models of processes that might well be altogether otherwise). And the postulated details were not really self-coherent; one can envisage clearly a small piece of Maxwell's dynamical system, and its behavior over a short time; but how these pieces could fit together, global-ly, over a large portion of space, and how they could be conceived to be-have over an extended time, is a baffling problem. Maxwell, therefore, in a truly fine piece of philosophical self-criticism, subjected his theory to an analysis, and showed that the essential contents of that theory—all of it that had a bearing upon known phenomena—could be preserved inde-pendently of any detailed account of the medium, if one only posited cer-tain functional dependencies (for which good evidence could be cited) of the kinetic and potential energies of the medium upon the electric and magnetic field variables themselves.

This situation presented three essentially different sorts of fundamental problem for further investigation: (a) to test the theory—for itself, and in comparison with competing theories—by experiments designed to realize the new phenomena it predicted, and to check any discrepancies among the predictions of the several theories; (b) to perfect the theory, by ex-tending it to those electromagnetic or optical processes which Maxwell did not deal with fully—namely, to the processes in ordinary material media, both at rest and in motion; (c) to perfect the theory in respect of its foundations, by finding out more about the characteristics of the under-lying medium than, according to Maxwell's analysis, is actually revealed in the electromagnetic processes that his theory treats. (Of course, (b) and (c) might well be expected to develop interdependently.) However, this account is a historical oversimplification. It contains no suggestion of the fact that for a generation following Maxwell's publications the Maxwell theory itself was very widely regarded as extremely obscure: that is to say, many competent people found it difficult to determine just what Max-well's theory said.[16]

[16] The historical facts seem to me to merit investigation. One repeatedly finds the statement that Maxwell's theory had no real influence in Germany, and, more gener-ally, on the continent of Europe, until Hertz's work of 1887. Yet Helmholtz, in Berlin,

Part of the difficulty stemmed from the notion of the "displacement current." This appears in Maxwell first as a consequence of a very special characteristic of his detailed model. In the refined version, the displacement current is retained, alongside—not as a consequence of—the general dynamical assumptions about the medium; and Maxwell's exposition (both in his definitive paper and in the later *Treatise on Electricity and Magnetism*) is very cryptic on the matter: one sees clearly neither what motivates nor what justifies the introduction of the displacement current; and the physical content of the hypothesis is obscure (because it is the one point of detail assumed about a medium that is otherwise left vague).[17]

2. The greatest merit for removing the obscurities of the subject belongs to Hertz. The essential general content of his contribution to this question is the following point: Maxwell's theory is independent of any solution to the problem I have stated above under (c)—the problem of a detailed account of the medium. This is the meaning of Hertz's famous aphorism, "The Maxwell theory is the system of Maxwell's equations." That remark, I think, has been treated by philosophers in a somewhat skew perspective. In Hertz, it has a good deal less to do with any quasi-positivist epistemology than with a concrete scientific judgment. The question "What is the Maxwell theory?" was an immediately serious one: scientists were having difficulty in deciding what the theory was, and in

took the theory very seriously as early as 1870, and encouraged a series of experimental tests of points related to it, in his laboratory and by his students; a series that culminated, of course, in the great discoveries of his student Hertz. And Gustav Wiedemann's encyclopedic treatise *Die Lehre von der Elektricität* (4th edition; vol. 4, part 2, Braunschweig: Friedrich Vieweg und Sohn, 1885) presents Maxwell's theory and its extensions by Helmholtz as the culmination of the theoretical development of the subject. It seems possible that local differences of perspective were important here, and that the tendency to suppress the Maxwell theory was most characteristic of Göttingen, where W. E. Weber's influence was dominant. The case of W. Thomson (Lord Kelvin) in England shows, on the other hand, that difficulty in comprehending Maxwell's theory was not restricted to the continent; and the testimony of Poincaré and of Hertz shows clearly that there was a real difficulty, not only for opponents of the theory.

[17] The difficulty appears to have been increased by the fact that Maxwell's *Treatise* was more widely studied than his paper. The *Treatise* is an admirable and fascinating work; but it is moderately challenging, even as a task of scholarship (with the help of good prior knowledge of Maxwell's theory), to locate the principles of that theory in this wide-ranging book. The paper, on the other hand, seems to me remarkably clear, and it is hard to believe that it could have occasioned the kind of bafflement that occurred. Is it possible that the *Philosophical Transactions* were, on the continent in contrast to England, less accessible than Maxwell's *Treatise*, and that this is partly responsible for the slower penetration of Maxwell's ideas on the continent? (Cf. note 16 above.)

particular in formulating what exactly the theory assumes about the ether. Hertz's answer does not exclude from the domain of scientific inquiry the problem of the ether; it says that this problem stands to Maxwell's theory of the electromagnetic field in the same relation as the problem of the cause of gravity to Newton's theory of gravitation, or the problem of the molecular nature of heat to the theory of thermodynamics. That Maxwell's theory is Maxwell's equations is even, in part, a biographical judgment: this is what Maxwell himself offered for us to believe, in contrast to what he left for our further investigation. As applied specifically to the displacement current, Hertz's dictum meant that to understand that hypothesis is to understand the role of the displacement-current term in Maxwell's equations; to test the hypothesis is to test the consequences of the theory that depend critically upon that term—and this, too, of course, it was Hertz's merit and glory to accomplish, thus solving the chief part of our problem (a).

3. Problem (b)—the extension of the theory to processes in material media (of which Maxwell gave only a tentative, preliminary, account)— was chiefly advanced by the electron theory of Lorentz. Passing over all details, I remark only that an ultimate radical consequence of this work was the elimination of the ether altogether as a material medium; that is, a thoroughly negative answer to what I have called problem (c).[18]

One reason why this result was radical is that it implied, for the first time in the history of physics since Newton, a failure of Newton's abductive scheme. The state of affairs is not, in this respect, parallel to the case of gravitation, where the nonexistence of a material medium propagating gravitational force has no such consequence—where, indeed, gravitation without a medium is a paradigm case for that scheme. The difference consists in the circumstance that, whereas Newtonian gravitational interaction is supposed to be instantaneous, Maxwell's fields are propagated with a finite speed; in the more abstract but closely related consideration, that

---

[18] This elimination really had two phases in the development of the Lorentz theory. In the first place, Lorentz assumed that bodies move freely through the ether—i.e., the ether is to no degree carried along by the motion of a body—i.e., that there is one fixed global spatial reference system with respect to which Maxwell's "equations for the free ether" hold everywhere, independently of the local state of motion. Since a material medium capable of internal motions, such as Maxwell envisaged, could hardly possess this absolute global rigidity, Lorentz's assumption seemed tantamount to rejection of such a medium. In the second place, the negative results of all experimental tests for effects of the velocity of bodies relative to the ether, and the modifications of the theory to take account of these results, which culminated in the special theory of

a Maxwellian system with a finite number of charged mass points (indeed, even one with no such points) has infinitely many degrees of freedom; and finally, in the physically fundamental conception of Maxwell's theory, that the dynamical quantities energy, momentum, and angular momentum are attributes of the field. Maxwell applied the dynamics of Lagrange, conceived as a mathematical transformation of the dynamics of Newton, to the field. When the field could no longer be conceived to be a Newtonian material structure, this application proved, in retrospect, to have amounted to a redefinition of the scope and the presuppositions of dynamics.[19]

With reluctance I must, for reasons of space, forgo a more thorough discussion of this whole development, and make only some general remarks that lead me to my philosophical summing-up. It has been suggested sometimes that the notion of the "reality" and "self-subsistence" of the field in the developed Maxwell theory can after all be regarded as a convenient fiction: the theory allows one to compute the (delayed) interactions between charges, and one may regard these as what is real, eliminating the field except as a possibly convenient device of calculation.[20] Now, in fact, there are technical reasons for objecting to this account of the theory; but I cannot go into these here. The point I wish to make is that such a proposed elimination of the field—in contrast to the successive eliminations of characteristics of the ether in Maxwell's self-criticism,

relativity, led to the conclusion that no absolutely distinguished reference system exists. So the ether, having first been deprived of susceptibility to changes in its state of motion, having thus its unique, unchanging state of motion as its one remaining "material" property, finally lost this property as well, and therewith its last hold upon existence.

[19] This redefinition, which was not contemplated by Maxwell, was what Einstein (justly, I think) referred to as "Maxwell's contribution to the idea of physical reality."

[20] Ernest Nagel, in The Structure of Science (New York: Harcourt, Brace and World, 1961), p. 396, makes such a remark, referring for support to the textbook of Mason and Weaver. Emil Wiechert, who contributed significantly to the theory (indeed, who discovered the main principles of the electron theory independently of Lorentz), emphasized at the end of his excellent review monograph Grundlagen der Elektrodynamik that "the electrodynamic phenomena can be viewed quite universally as superposition of interactions between the single material particles, which occur for each pair as if it alone were present." But this remark was not intended by Wiechert to imply an elimination of the field. It occurs in the context of a graceful historical appreciation of Weber's work, as a demonstration that there is no impassable chasm between the conceptions of Weber and those of the new electrodynamics (it should be noted that Wiechert's work appeared as the second part of the Festschrift zur Feier der Enthüllung des Gauss-Weber Denkmals in Göttingen, Leipzig: B. G. Teubner, 1899); and it is followed by the comment that "today we know that the mediation of the intervening medium requires time."

Hertz's subsequent clarification, and Lorentz's and Einstein's development of the theory—constitutes what I should call a philosophically specious quasi-positivist reduction. Philosophically specious because we know that epistemological positivism can eliminate anything: there is no intellectual gain in eliminating everything indiscriminately; and it is unphilosophical to discriminate in some arbitrary way, and then, retaining A, to eliminate B on grounds which if applied to A would dispose of it as well. We have as good reason to believe in the *fields* of the electron theory as we have to believe in the *electrons*. It is for reasons related to this point that I have called Newton's metaphysics deeper philosophy than Berkeley's or Hume's. Like the latter two, Newton rejects the obscure metaphysics of "corporeal substance." But Newton does not conclude that bodies are "unreal," or that their reality consists in our perceptions of them. He concludes rather that their "substantial reality"—in the only sense that truly has significance for us—*consists* in those combinations of properties which we have (gradually) come to know, through experience and the analysis of experience; or rather, more accurately, that what we know of their reality consists in this. Therefore in the case of mind Newton concludes neither, with Berkeley, that minds are the substance in which perceptions subsist (which is obscure metaphysics); nor, with Hume, that minds are congeries of perceptions (which is specious positivism); but far more modestly, with a skepticism that in my view is genuinely philosophical, that our knowledge here is so meager that "it would be rash to say what may be the substantial basis of mind."

I have said that the development of Maxwell's theory broke the Newtonian abductive scheme for natural philosophy. I think it is clear that this breach was made by methods quite in the spirit of Newton; and it is certainly clear that the new developments still fit the wider scheme that I have identified as Newton's metaphysics. The field distributions of the dynamical variables have been liberated from their former dependence upon the field of impenetrability, and even the bond between momentum and velocity has been in a certain sense dissolved; but despite these profound revisions of the fundamental physics, the basic conceptual structure remains the same. That cannot be said, however, in the case of the quantum theory: here, I think, even the metaphysics fails, and has to be replaced by a conceptual structure of a thoroughly new order. One of the points about quantum field theory is that, in its domain, the necessary conceptual structure has in fact not yet been found; I am tempted to say that the quantum theory of fields is the contemporary locus of metaphysical research. A second point is that the generic notion of a field, as (despite my opening remarks) a well-defined physical subject matter—but a kind of structure and process, rather than a well-defined class of phenomena—is itself one important example, or product, of what I think deserves to be called the discovery of the structure of reality. I remind you that Maxwell did already apply dynamics to the electromagnetic field (although he thought this depended upon there being a body there); and that according to the general theory of relativity, the electromagnetic field exercises gravitational attraction. I wish I had had space, in the more detailed and historical portion of this paper, to discuss the work on the Maxwell-Lorentz theory done by Poincaré. That work was of high quality and great value. Poincaré was a very great mathematician—one of the most creative in history; his interest in physical theory was intense; and he was, I think, not inferior in philosophical acumen to Hume. But his work on these subjects is nevertheless replete with failures to make the correct step, or the correct connection, in an intricate nexus of relationships: he clarifies the relationships with great brilliance, and fails to draw the right conclusion. These failures seem to me intimately connected with Poincaré's view of the relationship of theory to reality: with his not having regarded the basic principles of dynamics, the law of the conservation of momentum for example, as principles whose *truth* or *falsity* constitutes a deep characteristic of the world.

III

In the recent literature, there has been a certain tension between views strongly influenced by historical considerations and views strongly influenced by the systematic methodological and epistemological analyses of the logical empiricists. Since my paper has been very much concerned with historical matters, and since I have great respect for the logical empiricist tradition (and consider myself as to some extent within it), I should like before I close to say a word on this matter. It will be convenient to refer to Carnap's notion of a "linguistic framework," and to his distinction between "internal" and "external" questions—a distinction he makes specifically for "questions of existence," but which I shall construe more generally.

I think there is philosophical merit in distinguishing between precise questions and vague questions; in making as many questions precise as

285

possible; in making a given question as precise as possible; and in not ruling out of court such questions as one is unable to deal with in a precise way. I think those notions of Carnap's are intended to facilitate these ends —and do so to an appreciable extent; I therefore think there is merit in them.

Since writing the paper in which he introduced those distinctions, Carnap has come to recognize far more clearly than before (or at least than he had stated before) that in a full account of a highly systematized theory, the principles of the theory are apt to be deeply imbedded in the linguistic structure itself. Without attempting to elaborate this in a technical way in connection with the theories we have been discussing, I shall say that to formulate a theory precisely involves (or may involve) the construction of an appropriate linguistic framework. Within such a framework, a very important philosophical task becomes possible: namely, the analysis of the empirical content of the theory—and beyond this, the analysis of the lines of connection of the theory with empirical evidence. (Of course, the possibility of such analysis depends upon the possibility, within the given framework, of distinguishing what is "empirical evidence.") I have cited elsewhere[21] Newton's analysis of the notions of space and time as affording (when slightly amended) a classic case of the analysis of the empirical content of a set of theoretical notions. I think the self-criticism of Maxwell, completed by the theoretical work of Hertz, can pass as a classic case of the analysis of the *empirically supported* content of a theory. The importance of this kind of analysis seems to me clear beyond reasonable dispute. It is also clear how the results of such analysis may lead to revision of the framework itself, by purgation of redundant elements.

Where Carnap's notions, placed in this context and in relation to the history of science, seem to me deficient is in the treatment of the large-scale evolution of theories. For Carnap, the theory of induction itself is to be developed *within* a linguistic framework; in fact, Carnap's view of the situation appears to be that *when the empiricist program has succeeded, a framework will have been chosen more or less for good*—then only internal questions will remain. Such a prospect is hard to reconcile with the mutual dependence of frameworks and theories, and with what I believe most will agree is the unlikelihood that we are going to have—soon, at any rate—a general theory that will no longer undergo change. If, on the other hand, it is agreed that the program for a definitive "language of science," what-

---

[21] See the paper cited in note 5 above; the remark occurs on p. 197.

ever its prospects, has at least *not yet* achieved its aim, and that new theories may require new frameworks, then there is a danger that the internal/external distinction may lead to the neglect of important large questions that span the development of theories—on the grounds that these are questions external to the frameworks, and that only within a framework are clear criteria of meaning and truth available. Such an outcome would converge in a curious way with the tendency among historically oriented commentators to find in the succession of theories something akin to diverse artistic genres, as between which meaningful critical comparisons are dubious. I have called this a danger, because I think this tendency is wrong; and I have no doubt that Carnap, and empiricists generally, would agree with me. The general (although unsystematic) point of view that I would urge as the correct one here I have already tried to suggest—in distinguishing philosophically specious positivist criticism from analyses of constructive value like those of Newton or Hertz. No attempt to delimit, systematically and globally, the procedures and notions that are empirically legitimate—from "Hypotheses are not to be regarded in experimental Philosophy" to the verifiability theory of meaning and beyond—has really succeeded. To say this is not to depreciate the efforts that have been and are being expended upon this task—which may yet succeed, and which have contributed much of value though short of success; but it is to deprecate the appeal to programmatic notions as if the program had been realized: this leads to specious criticism. On the other hand, "hypotheses non fingo" and the verifiability theory of meaning both had a valid core; this I earnestly hope we do not forget. It has been possible for scientists, in creating, criticizing, modifying, and revolutionizing their theories, to apply what is valid in these principles, despite the lack of an adequate precise general formulation. There is no obvious reason why philosophers of science cannot do the same.[22]

COMMENT BY GERD BUCHDAHL

Comparing the papers by Professors Stuewer and Stein, and including in this comparison the contributions from Professor Schaffner as well as

---

[22] The author, heartily acknowledging the courtesy and consideration of the editor of this volume, feels constrained to record his objection to the circumstance that the conditions of publication have deprived him of final authority over the stylistic details of his paper. The author therefore reserves the right to publish elsewhere a version more fully to his own satisfaction.

my own, one notices a remarkable degree of unanimity concerning a certain thesis: It is insufficient to characterize scientific theories and their development, scientific thinking and its logic, by reference to certain "global" frameworks alone, be such frameworks conceived as "linguistic" (as in Stein's criticism of the Carnapian approach), or regarded as approaches toward generalized theories of confirmation and falsification. Instead of this, all the papers under review, emphasizing as they do the controversies that enter into the actual development of scientific theories, and the intellectual as well as personal or social stresses to which such a development is often exposed, draw attention to the need for an analysis of scientific theories to exhibit a much "finer structure" than is taken into account when one concentrates—as is usual—on the relative successes of theories to "account for the data."

Partly, this is due to a new recognition of the interplay between "data" and "conceptualization." As Stein affirms (and my paper implies and illustrates): an "empiricist program" and a "conceptual framework" cannot be kept rigidly apart; "the principles of a theory are apt to be deeply imbedded in the linguistic structure itself." More particularly, all these papers emphasize the presence of additional criteria of appraisal and of additional layers internal to the theories; they lay stress on how scientists do in fact argue and appraise their developing ideas at the time of their growth. And we need not for the present ask the (admittedly) important question how such additional criteria and considerations, which undoubtedly appear to such scientists rational *at the time*, can be shown to *be* rational when subjected to a philosophical metacritique.

In my own paper I had distinguished three groups of criteria which, respectively, relate to the conceptual articulation, constitutive (empirical) foundation, and "architectonic" determination of theories. Stuewer and Schaffner both emphasize the last of these, involving the criteria of consilience ("theoretical context adequacy") as well as relative simplicity and preferred explanation types, all requiring balance against the "empirical" component (Schaffner's "experimental adequacy"). Stein, on the other hand, pays particular attention to what he calls Newton's "metaphysical level of thought" (equivalent to my own "conceptual explication") with respect to the concepts of matter, force, and field. And we all agree that mere attention to absence of falsification, power of prediction, and similar global criteria—while necessary—is insufficient to make intelligible the nature of the controversies (mentioned as illustrations in these various papers) which those scientific episodes involved.

I will first make a few comments on the examples from the architectonic component. Stuewer's example of Richardson's theory of the photoelectric effect is interesting because it presents us with an explicit case of preference for a specific explanation type: "macro-theory" is preferred to "micro-theory"; though no doubt the latter is rejected not only because of a distaste by the adherents of macro-theory (as in the days of Ostwald and Mach) for "unobservable entities," but also because the particular unobservables in question here involve "paradoxical properties."

Stuewer uses this also to illustrate the deeply conservative tendency of most, including some of the greatest, scientists. If some philosophers of science therefore characterize science as the need to strike out with "bold hypotheses," it is important to be aware of a contrary tendency, not to say "need." Nor is such conservatism at all irrational. Evidently it links with the criterion of "theoretical context adequacy." As Stuewer puts it: though the photon concept is not entailed by the phenomenon of the photoelectric effect, and the latter (taken by itself) is capable of alternative explanations of which Stuewer's paper mentions a great many, nevertheless, the photon concept would at present be regarded as a "valid" one because it "derives its validity from an interlocking theoretical and experimental matrix."

It seems clear that such a criterion cannot be absolute; that it is neither necessary nor sufficient, if it is regarded as bestowing "conclusive validity" on some scientific hypothesis. Nevertheless, the opposition to Einstein's hypothesis on the part of Lorentz, Thomson, and Sommerfeld, though in the event it proved unavailing, was clearly a perfectly proper expression of paying regard to theoretical context criteria. So one should not be tempted to speak of an "invalid opposition" (or some equivalent pejorative expression) merely because of this knowledge of hindsight. And it should be added that it is not easy to appraise at any particular time whether such controversies as those discussed in Stuewer's paper will or will not yield additional predictions and thus eventually turn out to be (to use Lakatos's terminology) examples of either "progressive" or "degenerate" research programs. The debate between the Cartesians, the Leibnizians, and the Newtonians may not in their day have yielded anything "progressive," but since the attempt to mediate between the conflicting parties yielded the explicit formulations of a field approach on the part of Boscovich and

Howard Stein

Kant (as shown in my paper), it seems clumsy formalism to exclude such controversies because they seem incapable of yielding something of immediate positive value.

Consider for instance Stuewer's report of Lorentz's refusal to accept the hypothesis of light quanta because this was ("seemed"?) inconsistent with the phenomena of interference and diffraction. This shows again that the context criterion is not conclusive, if only because—in accordance with the Duhemian thesis—it is always possible that additional alteration in background theory may result in reconciling recalcitrant "old" theory (here: the interpretation of the diffraction phenomena) with a newcomer among theories. But none of this shows that the context criterion does not possess prima facie relevance. Indeed, in the tussle between the old and the new there are always a number of possibilities, a fact which is usually concealed from those who frame their methodologies "from the outside," bent on considering only the successful outcome of some theoretical or practical dispute. Thus Stuewer's paper shows that scientists usually try to reconcile a new "experimental" result (such as the photoelectric effect) with traditional existing theory (context criterion being used again) by making modifications to supplementary principles (such as the precise nature of atomic structure) which to them *at that time* were less well understood, rather than incur a clash between a well-entrenched theory and a new phenomenon. And these are "fine structure" activities, descriptive of the rationale of scientific thinking, that need somehow to be incorporated in our appraisal of the logic of science.

Stuewer's example of Richardson's theory illustrates a further important point: the very possibility of formulating a theory in terms of a preferred explanation type is often *felt* by scientists to bestow explanatory value. ("Feelings" may here again be rejected as having no bearing on the question of "rationality," but one may remark that the existence of such "feelings" of "explanatory power," attaching to some explanation type, has sometimes been used to define the notion of theoretical explanation on the part of philosophers of science, e.g., N. R. Campbell.) Now if we grasp that some particular explanation type may as such bestow explanatory power we shall be more sensitive in our reactions to Newton's "solution" of the problem of gravitational attraction at a distance. Stein, in his paper, calls Newton's opinion that gravitation occurs in virtue of an immediate action of God "disconcerting," and he intimates that no historian can be feeling anything but "puzzled," not to say "annoyed." Now as I

290

point out in my own paper, this reference to God's "action" should really be understood as being to "final causes," and thus to a physicotheological context. Two conclusions at once follow: First, Newton's solution is far from being so puzzling, since it is simply an actual fact of scientific development that it seeks to express its theories in terms of preferred explanation types. (Descartes opposed mechanical theories to "sympathetic relations," "formate souls," and "animate form." Fermat opposed a teleologically conceived concept of "least time" to Descartes's "incomprehensible" hypothesis of unobservable light particles moving through media with speeds proportional to their densities.) Second: "Bodies attracting by an immediate law of God" does not mean the same thing for Newton as "direct action at a distance" in the physical sense, contrary to what is claimed by Stein. For as I suggest in my paper, the teleological foundation which Newton postulates is meant as a "third possibility" mediating between the scylla of action by matter on matter at a distance and the charybdis of an ethereal mechanism whose modus operandi had proved inexplicable for twenty to thirty years—always assuming my interpretation of Newton's "metaphysical view" (contrary to Stein's) of the conceptual impossibility of distance action. Newton's solution may not add to our nomothetic knowledge of gravitation; it does tell us something about the forms and functions of scientific theories.

I shall return to Newton in a moment. There is, however, a problem to which I have already alluded; it is one that is brought into sharp focus in connection with the criterion of preferential explanation types. Here, if anywhere, it seems oppressively obvious that these "fine-structure" criteria have to be deeply historicist in kind. To be sure, it is trivially true that all criteria have a historical ingredient in that they are employed and applied at a given moment in time by putatively rational beings. But it seems blatantly obvious that what is an acceptable explanation type (e.g., teleology) to one scientist may be otiose to another. And as already mentioned, the fact that they appear to the actors involved to be part of a rational activity does not prove that they express in fact a feature of rational activity, understood in the normative sense. All this appears to introduce a degree of relativism into our account of scientific thinking.

In reply, I can urge only this much: First, that none of these criteria yield conclusive results. Second, that they can at best be no stronger than any inductive criteria, by which I understand the fact that such criteria validate an inference or "projection" on the basis of existing theory and

Howard Stein

evidence; they do not *entail* a result that has to stand, come what may. Third, that none of our criteria can be employed independently of the others, and that they must be given different weightings at different times, sometimes ignored, or suppressed, sometimes powerfully emphasized.

But one might go further; one might admit that there is a kind of "spectrum" of different "degrees of depth" enjoyed by these various criteria. If one for instance posits a causal hypothesis, then such a hypothesis will have to satisfy certain obvious criteria that have been elucidated by writers on inductive logic, from Bacon through Hume to Mill and our own time. Similarly, putative falsification clearly demands some modification to a hypothesis or its branching background theories, or even a reconsideration of the data given by the experiment (as in the case of Richardson and Compton, reported by Stuewer, page 259). Further along the scale would be considerations of concept formation: Leibniz and Huygens constructed gravitational theories rivaling those of Newton in response to a different conceptual evaluation of terms like matter, force, causation. Still further along there would be architectonic criteria like those of relative simplicity, theoretical context adequacy, all the way down to preferred explanation types, existence of important analogies, and so on. In general, one might say that it does not follow from the fact that none of these criteria are (a) conclusive or (b) essential that they should not play a *critical* part in the evaluation of theories, as members of a "family" of criteria.

I now turn to Professor Stein's paper, and as I have already indicated, I very much sympathize with his method of supplying a "finer structure" of Newton's physical thought, though I am not entirely clear about the denotation of his distinction between "heuristic" (or "research"), the "fundamental" and the "metaphysical" levels which he proposes. From what is said on page 278, it appears that "heuristics" corresponds to the formal mode of presentation of the *Principia* together with its explicit definitions. As I understand it, Stein further claims, however, that a proper *interpretation*—and to what "level" does such an "interpretation" belong?—of the procedures at the "research" level can be shown to involve the concept of the "field" as a "fundamental category for scientific explanation," i.e., of gravitational phenomena. The chief ground for this contention seems to be the fact that Newton's various formulations and laws involve "the notion of a disposition" and of "a contrary-to-fact conditional," as well as the notion of a single "order" of the phenomena involved, such that we can speak of "one field" surrounding a given body.

292

I am not certain of the demonstrative force of this contention. The dispositional and contrary-to-fact nature of laws is usually one of their universal characteristics, and thus seems too general a feature to permit of Stein's special conclusion. Perhaps before making this inference, one ought to explore additional criteria of different kinds of fields that have from time to time been suggested as necessary and sufficient conditions for granting the "reality" of a field. (Cf. the special criteria attached to a Faraday field in Mary Hesse's *Forces and Fields*.)

However, shelving this point, it is interesting to find Stein contending that at Newton's "metaphysical level of thought" we meet a similar obliteration of the dualism between matter and force (i.e., the *reduction of* matter to the field), and the emergence and generalization of a corresponding field concept—thus evidently supporting Stein's *interpretation* of the implications of the thought of Newton at the "research level." On the other hand, according to Stein, "fundamental theory" surprisingly does retain the dualist scheme, awarding an independent and prior place to "impenetrability" (and I would add: to inertia also), while requiring a special supplementary divine action to establish gravity, or alternatively, the intervention of the various ether mechanisms proposed by Newton from time to time. Moreover, Stein concludes, "fundamental theory" expresses the dualist viewpoint which is "the system of physics that Newton was led to develop," and which his immediate successors followed, whereas encapsuled in the bosom of the theory there lay always the unitary scheme, later to unfold itself in the approaches of Maxwell.

To find my way around these classifications I must try and compare them with my own "fine-structure" composition of Newton's thought. My "constitutive level" conflates Stein's "research" as well as "fundamental" levels and implies, as does his, a dualism of matter and force, with the merely de facto introduction of gravitational attraction. But I lack an "interpretation" of the "research level" as implying a monistic field action. All I say is that "interpretations" of formal structures are open-ended. On the other hand, my theory of Newton's "conceptual explication" (Stein's "metaphysical level") is that it supports the *dualist* and not the *monist* viewpoint. This being the case, one requires (according to my reading of Newton) a further contribution (my "architectonic component") in order to establish a rationale for gravitational attraction. The need for such a component in Stein's view is not very pressing since his "interpretation" of Newton's "research" as well as of his "metaphysics" implies a reduction of matter to the field. Therefore Stein is not inclined to take very seriously Newton's theological (or teleological) escape route as a means of relieving what he himself acknowledges nevertheless to be a "tension" in Newton's thinking.

Now I think that the "interpretation" of the "research level" that is here advocated might have seemed less plausible if Newton's "metaphysical thought" were not likewise construed in a monistic vein. Stein claims, however, to have a number of arguments which supposedly back up this version of Newton's thought. Perhaps the most global is the one noted on pages 272 and 277: Newton is there shown as holding that any knowledge of the action of matter (whether through attraction or impact) is based on experience. The sting in this is supposed to lie in the exclusion of an alternative possibility, viz., that we might gain the requisite knowledge by way of a kind of rational or intuitive insight. From this it would then follow that the affirmation of the empirical basis of physical knowledge entails the *universal* rejection of any possibility of grasping or conceiving the action of matter on matter, whether this be by impact or action at a distance. And it would then also follow that where Newton says that action without contact is inconceivable he would not be meaning to say that it does not happen, any more than he would want to argue for the physical impossibility of impact because *that* is inconceivable.

Now it is true that some writers did hold this view. In my paper for this volume I suggest that it is one of Locke's views, and that it may be found also in Maupertuis and Mill, among others. But it is not at all clear whether this was Newton's own understanding. As Stein admits, the Bentley letters ostensively contradict this: Why should Newton have *singled out* action by matter on matter "without an intervening material medium" as being "inconceivable"? (See also my comments on the Bentley passage in note 30 to my paper.) On Stein's reading of Newton it becomes only too clear why—as he notes on page 273—he cannot understand how Newton, "despite his principles, as I have explained them" (i.e., the doctrine of the *universal* unintelligibility of action, here attributed to Newton), can want to introduce ether hypotheses, let alone the "puzzling" and even "annoying" teleological escape! To overcome the obvious contradiction, Stein goes further: he suggests that we may safely interpret Newton's complaint in the Bentley letters at the inconceivability of action at a distance to be a universal one by "leaving out the words 'without mutual contact.'" I must say that by such methods almost any conclusion could be established!

At any rate, it seems to me that Newton's true views resemble more those of the Locke quoted on page 215 of my paper (not altogether consistent, as is usual with Locke, with the views mentioned on page 224). That is to say, Newton simply *had* no clear views on the notion of action on impact, and certainly did not appreciate the bearing of the empiricist basis of our knowledge of impact on this case. However, insofar as he had any views they seem more like the ones Mr. Stein attributes to Newton's "fundamental theory," viz., that impenetrability was the "first basic property of bodies," in the sense that "interaction by contact . . . is a *necessity*," a view echoing my Locke passage just mentioned. So Newton has not as yet grasped the implication of the empiricist doctrine that "knowledge by way of experience" excludes "rational insight." It is not surprising therefore (as Stein notes) that Newton nowhere calls "impenetrability" a force, since it is beginning to look as if he *was* a dualist at the "metaphysical level," espousing a dualism of matter and force, moreover, that seemed inimical to the "addition" of attraction *except* as an experiential induction. (For which reason, we *need* of course the ether or the teleological escape; neither of these would make any sense if it were the case that Newton's "metaphysical views" were those of Stein's construction.)

Stein, in his supplementary attempts to establish his viewpoint of Newton's field metaphysics further speaks of "impenetrability fields"; and this "activist" interpretation is then supported also by an earlier argumentation that *all* forces, as capable of *producing* change of motion, are "active principles." Does he want to say that the force of inertia falls under that description? His argument seems to leave this possibility open, but once again it is important to recall that Newton emphasizes that the three laws of motion (which, as Stein rightly points out, *jointly* define "inertia") are "passive principles" (so called by Newton), as contrasted with the "active principles" of gravity, fermentation, and the like. So again it appears that impenetrability as well as inertia is placed precisely at a level different from that of gravity, in line with the "fundamental view" here ascribed to Newton.

The final argument that Stein advances is Newton's view (found in the early *De gravitatione*) that those regions of space that God has rendered "impenetrable" as well as "movable" also have the "property, or power, of affecting our minds" when copresent with our "brain." I cannot see the force of this point. First of all, as Stein rightly notes, Newton observes that this "interaction" is once again "obscure"—a technical way in those days of saying that it is nonintuitive, like gravitational interaction. Second, and like the latter, it is not sufficient to find places where Newton *speaks* of interaction, for this language could quite well belong to the levels of "research" or "fundamental theory" (my "constitutive level"). After all, philosophers like Leibniz, who explicitly deny interaction at (what they call) the "metaphysical level," nevertheless claim the right to employ this language at the level of their *phenomena bene fundata*. But language employed at the "fundamental" (or theoretical or constitutive) level cannot be used to explicate the "metaphysical" views of an author, and I therefore very much doubt whether "Newton's metaphysics of body reduces the notion of matter to the notion of field."

Still, these are small-scale disagreements compared with my fundamental agreement which the former really imply, namely that scientists' thinking proceeds at a number of different levels, has different sets of components. If the views Stein attributes to Newton's "metaphysics" are really those of the later Locke, of Maupertuis and Mill (as regards the question of "unintelligibility"), and of Leibniz, Boscovich, and Kant (as regards the interpretation of matter as force), this is relatively unimportant. What is important is to realize that beneath the surface of the "research level," with its foundation upon the "data" or "observations," with its attempted verifications or falsifications, there are other levels which tend to articulate a given theory and infuse it with a kind of historical dynamic which at one moment may incorporate traditional, at another revolutionary, features, and which at any time may come to affect the conceptual articulation of the ostensive theory at the "research level."

I want to conclude with one or two remarks on Stein's general philosophical appraisal of Newton vis-à-vis that of other British philosophers such as Berkeley and Hume, and their modern descendents, positivists and instrumentalists. I cannot see much foundation for the praise of the youthful Newton's ascription of "substantial reality" to the "attributes" of bodies, unless a philosophical articulation of what all this implies had been given also. Now this is precisely what was done in the philosophies of Locke, Berkeley, Hume, and Kant. The story is a long one (I have tried to tell a little of it in my recent volume, *Metaphysics and the Philosophy of Science: Descartes to Kant*), so I will only note one single point. The difficulty as it appears to these philosophers is just how to preserve some kind of "reality" for entities which (and here, perhaps wrongly, they agreed with Newton) were being characterized by the logical label of "attributes." So they construe them as "existing" relative to a mind, whether they regard them as "ideas" (Locke, Berkeley), or "impressions" (Hume), or "appearances" (Kant). Now it is evident from the Queries to the Opticks that Newton himself was not a realist but rather something of a "subjective idealist": we know things by the "images" through which they appear to us—and it is not difficult to think of these "images" as representatives of the "qualitative" aspect of things rather than their "obscure substantial" being. Now such a viewpoint can hardly be appraised as "deeper philosophy" when compared with that of Berkeley and Hume. It is just simply no more than setting out the problems and questions and leaving later philosophers (as is their proper job) to get on with them.

Nor should we be deceived by the language of these philosophers into believing that their dispute is about the "reality" of things in the ordinary, nonmetaphysical sense of this term. (Here, too, there are many "levels" at which the discussion proceeds!) I agree with Stein that global attacks on "reality" remove too much so as to provide us with plausible analyses of the logical status of problematic scientific entities, such as the electron, or the electron field. But it is not even certain whether such radically phenomenalist thinkers as Berkeley and Hume ever contended for such conclusions. I need only refer to Hume's praise for Newton's hypothesis of an ether (mentioned on page 225 of my paper) and his frequent sympathetic allusions to "force and energy" as exemplifying a much greater largesse on the part of such philosophers than is usually accredited. And in my work referred to above I have shown that Berkeley certainly believes very frequently that he is not denying the existence of hidden structures, and even (sometimes) "insensible corpuscles." What he (and Hume) for the most part are concerned with is the therapeutic exercise of persuading us that the reality of a thing is not impugned just because there is no hidden metaphysical foundation for it to be found anywhere.

But, in conclusion, it must also be said that just because, say, a positivist analysis can be too radical in its rationing of the ontological population of the universe, and must hence be construed as really being neutral to the question of the existence of scientific theoretical entities, it does not follow that such entities do exist, any more than that they do not. I mean: the falsehood of the contentions of the positivists would by itself be insufficient to demonstrate the correctness of existentialist claims on the part of the philosophers of science. Different kinds of analyses would seem to be required in order to achieve this end, analyses of which the technical portions of Stein's paper give such a brilliant example.

COMMENT BY MARY HESSE

Professor Stein has given an excellent example of the marriage of historical and philosophical considerations in his analysis of the concept of "field." I am not competent to judge the validity of his interpretation of Newton in relation to the more "archaic" versions of Newton's metaphysics coming to light in the manuscripts (see, for example, the papers of J. E. McGuire and P. M. Rattansi referred to in my contribution to this volume). But I do wonder whether, even from the perspective of later physical categories, he has not blurred unduly the differences between Newton's conception of field and that of the nineteenth century.

Stein takes the fundamental notion of field to involve the dispositional, or counterfactual, power inhering in spatial points to exert a force on a test particle if one were present at the point (though it is not). Under this analysis Newton's and Maxwell's theories are indeed both field theories. But it is not clear that either theory *must* be interpreted in these dispositional terms. Stein asserts that without dispositions there is no induction, but this conclusion is controversial in philosophy of science, and there are cogent arguments to the effect that induction from factual instance to factual instance does not involve necessary reference to laws understood as quantified over counterfactual as well as factual instances. On Stein's criterion it seems that it would be impossible to describe any lawlike action at a distance without the intervention of a "medium of potential."

There are, however, alternative definitions of "field" which probe more deeply into their physical character, and serve to make important distinctions between Newton's action at a distance and Maxwell's (and Einstein's) fields. Stein does not mention Faraday, but it was he who gave the first and most subtle analysis of the difference between actions at a distance and actions through continuous media. Faraday had broadly speaking three tests for continuous action through space: (1) Can transmission of action be affected by changes in the intervening medium, as for example magnetic action by iron or electric action by dielectrics? (2) Are the lines of force curved? (3) Does the action take time to cross space? For gravitation Faraday thought the answer to all three questions is no; for all other physical actions the answer to (1) and (2) is yes, and, although Faraday did not know the answer to (3) in respect to electric and magnetic induction, it turned out in the light of Maxwell's theory to be yes in all cases except Newtonian gravitation, and even this exception was removed in Einstein's theory of gravitation. Thus instantaneous transmission became the primary criterion for action at a distance, and a finite velocity the criterion for field action. Indeed finite velocity *demanded* the presence of energy in the medium in a more than dispositional sense, in order to remain consistent with the conservation of energy. As Maxwell put it, "If something is transmitted from one particle to another at a distance, what is its condition after it has left the one particle and before it has reached the other?" His answer was, it becomes the energy of the intervening medium, that is, the field.

REPLY BY HOWARD STEIN

The striking affinity of theme noted by Dr. Buchdahl extends, in the case of his paper and mine, to a considerable part of the particular historical matter treated. I am grateful to him for the discussion he gives in his own paper, and for his comments on mine. That we agree on some important matters, both historical and philosophic, and disagree on others, will be apparent to all readers. It seems neither necessary nor desirable to attempt here to elucidate all points of disagreement; it may be left to our readers to weigh the considerations involved, and to ourselves to reflect further on the issues. I shall confine myself to a few points on which Buchdahl's comments seem to me to show either that my exposition has been insufficiently full and clear, or that a philosophical issue exists between us which requires sharper definition.

First let me say—not by way of argument, but statement of where I stand—that my general philosophic position is significantly closer to the empiricism of the Vienna Circle than is Buchdahl's (see part III of my paper). I should put the matter this way: Adopting Peirce's distinction of "abduction" (the process of finding hypotheses) and "induction" (the process of testing them), I think that the array of considerations that Buchdahl refers to as constituting a "fine structure" of inquiry is chiefly relevant to the former. I also think that this abductive phase of science is more important than the Vienna Circle, in its practice, seemed to hold; so I agree with Buchdahl's thesis of the importance of this fine structure. On the other hand, of the inductive phase, despite all the difficulties (the lack of a global theory of inductive inference, and the complexity of factors that have to be taken into account), I believe that the essential point does remain "the relative success of theories to 'account for the data'" (Buchdahl, page 288). "Fine-structural" grounds of dissatisfaction with a theory that accounts best (of all known theories) for (all known) data can be a significant motive for the further abductive process; but it is not a valid ground for *rejection* of the theory before a better one is found. This is what I think Newton meant by his statement that "Hypotheses are not to be regarded in experimental Philosophy," and by his Rule IV: "*In experimental philosophy we are to look upon propositions inferred by general induction from phenomena as accurately or very nearly true, notwithstanding any contrary hypotheses that may be imagined, till such time as other phenomena occur, by which they may either be made more accurate, or liable to exceptions. This rule we must follow, that the argument of induction may not be evaded by hypotheses.*" What history very remarkably shows, in relation to this issue, is that although no one has succeeded in formulating clear and acceptable general rules of inductive inference, in practice serious difficulty in deciding which of the actual theories does best account for the data has never persisted long. (A fairly recent and very instructive example is provided by Einstein's view of quantum mechanics: he greatly disliked the theory, and believed that it must eventually be superseded; but he fully acknowledged its current supremacy in accounting for the data.) One reason for insisting upon this point is my belief that, over several centuries now, the history of the inductive success of theories has occasioned very significant changes in the ends aimed at in abductive inquiry. This seems to me an important part of what has been called "the structure of scientific revolutions"; and I believe that a considerably higher degree of objective rational control has obtained in this process than has sometimes been claimed—essentially because of that feedback from induction to abduction.

Turning now to points of detail, I must amplify my evidently too compressed remarks on gravitation as "an immediate Law of the Creator of the Universe," which according to me (pages 273–274) does, and according to Buchdahl (page 291) does not, mean the same thing for Newton as direct action at a distance. I do not at all intend by my claim to impugn Newton's seriousness as a theologian; the role of God is undeniably, I think, from Newton's point of view, a real one (although *not* one that is to be appealed to as a premise in physical reasoning—i.e., in what Newton calls "experimental Philosophy"). The claimed equivalence of "immediate Law of the Creator" with "direct action" consists, rather, in the fact that *all interactions of bodies are, for Newton, ruled by "Laws of the Creator of the Universe"*; hence "direct action" is either *an entirely vacuous notion*, or else—and this seems the more convenient idiom—ought to mean nothing but *action governed by an "immediate" Law of the Creator*. Thus, I take the actual collision of ultimate particles (which according to Newton is a rare event) to be governed by "an immediate Law of the Creator"; but impact of elastic bodies (which, Newton holds, always derive their elasticity from an internal dynamical structure)—or gravitation, if it should prove to be caused by an ether or by streaming particles—I take to be governed by a *mediate* law: i.e., one that is reducible to, or derivable from, more fundamental laws.

What evidence is there that Newton's view of these matters in general, and of the "directness" of the process of impact in particular, is what I have said it is? For one thing, we have Newton's account of the nature of body in the *De gravitatione*, which I hope I have sufficiently described. What is crucial for our present question is not the reference to interaction with mind (Buchdahl, page 295), but the fact that the establishment of mobile regions of impenetrability, and the establishment of *all* their laws of motion and interaction, are represented equally as fiats of God. (According to Leibniz, gravitational attraction at a distance, resulting in the deflection of a body from uniform rectilinear motion, would amount to "a perpetual miracle." It follows from Newton's view that the word "miracle" is no more applicable to gravitational than to inertial motion.) That *De gravitatione* is an early writing is a point I have considered in my note 12; but let us turn to the quite late statement in Question 31 of the *Opticks* (Dover edition, p. 400): "All these things being considered, it seems probable to me, that God in the Beginning form'd Matter in solid, massy, hard, impenetrable, moveable Particles, of such Sizes and Figures, and with such other Properties . . . as most conduced to the End for which he form'd them." Not, then—I think I have warrant for saying, contra Buchdahl— "the ether or the teleological escape": at the point where Newton, whether early or late, refers to teleology, it affects equally hardness, impenetrability, inertia, and all. Again: does Newton reject "any possibility of grasping or conceiving the action of matter on matter, whether this be by impact or action at a distance" (Buchdahl)—or more precisely, reject such a possibility, as I claim, "in any other sense than that we have observed interactions of [these] type[s] to have occurred"? See *Opticks* (Dover edition), page 389: "All Bodies seem to be composed of hard Particles: For otherwise Fluids would not congeal; [etc., etc.]. Even the Rays of Light *seem to be* hard Bodies; for otherwise they would not retain different Properties in their different Sides. And therefore Hardness may be reckon'd the Property of all uncompounded Matter. At least, *this seems to be as evident as the universal Impenetrability of Matter.* For all Bodies, so far as Experience reaches, are either hard, or may be harden'd; *and we have no other Evidence of universal Impenetrability, besides a large Experience without an experimental Exception.*" In his explication of the third rule of philosophizing, in book III of the *Principia*, Newton makes an analogous statement, not only about hardness and impenetrability, but about the mobility and inertia and even the extension of bodies.

I think this, taken all together, provides a rather strong case for my conclusion that Newton's statement to Bentley, while *strictly accurate*, would remain accurate if the words "without mutual contact" were deleted. I may nevertheless be wrong; but I do not think I am irresponsible. Buchdahl's comment that "by such methods almost any conclusion could be established" seems to me a little unfair. And to his preceding remark, "On Stein's reading . . . it becomes only too clear why—as he notes on page 273—he cannot understand how Newton . . . can want to introduce ether hypotheses," I have to say that in the passage referred to I do not at all "note" such a thing, and I do not believe it to be the case; I rather claim to elucidate just this question, on the basis partly of Newton's fundamental conceptions (where impenetrability is given—as I have said, page 278—a somewhat special status), and partly of his whole program for understanding nature (which demands that *all* possibilities of "Tieferlegung der Fundamente" be explored).

Two somewhat isolated points call for brief comment: (1) I find surprising Buchdahl's characterization of Newton as "not a realist but rather something of a 'subjective idealist.' " He bases this upon Newton's account of sensation in the *Opticks*: namely, that we "know" things, in sensation, by the "images" generated, in our minds, by motions in our "sensoria"; which motions are conveyed to the sensoria by impulses propagated along the nerves from the sense organs (themselves stimulated by interaction with external bodies). Now, the language of traditional epistemology and metaphysics is far from precise, and traditional usage far from stable; but I think it quite unusual, even within the wide variability of that usage, to call such a view "subjective idealism." (2) Buchdahl refers to my characterization of "forces capable of producing change of motion" as "active principles," and asks: "Does he want to say that the force of inertia falls under that description?" The answer is no: as I have said, "One . . . force is both universal and intrinsic to all bodies: the *passive* principle of motion, or force of inactivity, *vis inertiae*. . . . But the object of greatest attention . . . is the class . . . of *active principles*. These are the forces that can produce, and more generally can change, motion. . . ." This is perhaps a rather scholastic point; but the distinction is Newton's, and I think I have been faithful to it; see *Opticks* (Dover edition), page 397: "The V*is inertiae* is a passive Principle. . . . By this Principle alone there never could have been any Motion in the World. Some other Principle was necessary for putting Bodies into Motion . . ."; *ibid.*, page 401: "It seems to me farther, that these Particles have not only a V*is inertiae*, accompanied with such passive Laws of Motion as naturally result from that Force, but also that they are moved by certain active Principles, such as is that of Gravity, and that which causes Fermentation, and the Cohesion of Bodies."

Buchdahl expresses doubt about my distinction between what I call "heuristic detail, or 'research,' " "fundamental theory," and "metaphysics," in the conceptions of Newton. The distinction is not one which I should like to make bear any great systematic burden; it is intended only ad hoc, for local convenience. I do not understand "heuristics" to imply (as Buchdahl suggests) "the formal mode of presentation of the *Principia* together with its explicit definitions"; I mean rather—in connection with gravitation—the considerations which were instrumental for Newton in *discovering and "proving"* (i.e., testing against experience) the theory of gravitation. By "fundamental theory" I mean the comprehensive conceptual scheme in which Newton places the discoveries he has made, and in which he hopes to place essentially all subsequent discoveries. In distinguishing a third level of Newton's thought as "metaphysical," I have in mind the circumstance that, although the conceptions of this third level can be characterized in the words I have just used about fundamental theory, these conceptions do not—in Newton's work—function in a cru-

Howard Stein

cial way in his physics. Bodies and forces are pretty much indispensable for Newton; but whether bodies are taken (to use a modern jargon) as values of individual variables, or as higher-order logical entities (in older language: as substances, or as attributes of some sort), is a question which he can and does set aside in his physical investigations. That Newton perceived the separability of such issues from his own physics is, in my judgment, an instance of critical philosophic acumen. That he addressed himself to those issues in the way *De gravitatione* shows he did is, in my judgment, testimony to his philosophic originality and depth. And that such issues, or such notions, can move from the (perhaps seemingly nugatory) "metaphysical" level, and can acquire a role in physics itself, is one of the fascinating lessons of the history of science and philosophy; this is the bearing of my remarks about the "discovery of the structure of reality."

I come now to an issue raised by Dr. Hesse as well as Dr. Buchdahl. They both question the appropriateness of my use of the word "field" in analyzing Newton's theory. To me, this is not a fundamental issue, as I have tried to make clear in the opening sentences of my paper. That the notion of field in modern physics has dimensions of significance not present in the notion that I refer to by that word in the case of Newton is quite true—and I have discussed this point on pages 282–283 and page 285. I certainly do not regard the notion of field as something merely "dispositional": even Newton's rudimentary "field of impenetrability" (as I call it) is "ideally testable," i.e., "dispositional," only "in a Pickwickian sense"; Maxwell's electromagnetic field is sometimes to be detected experimentally by an eye, rather than by a test charge or current!

Buchdahl's and Mary Hesse's references to Faraday's criteria, however, seem to me unhappy. Faraday (though I did not mention him in my paper) is one of my personal heroes; but I do not think his criteria for "continuous action through space" are very good. The second criterion in Dr. Hesse's list—"Are the lines of force curved?"—is clearly nugatory: except in the most special and trivial cases, lines of force *will* be curved. (That Faraday thought lines of gravitational force always straight is one of the rare instances in which his lack of formal mathematics betrayed him into outright error.) The first criterion—"Can . . . action be affected by changes in the intervening medium?"—played a strong positive motivating role in Faraday's work; but as a *criterion* it is deficient, because it is unclear. Newtonian gravitational action *is* affected by disposing new masses

304

through the "intervening medium"; and both in Faraday's time and our own, the effect of iron or of dielectrics on the magnetic or electric field is accounted for in an analogous fashion, as due to the disposition of magnets (currents) or charges in the medium. This leaves only the third criterion as having real substance. Passing from Faraday to Maxwell, this criterion—finite velocity of propagation—comes into significant play; and is related (as Dr. Hesse justly remarks) to the attribution of energy to the field (cf. page 283 of my paper). It is worth emphasizing once more that, for Maxwell, this feature was connected essentially with the notion of a *material* medium. The next-to-the-last sentence of Maxwell's *Treatise on Electricity and Magnetism* reads: ". . . whenever energy is transmitted from one body to another in time, there must be a medium or substance in which the energy exists after it leaves one body and before it reaches the other, for energy, as Torricelli remarked, 'is a quintessence of so subtile a nature that it cannot be contained in any vessel except the inmost substance of material things.' "

I do not, as I have said, consider the question how we should use the word "field" a fundamental issue; but how the word is used by the scientists we study is a point to be treated with due care. For Maxwell, attribution of energy (and other dynamical quantities) to the field is not a criterion of "a field theory"; he says (cf. my page 279) that his theory may be called a theory of the electromagnetic field *because it has to do with the space around the electric or magnetic bodies.* That *dynamical properties* are attributed to this space is rather what makes his theory "*a dynamical theory* of the electromagnetic field." This, to Maxwell, implied the existence, in the space concerned, of "matter in motion, by which the observed electromagnetic phenomena are produced." As I have briefly explained, it was only post-Maxwell that the idea of *a dynamical field theory without a material substratum* came to be seriously entertained (an idea outside the scope of what I have called Newton's "fundamental physics," but within the scope of his "metaphysics").

It may be of some (at least philological) interest to pursue the question of when—and with what precise meaning—the word "field" was first used in something like its present sense in physics. Felix Klein (*Vorlesungen über die Entwicklung der Mathematik im 19ten Jahrhundert*, volume II, page 39) says that the first such use known to him is by William Thomson (Lord Kelvin) in 1851; it occurs in Thomson's paper "On the Theory of Magnetic Induction in Crystalline and Non-Crystalline Substances,"

Howard Stein

Philosophical Magazine, March 1851—reprinted in his *Reprint of Papers on Electrostatics and Magnetism* (London, 1884), Article XXX; see section 605 of the latter book, pages 472–473:

Definition.—The *total magnetic force* at any point is the force which the north pole of a unit bar magnet would experience from all magnets which exert any sensible action on it, if it produced no inductive action on any magnet or other body. . . .

Definition.—Any space at every point of which there is a finite magnetic force is called "a field of magnetic force;" or, *magnetic* being understood, simply "a field of force;" or, sometimes, "a magnetic field."

The question of the role played in Newton's positive work by the notions I have emphasized—their role at the "heuristic" or investigative, and especially the *inductive*, level—may be divided into two parts. The first (easier, and less important) is historical: did Newton use these notions? I have met some skepticism from historians, especially about my construction of "the accelerative quantity of a centripetal force"; let me therefore review what seem to me decisive texts. First, Definition VI: "The absolute quantity of a centripetal force is the measure thereof that is greater or less *according to the efficacy of the cause propagating it from the center through the surrounding regions.*" (Vis centripetae quantitas absoluta est mensura ejusdem major vel minor pro efficacia causae eam propagantis a centro per regiones in circuito.) This does not (for Newton) imply, necessarily, finite velocity of propagation; but it does imply affection of "the surrounding regions." Of course, it is possible that only the surrounding *bodies* are meant; however, that possibility is weakened or destroyed by the following (in the second paragraph after Definition VIII): "We may . . . refer . . . the accelerative force to the *place* of the body [acted upon], as a *certain efficacy* [Motte renders: "a certain power or energy"], *spread from the center to the several places round it, to move bodies that are in them.*" (. . . licet . . . referre . . . vim acceleratricem ad locum corporis, tanquam efficaciam quandam, de centro per loca singula in circuitu diffusam, ad movenda corpora quae in ipsis sunt.)

The more important part of the question—raised with particular clarity by Dr. Hesse—is whether Newton *had to* use such notions. That he did have to use them (in some form or other) I have argued on pages 267–269 of my paper, and in note 4. I am puzzled to read in Dr. Hesse's comment that "Stein asserts that without dispositions there is no induction"; and that "this conclusion is controversial in philosophy of science." For the first part, I do not find that I have said this; I have said that Newton's inductive argument—which is, it seems to me, a clear test case for any theory of induction—makes essential use of a dispositional term or a contrary-to-fact conditional. For the second part: the claim I make, I argue for. Now, when one offers arguments or evidence, either they are supererogatory, or their bearings are upon a controversial issue: the aim is, at best, to settle a controversy, but at least to clarify one. A premise may fairly be challenged as controversial; to admit this as an argument against a conclusion is to preclude any progress in philosophical discussion.

A really careful analysis of Newton's whole argument for universal gravitation is more than I could undertake in the paper, or can here; but a review and amplification of some salient points seems in order. It is well known that Newton's treatment of the force on the moon is a critical juncture in his argument; for this provides the only link connecting the planetary motions with the terrestrial force of weight. To establish the link was essential, both because (1) a major part of Newton's achievement was precisely this classification of the astronomical motions under the same head as the phenomena of weight—their exhibition as effects of the same "natural power" (this conclusion was accepted, for example, by Huygens, and regarded by him as a very great discovery: one of its implications was that any mechanical theory of weight must ipso facto be a mechanical theory of planetary motion as well; and must incorporate the law of the inverse square of the distance); and because (2) Newton's more far-reaching conclusion—that all bodies have a power of gravity proportional to their mass, affecting other bodies with centripetal accelerations inversely proportional to the squares of their distances—was arrived at only through considerations based upon that classification.

Now, the proposition that the impressed force giving rise to the moon's orbital acceleration is its weight—Proposition IV of book III of the Principia—is established by comparing the moon's acceleration with the acceleration of terrestrial falling bodies. The former is to the latter as 1 to 3600. The radius of the moon's orbit is to the radius of the earth as 60 to 1. How do these data allow us to infer anything of interest? We see that the ratio 1:3600 is the inverse square of the ratio 60:1; but it is certain a priori that the one ratio must be some power of the other: what follows from this special power? The answer, of course, is that this relation is significant because it coincides with the relations found for accelerations and radial distances among the planets, and among the satellites of Jupiter (later also among the satellites of Saturn; but only one satellite of Saturn was known at the time of the first edition of the *Principia*). Nevertheless, the question remains: what follows from this agreement of relationships? That x is to y as Y^2 is to X^2 is a propositional schema that has innumerable true instances; we cannot ordinarily draw strong inductive conclusions from them. For instance, the fact that the intensity of illumination by a point light source varies inversely with the square of the distance has no bearing upon the relation of the moon to earthly bodies.

There is more than one technical way to present the point that is crucial here. Indeed, having given one argument in the first edition of the *Principia*, Newton himself later added, in a scholium, an alternative version. One can put it that *if the moon* (or a small piece of the moon) *were brought down to the region of the earth's surface, the force upon it would increase continually in the same proportion as that by which the square of its distance from the earth's center diminishes;* and hence, by the data already cited, would—when the body reaches the earth—be just the force that we should then call its "weight." The proposition in italics is the conclusion that Newton arrives at, by induction, from the astronomical evidence: partly from that about the moon in its orbit; but, as I have explained, chiefly from that about the other bodies in the solar system. I have formulated the proposition here using a contrafactual conditional. Newton, in his first version, says rather: "Since *that force,* in approaching the earth, *increases* in the inverse duplicate ratio of the distance . . . a body in our regions *falling with that force* must describe . . . in the space of one second [the same distance that freely falling bodies do in fact describe in one second]." What, in this formulation, is meant by "that force"? Either—I claim—"the force that would be experienced by the moon if brought down," i.e., what I have said above, or "the field (whose intensity is an 'accelerative force') which the astronomical evidence shows to exist about all central bodies of the solar system." Clearly, there is no substantive difference between the two constructions of the phrase. That the second is perhaps closer to Newton's own way of conceiving his argument is suggested not only by his account of "accelerative force," which I have cited, and by his use here of the indicative mood ("that force . . . increases"), but still more by the scholium he added as a more ample explication of the argument. What he does in that scholium is to imagine *many new test bodies* introduced to (as I put it) "explore the field." He says, "Suppose several moons to revolve about the earth, as in the system

308

of Jupiter or Saturn; the periodic times of these moons (by the argument of induction) would observe the same law which Kepler found to obtain among the planets; [etc.]" Again we have, as in my first formulation, a contrary-to-fact conditional, whereas in Newton's first version, on my reading, a dispositional property is implicit. Point (1) of my note 4 is that one or the other of these (equivalent) devices—or, perhaps I should have said, something else, also equivalent to them!—is needed in order to formulate the proposition in question.

This conclusion may seem to lack demonstrative force; but, for me, it is based upon my conception of what the proposition at issue is: I see it as intrinsically involving such a component. If anyone is of another opinion, however, he ought to defend that opinion by showing how else to express the proposition.

What appears to me a more serious possibility is that the proposition we have been discussing might be bypassed: that is, that Newton's main conclusions might somehow be justified, on the basis of the data, through a train of inductive arguments of a more subtle structure, making no appeal to notions like field or disposition or "nomological connection." I do not see how to do this; but I don't pretend to settle deep questions about the logic of induction by that criterion—I have tried to make this clear in note 4. What I do maintain as of some significance is that, however the difficulties about inductive inference are resolved, the resolution must take account of, or be compatible with, the three points set forth in note 4. The first of these I have now defended at some length. The second, I think, is noncontroversial, and also presents no real difficulty. For the theory of inductive inference, it is point (3) that I consider most important. But this point has not been challenged; and I believe that what I have said bearing upon it in the paper and the note (and in some of the considerations advanced in the preceding portions of these remarks) is sufficient.

One further remark should still be made, however, concerning the relationship of point (1) to point (3). I have said that the possibility may exist of bypassing Newton's proposition that if (a piece of) the moon were brought to earth, its orbital acceleration would become the acceleration due to gravity, as a premise of his further inductive argument. I have given, in outline, the inductive argument which leads to that proposition as a conclusion; and have referred to the proposition as justifying, in Newton's logical account, the further conclusion that (to put it so for once) gravity is the cause of the moon's orbital acceleration. What I now want to point

309

out, as illustrating what the theory of induction has to show how to deal with, is that this proposition of Newton's serves for him in an extremely straightforward way as *a premise to a further inductive conclusion* (of undoubted empirical value). In his argument for Proposition IV, Newton can assert that a piece of the moon, if brought to the earth, would experience a force increasing as the inverse square of the distance, for his astronomical evidence supports this. He *cannot*, at this stage, assert the symmetrical proposition: that *a terrestrial body, raised to the height of the moon, would diminish in weight* in that same ratio; for he has no evidence at all to support this. Once Proposition IV is established, however, and the acceleration field of Galileo has been identified with the acceleration field of Kepler, this new proposition *can* be asserted. Of course it has now become possible—three centuries later—to put both propositions to experimental test (in the opposite order): we have lifted terrestrial bodies to the moon, and have brought lunar bodies down to earth; and in doing so have confirmed both propositions directly.

KENNETH F. SCHAFFNER

Outlines of a Logic of Comparative Theory Evaluation with Special Attention to Pre- and Post-Relativistic Electrodynamics

I. Introduction

It would be false to say that case studies drawn from the history of science have had no influence on the philosophy of science in the past ten years. The contributions of the late N. R. Hanson (1958), and S. Toulmin (1961), P. K. Feyerabend (1962, 1965), and T. S. Kuhn (1962) utilize examples drawn from the science of Aristotle, Buridan and Oresme, Galileo, Newton, Lavoisier, Dalton, Maxwell, and Einstein. On the basis of such examples these current authors develop their views of the nature of scientific thought, and though they by no means agree in all particulars with one another, their general "historical" approach has raised some perplexing questions for philosophers of science who have based their views on the more "logical" analyses of, say, Carnap, Hempel, and Nagel.[1]

By focusing attention on the richness and adaptability of historically discarded scientific theories, and on the many instances of theory competition that exist in the historical record, these "historical" philosophers of science have called into serious question many of the central doctrines of earlier philosophers of science.[2] In varying ways they have suggested that:

AUTHOR'S NOTE: I am indebted to Professor Dudley Shapere of the University of Chicago for very helpful discussions on many of the points discussed in this paper. I should also like to thank Professor Ernest Nagel of Columbia University and Professor Manley Thompson of the University of Chicago for reading a version of this paper and for making most useful comments. Grateful acknowledgment is made to the National Science Foundation for support of research.
    [1] The distinction between the "logical" and the "historical" approaches to the philosophy of science is made in Shapere (1965). For representative selections of the former approach see Carnap (1956), Hempel (1965), and Nagel (1961).
    [2] These central doctrines had not of course gone uncriticized before the last decade. In fact some of the work of the earlier critics of the logical empiricist approach, such

311

Kenneth F. Schaffner

1. Experiments in science are not theory neutral but have interpretations placed on them which are theory dependent. This claim immediately gives rise to the question whether a "crucial experiment" which would choose between conflicting theories is ever possible, as the experiments are, on this view, interpreted in radically different ways within the different competing theories.

2. There is no common "observation language" as many of the logical empiricists believed there was for different theories applying to the same domain of inquiry. Rather, observational terms do not possess a meaning per se, but are only meaningful in connection with a theory. The traditional position has accordingly been neatly inverted: theories are understandable per se, observational terms only admit of a partial interpretation through theories.[3]

3. Recalcitrant experiments, or experiments which prima facie falsify a theory and resist reinterpretation which might save the theory, do not cause a theory's rejection (or, it is implied, even its modification—cf. Kuhn, 1962). All theories are considered at all times to be in some difficulty. When conditions are right a theory is rejected in toto and replaced, apparently for irrational reasons, by an alternative theory which is either inconsistent or "incommensurable" with the former. Scientific change is viewed as fundamentally cataclysmic and discontinuous, as opposed to the traditional account of an uneven but cumulative progress of science. Though one of the conditions for change is that an alternative theory be available which in some notoriously obscure sense is "about" or "associated" with the "same" experimental domain as its less fit predecessor, it is clear that this claim is in conflict with the comments on the discontinuity and incommensurability of theories and observations cited above. In spite of the difficulties of unsatisfactory formulation, however, this claim does contain plausible arguments against any general "falsifiability" approach to science.[4]

4. The historical school, if I may be permitted to call it a "school," has also suggested that one has to give the standard term "theory" a new and broader meaning so as to be faithful to the way it functions in the history

as Wittgenstein (1953) and Popper (1934, but hereafter cited as 1959), served as the source and stimulants of the historical school's critiques.

[3] Cf. Feyerabend (1965).

[4] One clear exception to this criticism of a "falsifiability" approach is the recent work of I. Lakatos (1968, 1970), who has proposed a "sophisticated falsification" approach which is to bridge the gap, as it were, between Popper and Kuhn.

312

of science. Feyerabend (1962) and Hanson (1958) have respectively proposed that a theory be understood as "a way of looking at the world" and "what makes it possible to observe phenomena as being of a certain sort . . ." Toulmin (1961) has introduced the phrase "Ideal of Natural Order" to articulate how he understands a scientific theory, and Kuhn (1962) has proposed the term "paradigm" to stand for that complex of theory, methodology, standards, and metaphysic which a scientist works with, implying it is a good deal more than a theory that is tested by an experiment. The logical empiricist analysis of a theory is treated as irrelevant, if not incorrect, for an adequate philosophy of science. Such a view of "theory," when taken in conjunction with the three earlier theses, indicates the extent to which the reaction of the "historical" school to logical empiricism has led them in the direction of a new "idealism."

Such theses as the four cited above present a very different picture of science from what most of us hold or once held. The cumulative view of science and science's "objectivity" seem to disappear. The control over speculation exercised by observation and experiment appears severely weakened, and the rationality of science and the progress of science are denied.[5]

Though the history of science has, then, exerted a significant influence on the philosophy of science, the claim of a contrary influence is easy to deny. With several exceptions, such as Joseph Clark's attempt to consider the implications of a logical empiricist's conception of a scientific theory for the history of science,[6] historians of science do not avail themselves of the contributions of the philosophy of science. Some historians do, of course, like L. Pearce Williams,[7] read Kant and Schelling—but this is hardly contemporary philosophy of science.

Nevertheless historians do face what can legitimately be called "philosophical" problems. When they are forced to assign a date to the discovery of oxygen, they must involve themselves in conceptual analysis to determine whether "dephlogisticated air" is to count.[8] When a historian is asked to determine whether it was Lorentz and Poincaré, or rather Einstein, who first articulated the special theory of relativity, certain questions about the nature of scientific theory and of alternative interpretations of

[5] See Scheffler (1967) and Shapere (1964, 1966) for general comments and critiques of these and other implications of the historical school.
[6] Cf. Clark (1962).
[7] Cf. Williams (1965, 1966).
[8] See Kuhn (1962), pp. 54ff on this problem.

313

Kenneth F. Schaffner

mathematical formalism must be considered. When a historian wishes to give an account of the relevant circumstances involved in the replacement of one scientific theory by another, he must know something about the logic of theory selection as it affects the behavior of the practicing scientist.

This paper is an attempt to construct a logic of comparative theory evaluation which will resolve some of the paradoxes associated with science's apparent rationality which were raised by the historical school of philosophy of science. It is also hoped that it will assist in solving some of the disputes that have arisen in historical circles over the priority of the discovery of the special theory of relativity. By absorbing certain aspects of the major theses of the historical school as outlined above, and by re-presenting some of the logical empiricists' conceptual tools in a new form, this paper will attempt to articulate a characterization of science which permits the elaboration of a logic of comparative theory evaluation.

II. An Analysis of Scientific Theory and Experiment

Before I begin an inquiry into the logic of comparative theory evaluation, it will be necessary to digress slightly in order to discuss some basic ideas which will occur in my analysis.

A. Antecedent Theoretical Meaning

The term "scientific theory" will be used in this paper to refer to a set of sentences of universal form. These sentences will contain a class of "primitive" nonlogical terms which are given what I call *antecedent theoretical meaning* by drawing on antecedently understood domains of discourse.[9] These domains are not intended to provide "models"—in any of the traditional philosophical senses—for the theory. An entity term, such as "gas molecule," has its meaning *created* by sentences providing an appropriate analogical description, usually drawn from several diverse domains including branches of mathematics. Such meaning-creating sentences, which are to be found in the text of scientific articles and monographs, are usually characterized as analytic. The antecedently meaningful terms are then interrelated in sentences which have the character of hypotheses, that is, which are usually considered to be synthetic.[10]

---

[9] See my (1969b) for more details on this notion of antecedent theoretical meaning.

[10] The distinction between analytic and synthetic sentences becomes somewhat relative in this analysis and will depend on the extent to which certain properties are considered definitional. There are also numerous instances in the history of science in which definitions change. See Putnam (1962) for a further discussion of relevant examples and issues.

James Clerk Maxwell (1867) gave an example of this procedure in connection with the term "gas molecule" when he wrote in one of his fundamental papers on kinetic theory: "In the present paper I propose to consider the molecules of a gas not as elastic spheres of definite radius, but as small bodies or groups of smaller molecules repelling one another with a force whose direction always passes very nearly through the centers of gravity of the molecules, and whose magnitude is represented very nearly by some function of the distance of the centers of gravity." Maxwell continued, further elaborating his notion of a "molecule," and also proposing a law or specific hypothesis for the interaction of these entities: "If we suppose the molecules hard elastic bodies, the numbers of collisions of a given kind will be proportional to the velocity, but if we suppose them centers of force, the angle of deflection will be smaller when the velocity is greater; and if the force is inversely as the fifth power of the distance, the number of deflections of a given kind will be independent of the velocity. Hence I have adopted this law in making my calculations."

We see in the quotations above how the meaning of the term "gas molecule" is carefully created, and how it is possible to postulate modes of behavior of the entity. Such meaning as is conferred on the term, of course, is not forever fixed, and could change if a different theory utilizing, say, a different notion of "gas molecule" were formulated. It should also be mentioned that insofar as the effects of the actions of the gas molecules come to be analytically associated with the notion of "gas molecule," the adjunction of certain types of correspondence rules to a theory about gas molecules can extend and modify the meaning of the term "gas molecule." I shall have more comments to make concerning correspondence rules below.

B. Correspondence Rules and Theory Interdependence

In order to provide experimental control and tests, as well as to enable theories to account for observationally accessible states of affairs, a theory must also have associated with it an additional set of sentences, which I shall term C-sentences; these state how the entities and/or processes described by the theory's axioms ultimately affect our sense organs. Let us call O-sentences those sentences which describe an intersubjectively testable experience, such as the "seeing of alternating light and dark bands or fringes through a telescope" or the "hearing of a tone." We shall also characterize as O-sentences those sentences which name an entity that is observationally accessible without the immediate use of artificial instruments. The second class of O-sentences covers such things as photographs with curved lines on them (bubble chamber *photographs*) and the numerical total, i.e., the readable number *sign*, on a particle counter.

The C-sentences then connect theoretical terms with "observational" terms, and are thus strongly analogous to what traditionally have been termed "correspondence rules." They differ from the usual analysis of correspondence rules in that these C-sentences are further analyzable into chains of sentences, each element of which characterizes a state of affairs, and which causally or nomically implies the next element in the chain. The causal or nomological implications are warranted by well-confirmed or well-corroborated auxiliary theories of science. Thus in the C-sentence of the Bohr atomic theory connecting the "theoretical" notion of an "electron transition between orbits" with the "observational" notion of a "spectral line," the theory of physical optics is appealed to in order to account for the behavior of light in prisms and telescopes which makes the "spectral line" associable with the "electron transition." Accordingly, the Bohr theory employs, as an auxiliary theory, the theory of physical optics. A view of the theoretical interdependence of science, not dissimilar to Duhem's (1906), is a consequence of such an account of C-sentences.[11]

The O-sentences mentioned above are not meant to be necessarily atheoretical. In many cases they may involve universal concepts, such as numbers, and are best understood as Popper understands his "basic statements"—as provisional but conventionally accepted reports of intersensually testable events.[12]

The concept of an O-sentence introduced here is thus characterized by several definientia which are possibly epistemologically distinct, e.g., such sentences were noted as "describing an intersubjectively testable experience," as "naming an entity which is observationally accessible," and as "reports of intersensually testable events . . ." Though I believe that these various notions could well profit by more detailed analysis, it seems that the characterization of O-sentences provided here is sufficiently clear so as to provide the loose type of experimental constraint discussed in the later philosophical and historical sections of this paper.

[11] I have discussed this analysis of correspondence rules more extensively in my (1969b), and have also applied it in the context of the reduction of biology to physics and chemistry in my (1969c).

[12] Cf. Popper (1959), especially chapter 5.

C. Theory and Experiment

On the view to be defended here, which is in many respects very much like both Popper's and the "historical" school's position, any experiment takes place within a theoretical context. A theory (or perhaps theories) is applied to a relatively specific situation and yields, in accordance with the antecedently meaningful sentences and associated C-sentences, O-sentences as described in the previous section. This means that it is the theory, by virtue of its antecedent theoretical meaning and its antecedently understandable hypotheses, and the associated theories which indicate which experiments are relevant to its claims about the world. This is the case whether the experiments have been done earlier, even in connection with a discarded theory (unless an important parameter is now claimed to have been overlooked in the earlier version of the experiment), or are yet to be performed.

We can see the features of this analysis exhibited in the Michelson-Morley interferometer experiment. Lorentz's (and also the relativists') analysis of the Michelson-Morley experiment will be discussed later, so suffice it for now to say that if the ether is assumed stationary in the universe, and if the Michelson interferometer is rotated, then there will be a slight displacement of the fringes viewed through the telescope of the interferometer. There were associated well-confirmed optical and cosmological theories which were taken to be beyond question in the context of this experiment: it was not doubted that a wave disturbance of light would travel in a straight line at a velocity approximately equal to 1.86×10^5 miles per second and obey the laws of reflection and refraction, nor was the "hypothesis" that the earth was moving about the sun at a velocity of 18 miles per second questioned.

The observation reports or O-sentences which are the output of this experiment are representative of what a careful observer, whether he be a proponent of Lorentz's theory, Stokes's theory, or Einstein's theory, would see through the interferometer's telescope as the interferometer is rotated. (How and why such agreement can occur will be discussed below in the "experimental adequacy" section.) The language of the O-sentences is, it is true, mathematical, for it mentions fringe shifts of so many tenths of a centimeter. Nevertheless the theoretical component of the O-sentences is clearly minimal, as is supported by the willingness of observers of different theoretical commitments to make the same observations. Furthermore, and this is most important, the O-sentences *need not agree* with the ex-

317

Kenneth F. Schaffner

pectations of the theorist, as the results of the Michelson-Morley experiment did not agree with the expectations of Lorentz,[13] nor may it be possible for the theorist to *easily* and nontrivially modify his theory so as to accommodate the unexpected O-sentence. Accordingly, experiments can exert some control over theory, even though the results of the experiments, the O-sentences, may have theoretical elements infused in them.[14] There are, of course, greater difficulties of control if the theoretician can easily modify his theory to accommodate the O-sentence(s), but this is an issue which will be taken up in the section in which ad hocness and simplicity are analyzed, where it will be argued that such considerations restrict the theoretician from making facile accommodating moves.

This analytical interlude which I have now completed was necessary to set the stage for later accounts of the interaction between theory and theory and between theory and experiment that will be developed in the next two sections. It is hoped that the rather schematic aspects of this logical analysis will become more distinct and complete in the context of the specific examples to be discussed later.

III. Hertz on the Foundations of Mechanics and the Logic of Comparative Theory Evaluation

In the long introduction to his posthumously published (1894) monograph, *The Principles of Mechanics*, the distinguished physicist Heinrich Hertz attempted to outline the reasons why he found both the old Newtonian account of mechanics and the newer energeticist formulation of mechanics inadequate. Hertz understood each formulation of mechanics to be an "image," or *Bild*, constructed from its own basic ideas or primitive terms by combining those ideas or terms into fundamental principles or propositions or axioms. Each set of ideas and principles had to be adequate for deriving the whole of mechanics, but the ideas and principles could vary: some primitive terms in one image being defined terms in others, some principles of one image becoming theorems in a different image.

In particular, Hertz considered three different images of mechanics: (1) the Newtonian-Lagrangian image which Hertz took to be based on the

---

[13] See my (1969a) and also chapter 6 of my (1970) for a detailed discussion of Lorentz's difficulties with the results of the Michelson-Morley experiment.

[14] This analysis, which permits disagreement between the consequences of a theory and experimental outcomes, partially accords with Feyerabend's account discussed in his (1965), pp. 214–215, but is, I think, a more intelligible analysis of the process representing the prelude to falsification.

ideas of space, time, mass, and force, and to employ as principles either Newton's laws of motion or Lagrange's generalization of d'Alembert's principle, (2) the energeticist image, which founded mechanics on the ideas of space, time, mass, and energy, and which Hertz understood to use Hamilton's integral principle of least action, and (3) Hertz's own image of mechanics, which had as its basic ideas only space, time, and mass, and which utilized a principle of minimal curvature of path of a mechanical system in motion.

It is not my intention to consider Hertz's mechanical investigations any further. What I wish to do, rather, is to examine the metascientific ideas which he constructed in order to assess the relative merits of these three competing images.

Hertz introduced three metascientific concepts: *Zulässigkeit* or permissibility, *Richtigkeit* or correctness, and *Zweckmässigkeit* or appropriateness. Hertz believed that only if we could obtain a clear conception of what properties were to be ascribed to the images for the sake of permissibility, correctness, and appropriateness could we "attain the possibility of modifying and improving our images."

The concept of permissibility reflects a type of Kantian bias on Hertz's part, for it involves an a priori and immutable assessment of an image or a scientific theory. In connection with permissibility Hertz wrote:[15]

We should at once denote as inadmissible all images which implicitly contradict the laws of our thought. Hence we postulate all our images shall be logically permissible [i.e., self-consistent]. . . .

What enters into the images in order that they be permissible is given by the nature of our mind. To the question whether an image is permissible or not, we can without ambiguity answer yes or no; and our decision will hold good for all time.

Correctness on the other hand is associated with experiments and observations:

We shall denote as incorrect any permissible image, if their essential relations contradict the relations of external things. . . .

What enters into the images for the sake of correctness is contained in the results of experience, from which the images are built up. . . . Without ambiguity we can decide whether an image is correct or not; but only according to the state of our present experience, and permitting an appeal to later and riper experience.

[15] All quotations from Hertz in these pages are from the Introduction to his (1894) monograph.

Kenneth F. Schaffner

Appropriateness is defined in the following terms:

Of two images of the same object that is the more appropriate which pictures more of the essential relations of the object—the one which we may call the more distinct. Of two images of equal distinctness the more appropriate is the one which contains in addition to the essential characteristics, the smaller number of superfluous or empty relations—the simpler of the two.

What is ascribed to the images for the sake of appropriateness is contained in the notations, definitions, abbreviations, and, in short, all that we can arbitrarily add or take away.

It is most difficult to determine the appropriateness of an image, as Hertz noted when he wrote: "We cannot decide without ambiguity whether an image is appropriate or not; as to this differences of opinion may arise."

IV. A Generalization of Hertz's Categories of Comparative Theory Evaluation

Though Hertz's three categories in terms of which to compare competing theories are suggestive, and were used by him with excellent results in the domain of mechanics,[16] I feel that they must be somewhat generalized if they are to accommodate the advances made by twentieth-century physicists, historians, and philosophers.

A. Theoretical Context Sufficiency

The notion of "permissibility," understood in the sense given above in the quotations from Hertz, lacks the flexibility that is needed to describe the scientific revolutions which physics has gone through in this century, as, for example, the revolution in our views of geometry and space associated with Einstein's general theory of relativity. There is, happily, a suggestion in Hertz which can be followed up in order to attain a more adequate characterization of this dimension of theory comparison.

In what was almost an aside before he commenced his critique of the permissibility of Hamilton's principle of least action, Hertz remarked: "In order that an image of certain external things may in our sense be permissible, not only must its characteristics be consistent amongst themselves, but they must not contradict the characteristics of other images already established in our knowledge." Now this expansion of the view held earlier indicates that one of the significant aspects of permissibility is the amount

[16] See Sommerfeld (1952) for a critique, though, of Hertz's account of the Newtonian-Lagrangian image.

320

of concordance between a theory to be assessed and the corpus of accepted scientific knowledge. If I may give this assessment category, as so generalized, a name, I shall label it *theoretical context sufficiency*, and shall accordingly be discussing the historical theoretical context within which a theory is to be judged.[17]

But we must do more for this dimension of theory comparison than simply broaden it to include the theoretical context of the time. It must also be stressed that a decision made in terms of the "theoretical context sufficiency" of a theory is *not* good for all time, for it may well turn out that a theory which is inconsistent with currently accepted theories is in fact correct, and that it is the various accepted theories constituting the theoretical context which require revision. This relativization of judgments made of the "theoretical context sufficiency" will not necessarily result in the type of relativity suggested by the historical school of philosophers of science, because of the joint manner in which the other categories of comparative theory evaluation function. We shall return to this difficulty in the context of a specific example below.

It must also be stressed that an assessment of the theoretical context sufficiency *will* vary from scientist to scientist, depending on his perspective and knowledge. Any philosophy of science must permit these "subjective or individual fluctuations" from the *consensus* of science at the time, though I think that it is a mistake to turn such fluctuations into a normative element of a philosophy of science as both Kuhn and Feyerabend seem to have done. We shall have an occasion to examine such "subjective fluctuations" in connection with Einstein's view of Lorentz's theory in 1905 and Lorentz's view of Einstein's in 1909.

B. Experimental Adequacy

Hertz's notion of "correctness," though it already has some flexibility and dynamism built into it, as it provides for revision in the assessment of correctness based on new and later experiments, is not yet adequate for our purposes. Recall the earlier account which I gave of the relation between theory and experiment. In that account experimental results or observation reports were admitted to be infused with theoretical meaning. Observation reports were, analogous to Popper's basic statements, taken to be observationally accessible provisional stopping points captured

---

[17] This "theoretical context" is similar to what other philosophers have termed the "background knowledge." Cf., for example, Shapere (1970).

in some language that is, by convention, admitted to be descriptive of those points.

Disagreement between a theory and an experimental outcome, then, is on this view a conflict between some complex of elements drawn from the corpus of accepted scientific knowledge, the theory under test, and sentences describing the interpreted sense experiences associated with the experiment. This claim is nothing new, for it was pointed out by Duhem (1906) early in this century. Nevertheless there are difficulties which are raised both by the Duhemian thesis and by my acceptance of some of the historical school's claims regarding the infusion of theoretical meaning into observational sentences, and it would be worthwhile to indicate how the view being developed here (and in section II above) can be fitted with a thesis of experimental control over the acceptance of a theory. Accordingly what I wish to do now is to consider how, given the apparent flexibility of theories and their relations with experiment, a category of anything like "correctness" can continue to function.

The difficulty of introducing a modified form of Hertz's category of correctness, a category which I shall term "experimental adequacy" inasmuch as "correctness" has unfortunately restrictive connotations, is twofold. First, as was discussed briefly above, the meaning of the O-sentence, since it is admitted to contain "theoretical elements," might well change from theory to theory, thus ruling out an intertheoretic common experimental base to which competing theories must conform. This difficulty was sidestepped earlier, when I simply noted that certain O-sentences, such as are the result of the Michelson-Morley experiment, seem to have a common element from any of the competing theories' points of view. But simply pointing out the prima facie common element is not sufficient; for we have a right to ask why there is this common element so as to be able to guard against the illusion of O-sentence meaning change. In other words, some general philosophical support is required for the apparent overlap of O-sentence meaning.

Second, we must also resolve a different difficulty which is connected with the issue of experimental control over theory acceptance. This is the issue of the *relevance* of an experiment (or observation report): does an experiment which is closely associated with theory T_1 in domain D also *have to be associated* with theory T_2 also applied in domain D?[18]

---

[18] Of these two difficulties, Feyerabend (1962, 1965) stresses the first, and Kuhn (1962) the second, though neither would disagree with either.

We shall attempt to make use of the distinctions developed in section II above to resolve these difficulties.

Suppose that we are given two competing theories, say a Newtonian-like corpuscular optics and a Young-Fresnel wave optics. Further, suppose that O_a represents an O-sentence describing a fringe pattern on a certain field of view, say on a screen, a, located behind two narrow parallel slits illuminated by a monochromatic light source. I shall assume that such a sentence contains (1) terms which are "ostensively" definable (to provide the "primary sense" noted below), (2) logical terms which serve as connectors, and (3) mathematical terms which are theory neutral. Our difficulties, then, will concern the meanings of the first set of terms, which we may call O-terms or O-predicates.

We now attempt to distinguish two general types of senses ascribable to the O-predicate in this situation. The first or primary sense is that associated with the referent, the bandlike appearance one experiences in learning the meaning of "optical fringe phenomena." The second sense is that which is associated with the O-term by virtue of the term's being embedded or hooked into a theory through correspondence rules or C-sentences such as were discussed earlier. Such embedment is achieved if the O-term appears in an O-sentence which is the terminal element of a C-sentence chain initiating in a theory. Only in the second sense do we talk of seeing the fringes *as evidence for wave optics* (or for corpuscular optics).

The primary sense of the O-term then is common to both the Newtonian theorist's use of scientific language and the wave theorist's use. Furthermore, it is common not by virtue of being part of some "higher-level background theory," as Feyerabend (1965) would suggest, but rather because both our higher-level theories of wave optics and corpuscular optics contain low-level theories (supposing such O-sentences to be theoretical) which are connected with them by C-sentences. This low-level theory uses the theoretically uninteresting language of spots, fringes, colors, and the like, in as precise a way as possible by measuring the fringes with scales and attempting to construct intensity scales of colors and brightness. More overlap than this, however, is not necessary, for overlap only need occur at the points where both competing theories make contact with experience. The terms, in their *primary* observational sense, simply do not change meaning with the ease which Feyerabend suggests they do. Feyerabend can only maintain his thesis by conflating the two senses of the O-terms which I have distinguished.

Such a thesis, then, contends that a "withdrawal" or "retrenchment" is possible when scientists of different and conflicting theoretical orientations are analyzing the output of an experiment. If disagreement on the meaning of the O-sentence, in its primary sense, is present, it is possible, such a thesis would maintain, that the sentence could be further analyzed so as to reach a common level. This new common level would then be the primary sense of the O-sentence vis-à-vis this theoretical conflict. Thus the distinction between primary and secondary senses is relative, but relative to theoretical comparison, and not to theories in isolation. Though it is logically possible that there be no common basis no matter how far the primary sense is sought for, this seems most unlikely, for it is not supported by any of the examples in the literature designed to exemplify differences in the meaning of observation terms relative to different theories. Though this thesis of withdrawal or retrenchment in the search for a common empirical base is perhaps put to new ends here, it is to be found in its essentials in Popper's (1959) discussion of the "relativity of basic statements."

An answer to the second difficulty mentioned above, that of the change in the relevance of an experiment due to a change in theory, is more complex and requires to a certain extent a more pragmatic argument. Again I refer to Popper (1959) who asserted: ". . . a theory which has been well corroborated can only be superseded by one of a higher level of universality; that is, by a theory which is better testable and which in addition, contains the old, well corroborated theory—or at least a good approximation to it."

Though it is precisely this and similar theses which are effectively attacked by Feyerabend (1962, 1965) and Kuhn (1962), there is still some merit to such an inclusionist claim, though it can be overstated and misinterpreted. The merit accrues to the claim because the purpose of science is both to introduce system into our knowledge and to enable us to anticipate the future course of events. Consequently, if a new theory has significantly less breadth than the presently accepted theory, it is not likely to be considered as a serious candidate to replace the older theory. The body of experimental evidence associated with and tied together by the old theory accordingly constitutes a body of knowledge which the new theory must either accommodate or give good reasons why it cannot and should not do so.[19] Such "good reasons" can be strictly internal to the new theory.

On the other hand, as I pointed out in section II, it is the new theory, by virtue of its antecedent theoretical meaning and its antecedently understandable hypothesis, and its associated theories, which will inform us whether a body of experimental knowledge should be relevant to its claims. It is this aspect of the theory which permits it to be extended into new domains, as well as to allow it to accommodate old knowledge in new ways. Thus there is the possibility of a tension between what "experimental" knowledge the old theory informs us should be accommodated by its potential replacement, and that to which the new theory says it should apply. Thus Newton's optics indicated that any new optical theory should relate and explicate the rectilinear propagation of light rays, reflection, refraction, and "Newton's rings." Fresnel's optics agreed, and also contended that it should also account for diffraction patterns.

In order for two theories to compete, as Lorentz's and Einstein's theories do, and as Lorentz's and Mendel's genetic theory do not, the competing theories must overlap in the sense of providing alternative accounts of the same experiments and observations. I indicated above how and why this was possible. But when the theories diverge, say as Lorentz's and Einstein's do inasmuch as Lorentz's theory gives an explanation of the Zeeman effect, which Einstein's does not, and Einstein's theory accounts for the inertia of energy, i.e., $E/c^2 = m$, whereas Lorentz's theory does not, the external constraint on the relevance of an experiment disappears, and only the (internal) antecedent theoretical meaning can determine the relevance of experiments. Nevertheless, there is still the possibility of conflict between the assertions of the theory which are testable by experiment and the O-sentences which are the outcome of the experiments. Accordingly, even where there is divergence between successive theories, the category of "experimental adequacy" may still be employed, though in less straightforward a manner than in cases where there is overlap between competing theories and a more obvious determination of the relative success of the competing theories can be made on a completely common field of battle.

[19] Shapere (1970) has formulated the technical notion of a "domain" in order to introduce in an extra-theoretical manner similar constraints to those which I cite here. Shapere's notion, however, is directed not so much at the problem of theory acceptance as at the problem of theory creation and modification.

Kenneth F. Schaffner

C. Simplicity

Hertz's notion of appropriateness is the most difficult to specify as he himself admitted. What is wanted here is something like the minimal number of concepts and principles necessary to account for a *collection* of experimental and theoretical results constituting a branch of science. As such, appropriateness is very much like sparseness or "simplicity," and I shall be using this latter term in place of Hertz's term "appropriateness." It will turn out, though, that there are several different senses of "simplicity," which it is wise to distinguish, and even though some of them have been so distinguished in recent writings on simplicity by philosophers of science,[20] not all of the senses of the term, as applied in Hertz's analysis, have been characterized.

Before I turn to an analysis of simplicity, however, it would be well to pause and to note one difficulty which was raised in the previous section on experimental adequacy and which is relevant here. This was the claim that different competing theories might diverge in their areas of application. A consequence of this possible divergence is the difficulty of giving any extra-theoretical characterization of what it is that a theory should account for, i.e., there may be a difference in the "collections" cited in the immediately preceding paragraph for two competing theories. This fact might be thought to lead to considerable difficulties if a broader theory were more complex, for one would be faced with estimating the relative merits of a simple narrow theory and those of a broad complex theory. Happily this difficulty does not seem to arise, insofar as simplicity considerations are normally outweighed by considerations of experimental adequacy in such cases of category assessment conflict. In general, then, a broader and more complex theory will be acceptable over a simpler narrower theory. (To a certain extent a case like this occurred when Maxwell's electromagnetic theory of light became, in 1885 in England, favored over a conjunction of the simpler Weberian theory of electrodynamics and the simpler optical ether theories. This occurred before Hertz's important experiments.)

My first task as regards a generalization of Hertz's category of appropriateness is to convey a clearer idea of his notion, or rather of his notions of appropriateness, as there are several intersecting ideas grouped under this term. Unfortunately the only way to do this is to examine the manner

[20] For example see Rudner's (1961) typology of simplicity, and also the collection of articles on simplicity in Foster and Martin's (1966).

in which Hertz applied the notion in its various senses in the context of his analysis of the relative merits of the different images of mechanics. We can then generalize from these examples.

In connection with the Newtonian-Lagrangian image of mechanics, Hertz noted that there were many situations permitted and characterizable by that image for which there was no experimental evidence. Hertz (1894) wrote:

All the motions of which the fundamental laws [of the Newtonian-Lagrangian image] admit, and which are treated of in mechanics as mathematical exercises do not occur in nature. Of natural motions, forces, and fixed connections, we can predicate more than the accepted fundamental laws do. Since the middle of this century we have been firmly convinced that no forces actually exist in nature which would involve a violation of the principle of the conservation of energy. . . . In short, then, so far as the forces, as well as the fixed relations, are concerned, our system of principles embraces all the natural motions; but it also includes very many motions which are not natural. A system which excludes the latter, or even a part of them, would picture more of the actual relations of things to each other, and would therefore in this sense be more appropriate.

I shall call this sense of appropriateness "fitness," for want of a better term. It corresponds to Hertz's earlier use of the term "distinctness" which does not seem apt. There is, as far as I am aware, no sense of simplicity in the current literature which touches on this issue in exactly the Hertzian sense, though some explications of "inductive simplicity" approximate it if generalized from the excessively formalistic "curve-fitting" orientation which infuses such explications. (Cf. Rudner, 1961.)

It should be noted that the determination of "fitness" will usually require some time lapse after the initial enunciation of a scientific theory. Since fitness is related to a minimization of the ultimately nontestable, and usually less easily initially checkable, experimental consequences of a scientific theory, it cannot be easily determined when a theory is first broached. Nevertheless it is sometimes fairly easy to make a relative or comparative determination of the fitness of two competing theories within a previously well-researched area of application, if the new theory does not go, or is not thought to go, extensively beyond the scope of the old theory.

Hertz also considered another dimension of appropriateness. He wrote in the same context as above:

We are next bound to inquire as to the appropriateness of our image in a second direction. Is our image simple? Is it sparing in unessential characteristics—ones added by ourselves, permissibly and yet arbitrarily, to the

327

Kenneth F. Schaffner

essential and natural ones? In answering this question our thoughts turn again to the idea of force. It cannot be denied that in very many cases the forces which are used in mechanics for treating physical problems are simply sleeping partners, which keep out of the business altogether when actual facts have to be represented. . . . [Now] we have felt sure from the beginning that unessential relations could not be altogether avoided in our images. All that we can ask is that these relations should, as far as possible, be restricted, and that a wise discretion should be observed in their use. But has physics always been sparing in the use of such relations? Has it not rather been compelled to fill the world to overflowing with forces of the most various kinds—with forces which never appeared in the phenomena, even with forces which only came into action in exceptional cases? . . . Now if we place these conceptions before some unprejudiced persons who will believe us? Whom shall we convince that we are speaking of actual things, not images of a riotous imagination? . . . Whether complications can be entirely avoided is questionable; but there can be no question that a system of mechanics which does avoid or exclude them is simpler, and in this sense more appropriate, than the one here considered; for the latter not only permits such conceptions, but directly obtrudes them upon us.

In these passages we can distinguish, though Hertz does not do this explicitly, two further senses of appropriateness or simplicity: one is a terminological and/or ontological simplicity; secondly there is a simplicity of system. The conjunction-disjunction of terminological ontological is used because for Hertz, as for myself, the fundamental terms which appear in a scientific theory are to be taken in a realistic sense. The notion of simplicity of system refers to the comparative simplicity of two axiom systems, in which it is assumed that each axiom represents a further irreducible physical effect. This restriction is imposed to eliminate the trivialization of this aspect of simplicity which would occur if all the axioms could be conjoined into one axiom.[21]

It is obvious that by making a virtue of simplicity we make a vice of

[21] It could happen that the second and third senses of simplicity distinguished here, i.e., the terminological/ontological and the simplicity of system, collapse into one basic sense. This would be Goodman's (1966) position. I believe, however, that such a reduction to the simplicity of predicate bases would not be very helpful. This belief is partly based on Goodman's approach, which assumes replaceability as a precondition for simplicity orderings. This is what our problem is all about. Secondly, Goodman's criticism, that the simplicity of a system cannot be determined by comparing the number of hypotheses because all axioms can be conjoined into one axiom, has been eliminated by the stipulation that each hypothesis represent an "irreducible physical effect" in the sense that two "irreducible effects" not be known to be derivable from one such effect, nor to be derivable by the detachment of one of the conjuncts from a conjunction. Such a restriction is perhaps ad hoc in the context of logic, but it represents a quite suitable restriction in the context of physics.

328

complexity. One particularly odious form of complexity is that which accrues to a theory which has many ad hoc hypotheses associated with it. Such complexity not only is contrary to our desire for sparseness and elegance in our scientific theories, but also permits a theory to outflank the controls imposed on theoretical speculation by the restrictions cited in connection with the category of "experimental adequacy." Therefore, from the point of view of the logic of comparative theory evaluation being sketched here, theories containing ad hoc hypotheses are doubly suspect.

The notion of an ad hoc hypothesis has been succinctly characterized by Grünbaum (1964) as a hypothesis which is added to a theory solely to enable it to outflank some unexpected and embarrassing result, an E-result, and which hypothesis (plus the theory under test) has no further additional testable consequences which differ from the E-result in an interesting and significant way. Though the use of the terms "interesting" and "significant" introduces pragmatic elements into the characterization, they are not necessarily subjective ones. This sense of ad hocness can be termed *intra*systemic ad hocness, to distinguish it from a different sense of ad hocness of an *inter*systemic type. In this form, a hypothesis H is ad hoc if it is conjoined to a theory T_1, to obviate the necessity of accepting a new theory T_2, which accounts for the E-result *without* H. H, as in the previous sense, is not supposed to entail, in conjunction with T_1, any additional "interesting and significant" results.

I believe that I can accept either or both of these senses of ad hocness within the logic of comparative theory evaluation which is being proposed here. In both cases, if an ad hoc H is accepted, there will be an increase in the system complexity, and possibly also an increase in the terminological/ ontological complexity. A decrease in fitness is also likely in such circumstances.

It now remains only to introduce some comments on the way in which a judgment of theoretical context sufficiency may outweigh and overpower the type of judgment made in connection with the relative simplicity of two competing theories.

The judgment whether one ontology, say, is simpler, in any useful sense, than another is contingent on the ability of the scientist to seriously consider ontological shifts. Thus a particle physics ontology based on "quarks" is generally thought to be simpler than one without the still observationally inaccessible particles. Suppose, however, that physicists could not accept the existence of quarks because their existence would violate some fundamental physical principle, say the conservation of energy or Lorentz covariance. If this were the case, it would seem that the "ontological simplicity" to be gained by the acceptance of a quark-based particle physics would be of questionable value. Later I shall consider a case quite similar to this which actually occurred in the history of science. The implication is that the theoretical context considerations may overwhelm any simplicity assessment.

The theoretical context also, and perhaps more directly, influences simplicity assessments in yet another way. Given strong theoretical reasons for wishing to retain or maintain some entity or principle, an additional hypothesis, which is conjoined to a theory to enable the theory to survive a prima facie falsifying experiment, will not appear as odiously ad hoc as it would if such strong reasons were lacking. Accordingly, the assessment of system simplicity vis-à-vis the problem of ad hoc modifications of the system is also potentially influenceable by determinations made in the first assessment category.

It should be clear that such a logic of comparative theory assessment as I have been outlining will not employ formal or "effective" concepts. It does not constitute a type of easily applicable schema which can result in an automatic decision for the person thinking in terms of its categories. Nevertheless I do think that the categories of theoretical context sufficiency, experimental adequacy, and relative simplicity do accurately characterize the process of comparative theory evaluation as it is practiced by scientists making the history of science. In order to show how the categories work in practice and in order to delineate the features of the categories more adequately I now turn to a specific and somewhat controversial case in the history of recent physics.

V. The Electrodynamics of Moving Bodies in the Early Twentieth Century

An examination of scientific articles and textbooks written in the years 1900–5 yields the unmistakable conclusion that the electron theory of H. A. Lorentz enjoyed supremacy among the various electromagnetic theories of moving bodies.[22] In a long monograph published in 1901 which was based on lectures given in 1899 at the Sorbonne, Henri Poincaré compared the theories of Maxwell, Hertz, Larmor, and Lorentz on the electrody-

[22] See Schaffner (1969a, 1970), Goldberg (1969b), and McCormmach (1970 and unpublished) for support for this claim.

namics of moving bodies. Although he had certain reservations having to do with an apparent failure of Newton's third law in Lorentz's theory, Poincaré concluded that the Lorentz theory offered the greatest promise of all the then current theories.[23]

P. Drude's important *Lehrbuch der Optik* which was published in Germany in 1900 and in English translation in 1902 and which became a widely used optics text, presented Lorentz's 1895 theory in a section on the optics of moving bodies. In the next few years, studies on the properties of the "empirical electron," and Lorentz's development of a more adequate second-order theory of the electrodynamics of moving bodies, only served to increase the influence of the Lorentz electron theory. A new world picture based on the electron theory and conceiving of all inertial mass as due to electromagnetic interactions was being developed by Wien and Abraham. Lorentz's theory fit quite well into this new *Weltbild* which suggested that traditional mechanics might be in need of revision if it was to conform to the new picture of the world as basically electromagnetic in nature.[24]

In September of 1905 Einstein published a brief paper on the electrodynamics of moving bodies which obtained what appeared to be many of the same results as had Lorentz, though Einstein reasoned to them in such a way as to throw a very different perspective on the nature of space and time and the laws of nature. By 1909–10, in only about four or five years' time, most of the German physicists had been converted to Einstein's theory from the Lorentz approach.[25] In other countries the pace of conversion was slower, but in general was not delayed beyond the middle teens.

Einstein's theory of the electrodynamics of moving bodies—the special theory of relativity—constituted in many ways a very abrupt departure from traditional ways of thought—a true scientific revolution. This, together with the fact that there are many diverse points at which the theory obtains experimental support and the fact that it appears so different from the Lorentz theory as regards its prima facie "simplicity," suggests that the study of the transition from pre- to post-relativistic electrodynamics might afford an excellent test area in which to apply the logic of comparative theory evaluation sketched out above.

[23] See Poincaré (1901), p. ii.
[24] See Jammer (1961), McCormmach (unpublished), and Goldberg (1970b) for discussions of the electromagnetic view of nature.
[25] See Goldberg (1969a) and also his (1970a) for indications of the difference in the rate of acceptance of relativity by country.

Kenneth F. Schaffner

VI. The Lorentz "Absolute" Theory

A. The Foundations of the Electron Theory

The genesis and development of H. A. Lorentz's theory of the electrodynamics of moving bodies cannot be examined in any significant detail in these pages.[26] What I propose to do is to outline the structure of the Lorentz electron theory, largely insofar as it applies to the electrodynamics of moving bodies, in its mature 1904–9 form.[27] Then I shall do the same for Einstein's theory, and conclude with a relative assessment of the two theories based on the modified tricategorical Hertzian analysis developed above.

Lorentz's electron theory has its roots in Maxwell's electromagnetic field theory. Lorentz proposed, as was expected for any field theory at the end of the nineteenth and in the beginning of the twentieth centuries, the existence of an ether or a medium which pervaded all space and, for Lorentz, even ponderable bodies. There were certain "states" of this medium which gave rise to electrical and magnetic forces acting on electrified bodies and magnetic poles. Contrary to Maxwell, though like Hertz, Lorentz refused to speculate more on the state of the medium which might account for these forces. It was sufficient to define the electrical and magnetic fields in terms of these "forces" and to relate the forces in partial differential equations. For Lorentz the ether was essentially motionless; bodies could not drag any ether with them as they moved through it and though the ether could act on electrified and magnetized bodies, these bodies could not react on the ether, as this would require setting it in motion.[28] Lorentz does introduce, as did Maxwell, a "dielectric displacement" though we are not told what, if anything, is being displaced, and in the free ether, anyway, the dielectric displacement and the electric force have the same magnitude and direction.

The basic equations which hold for the electromagnetic field in the (charge) free ether are:

(1) $\operatorname{div} \mathbf{d} = 0$

(2) $\operatorname{div} \mathbf{h} = 0$

(3) $\operatorname{curl} \mathbf{h} = \dfrac{1}{c} \dfrac{\partial \mathbf{d}}{\partial t}$

[26] See Schaffner (1969a, 1970), Goldberg (1969b), Hirosige (1962, 1966), and McCormmach (1970 and unpublished) for more details on the genesis of the electron theory of Lorentz.

[27] This analysis is based on Lorentz's (1904b, 1909) works.

[28] Lorentz subsequently relaxed his position on this point by admitting the existence of an "electromagnetic momentum" in the ether. This also resolved difficulties over the

332

$$(4) \quad \text{curl } \mathbf{d} \ = -\frac{1}{c}\frac{\partial \mathbf{h}}{\partial t}$$

where **d** is the dielectric displacement, **h** is the magnetic force, and c is a constant depending on the properties of the ether (c, of course, will turn out to be the velocity of a disturbance propagated through the ether which, for a short range of frequencies, is visible light).

Lorentz also introduced the notion of extremely small charged particles which he at first termed "ions" but later, after the analysis of cathode and β rays had been carried out and Stoney's term adopted, called them "electrons." These electrons have a specific structure for Lorentz. They are essentially spherical but he also noted: "We shall consider the volume density ρ [of the electron's charge] as a continuous function of the coordinates so that the charged particle has no sharp boundary, but is surrounded by a thin layer in which the density gradually sinks from the value it has within the electron to 0."[29] Some of these electrons are free to move, especially electrons found in the interior of conductors; others in dielectrics, for example, are bound loosely to atoms in the ponderable bodies and can vibrate.

The equations for the state of the ether within the electrons require only the slightest modification of the equations already given for the (charge) free ether. In place of (1) we write:

$$(1)^* \quad \text{div } \mathbf{d} = \rho$$

and in place of (3) we substitute:

$$(3)^* \quad \text{curl } \mathbf{h} = \frac{1}{c}\left(\frac{\partial \mathbf{d}}{\partial t} + \rho \mathbf{v}\right)$$

where **v** is the absolute velocity of the charge. Lorentz also adds a fifth equation:

$$(5)^* \quad \mathbf{f} = \mathbf{d} + \frac{1}{c}(\mathbf{v} \times \mathbf{h})$$

which represents the dynamical force per unit charge produced by the ether on a charge. This equation can be derived from the others by adding vectorially the two forces experienced in a field by a moving charge.

From the above the reader should be able to see how the antecedent theoretical meaning of the theory of electrons is developed.

apparent failure of Newton's third law in his electron theory. See Lorentz (1904a, b, and 1909) for discussion of electromagnetic momentum.

[29] Lorentz (1909), p. 11.

Kenneth F. Schaffner

By applying these equations to free electrons moving in metals Lorentz obtained causal explanations for many empirically established formulas, such as had been established by the experimental work of Drude and others.[30] Lorentz also developed an explanation of the splitting of spectral lines in a magnetic field—the well-known Zeeman effect—by utilizing Newton's laws of motion together with his own equations and applying them to the loosely bound electrons in atoms. It is in this and similar accounts that we see the exemplification of what I referred to as the establishment of correspondence rules or C-sentences: connections between theoretical processes and experimental or observationally accessible states of affairs.

B. Lorentz's Electron Theory Applied to Moving Bodies

From the birth of his electron theory in 1892 on, Lorentz had been concerned to give explanations of various experiments made on systems of moving light waves. The well-known Fresnel partial dragging coefficient,

$$\left\{ 1 - \frac{1}{n^2} \right\},$$

where n is the index of refraction of a transparent body, which is needed to account for many experiments in the field of the optics of moving bodies, had been derived by Lorentz in 1892.[31] The Michelson-Morley experiment received a sort of ad hoc explanation when Lorentz suggested that objects moving through the ether contracted just enough to compensate for the difference in the path of the light waves in the moving interferometer.[32] In 1895 Lorentz proved a general theorem from his equations —known as his theorem of corresponding states—which showed that if a new time variable, a "local time," could be substituted for the universal time in the moving system, no first-order effect of the earth's motion on an electromagnetic system would be detectable. It is most important to realize, however, that the "local time" for Lorentz was, as he later termed it, "no more than an auxiliary mathematical quantity." Pauli (1958) explicates the sense of this by noting that such a "local time" means that "the origin of t' was taken to be a linear function of the space coordinates, while

[30] Cf. Lorentz (1909), pp. 62ff.
[31] See Schaffner (1969a) and especially (1970), chapters 3 and 6, for a discussion of the Fresnel partial dragging or convection coefficient.
[32] See same references as cited in note 31 for a discussion of the Michelson-Morley experiment

334

the time scale was assumed to be unchanged."[33] Lorentz did not invest his "local time" with any physical significance, other than to associate it with the ether wind, and in particular he did *not* identify it with the time of the moving system, until after he had read Einstein's papers on relativity.

Lorentz's theory of 1895, then, gave a good account of experiments in the domain of the electrodynamics of moving bodies known at that time. But the field was not static, and by 1904 new second-order experiments had been performed by Trouton and Noble and by Rayleigh and Brace which raised additional problems for Lorentz, and which caused him to reformulate his theory of moving bodies and to generalize it to second- and higher-order experiments.

Lorentz, in developing his 1904 theory, began with the fundamental equations of his electron theory given above as $(1)^*$, (2), $(3)^*$, (4), and $(5)^*$. He then applied a "Galilean velocity transformation," $x_r = x - vt$, to these fundamental equations, and then superadded another change of variables:

(6) $x' = \beta l x_r$
(7) $y' = l y_r$
(8) $z' = l z_r$

(9) $t' = \dfrac{1}{\beta} t - \beta l \dfrac{v}{c^2} x_r$

where

$$\beta = \sqrt{\frac{c^2}{c^2 - v^2}}$$

and l is an as yet unknown function of v. Equations $(6)-(9)$ are the well-known Lorentz transformation equations, but it would be well to pause for a moment and recall their meaning for Lorentz.[34]

Equation (6) represents the transformation of length that captures the Lorentz-Fitzgerald contraction. This is a shrinkage caused by the motion of bodies through the ether with an *absolute* velocity equal to the v appearing in the transformation. The t' in equation (9) represents the latest form of the "local time" which holds in *moving* electromagnetic systems.

---

[33] The partial quotation from Lorentz is from his (1915), p. 321, the Pauli quote is from his (1958; originally published in 1921), p. 1.
[34] The difference in meaning between Lorentz's and Einstein's transformation equation has been discussed by d'Abro (1927), Grünbaum (1961, 1963), Popper (1966), and myself (1969a, 1970).

335

It is still to be distinguished from the "true" Newtonian time holding in bodies at rest in the ether.

Transformations for **d** and **h** were introduced by postulation, but they vary only slightly from what would have been expected from a standard referral of a system of charges to a moving coordinate system. Lorentz also postulated transformations for charge density:

(10) $\rho' = \rho/\beta\, l^3$

and he also introduced a modified velocity addition formula, claiming that if an electron had a velocity u in addition to the velocity v of the system of charges to which it belonged, the velocity in the primed system of coordinates would be

(11) $u_x' = u_x\beta^2 \quad u_y' = \beta u_y \quad u_z' = \beta u_z$

In a later section of his (1904b) paper, Lorentz also introduced a new vector representing electric moment (due to polarization), symbolized by **P**, and also postulated its transformation properties.

From his fundamental equations plus the transformations which he had simply postulated in addition to them, Lorentz was able to derive expressions for the dielectric displacement and Lorentz force acting on electrons as referred to the moving system. He was also able to calculate the electromagnetic momentum of such a system, something he had to do to obtain results that would enable him to account for the outcome of the Trouton-Noble experiment.

This rather complex system of equations, however, had to be complicated even more, before Lorentz could attain to his goal of proving a theorem of corresponding states for many electromagnetic phenomena to all orders of v/c. He also found it necessary to add: (1) a special hypothesis postulating that the electron contracted in accordance with the Lorentz-Fitzgerald contraction and (2) still another hypothesis concerning the transformation properties of any forces holding between particles and (3) yet another hypothesis proposing that all the electron's mass be taken as electromagnetic mass.[35] The last hypothesis had some experimental evidence in its favor, however, and was then a current belief of all those who were working out electromagnetic *Weltbilder*.

By applying these various hypotheses to the motion of electrons, and, in addition, by making use of Newton's second law of motion, Lorentz was

[35] Holton (1960) lists 11 ad hoc hypotheses in the (1904b) Lorentz paper, and Goldberg (1969b) concurs in this assessment.

able to obtain the value of the undetermined l function which appeared in his transformation equations; it becomes equal to unity.

From here Lorentz went on to prove a generalized theorem of corresponding states. Lorentz wrote (1904b):

We may sum up by saying: If in the system without translation, there is a state of motion in which, at a definite place, the components of **P**, **d**, and **h** are certain functions of the time, then the same system after it has been put in motion (and thereby deformed) can be the seat of a state of motion in which, at the corresponding place, the components of **P'**, **d'** and **h'** are the same functions of the local time.

Lorentz concluded that these new time variables, contractions, mass transformations, and the like, so interacted that many experiments, and in particular the Michelson-Morley, Trouton-Noble, and Rayleigh-Brace experiments, performed in optics and electromagnetic theory would produce null results.

Lorentz's theory does not constitute a theory of relativity nor does the theorem of corresponding states constitute a principle of relativity. In Lorentz's own words, he was attempting to show that "many electromagnetic actions are entirely independent of the motion of the [moving] system" (my emphasis).

I do not have the space in this paper to comment on Poincaré's extension of Lorentz's approach, and his publication, independently of Einstein, of a principle of relativity which he, Poincaré, actually used to modify Lorentz's charge density transformations, and the velocity transformation, so as to obtain complete invariance of the fundamental equations.

VII. Einstein's Special Theory of Relativity

Einstein's special theory of relativity constituted a significant departure from the Lorentz theory. Einstein's assessment of the theoretical context sufficiency of Lorentz's 1895 theory was somewhat different from Lorentz's assessment which motivated him to produce his 1904 version.[36] By early 1905 Einstein had already satisfied himself as to the microscopic inadequacy of the Maxwell and the Lorentz theories in his paper on light quanta and the photoelectric effect.[37] Einstein also felt that Lorentz's theory, with its ether rest frame, implied the existence of asymmetries in

[36] Einstein knew of Lorentz's (1892a) and (1895) papers, but not of his later modifications. See Born (1956) for a letter from Einstein stating this.
[37] Cf. Einstein (1905a).

optical and electromagnetic phenomena for which there was no experimental foundation.

In Einstein's assessment of the adequacy of the Lorentz theory we see an example of a "subjective fluctuation" from the consensus of science. Part of Einstein's distrust of the Lorentz theory was shared by others, in particular Poincaré, but part was uniquely his own. This is especially true of Einstein's rejection of the Lorentz "constructive" approach because of his own work on light quanta. Perusal of Professor Stuewer's paper in this volume will indicate how alone Einstein stood on this ground.

Though Einstein was not able to formulate what he would call a "constructive" type of theory of electrodynamics, basically because of the difficulties introduced by the light quantum hypothesis, he did discover that a theory of principle—on the analogy with the second law of thermodynamics, and as distinguished from a "constructive" type of theory such as statistical mechanics—could be formulated. In order to do so, however, a new *kinematics* would be necessary: only with the aid of new conceptions of space and time could a simple and consistent theory of the electrodynamics of moving bodies be developed.

Einstein's theory is essentially a new kinematics based on the principles of the constancy of the velocity of light and the principle of relativity. The first principle is the heritage of the wave theory of light and asserts only, contrary to the beliefs of some students of relativity, that the velocity of an emitting body has no effect on the velocity with which the light is propagated. This, when conjoined with the principle of relativity, that is, that "the laws by which states of physical systems undergo change are not affected whether these changes of state be referred to the one or the other of two systems of coordinates in uniform translating motion,"[38] resulted in a paradox. The paradox was resolved by Einstein by a brilliant reanalysis of the notion of time or of simultaneous measurements in which he exposed the conventional character of some of the fundamental notions of physics.[39]

Einstein's kinematics led to what formally appeared to be the same space and time transformations as Lorentz had postulated. There were, however, several significant differences. Central to the antecedent theoretical meaning of the special theory of relativity is the thesis of the equivalence of all unaccelerated reference systems and of the *relativity* of veloc-

[38] Einstein (1905b).
[39] See Reichenbach (1957) and Grünbaum (1963) for in-depth analyses of the philosophical aspects of Einstein's reanalysis of simultaneity.

ities which appear in the transformation equations. For Lorentz the v term appearing in his equations represented absolute velocity, and the transformations between moving and rest frames are not reciprocal, for in shifting from a moving frame of reference to one at rest in the ether one obtains an "expansion" effect which is built into Lorentz's formalism. For Einstein there were reciprocal contractions which are due to reciprocal failures of the identity of simultaneity and not to alterations in the molecular forces due to the translation of a body through the ether.[40]

It should also be pointed out that Einstein's derivation of the transformations transcended the electron theoretical foundations: it was a "kinematical" derivation based only on the two principles cited above, and such generality immediately suggested that the new kinematics might well apply in other areas outside of electrodynamics.

Einstein's principle of relativity and his new kinematics were, in the second part of his 1905 paper, applied to electrodynamics, i.e., to the Maxwell-Hertz equations for bodies at rest and also to the fundamental equations of the Lorentz electron theory. Einstein obtained the same transformations for electric force (or dielectric displacement) and magnetic force as Lorentz was forced to postulate, and Einstein also derived a different charge density transformation, which, taken together with the Einsteinian velocity addition formula proven in the kinematical part of the paper, yielded complete invariance of the Lorentz equations under the Lorentz space and time transformation equations. Einstein also obtained, deductively, the Lorentz expressions for the velocity dependence of the mass of an electron. There are other important results developed in this second part of Einstein's paper, but they do not concern our present task.

VIII. A Comparison of Lorentz's and Einstein's Theories from the Point of View of the Logic of Comparative Theory Evaluation

A. Theoretical Context Sufficiency

The theoretical context in which the Einstein special theory of relativity first appeared was still dominated by Newtonian mechanics and its later reformulations, though the mechanical approach was considerably weaker than it had been, say, in the middle years of the nineteenth century. Maxwell's electromagnetic theory, owing to the experimental work of Hertz, had, about seventeen years previously, triumphed over various action-at-a-distance theories of electromagnetism, and had in the 1890's begun to

---

[40] See my (1969a) for more details.

Kenneth F. Schaffner

replace the older theories of the mechanical optical ether. Maxwell's theory was still conceived of as ultimately explicable by mechanics,[41] though in the early years of the twentieth century several physicists, building on the considerable successes of Lorentz's electron theory, began developing a new electromagnetic view of nature which would serve to compete, as regards fundamentality, with the Newtonian mechanical view.[42] The success of thermodynamics as a nonmechanical theory had engendered a vigorous, if rather polemical, energeticist school, and Ernst Mach's important philosophical-historical critiques fanned the flames of the antimechanist approach to physics. Max Planck had, in 1900, discovered the quantum of action, but was still attempting to accommodate it within classical physics.[43]

In spite of the weakening of the classical Newtonian ideal, Max Born, in commenting on the scientific atmosphere at the time he was a student of physics during 1901–7, indicated that the influence of the Newtonian ideas was still significant. Born (1956) wrote: "Newton's mechanics still dominated the field completely, in spite of the revolutionary discoveries made during the preceding decade, X-rays, radioactivity, the electron, the radiation formula and the quantum of energy, etc. The student was still taught—and I think not only in Germany, but everywhere—that the aim of physics was to reduce all phenomena to the motion of particles according to Newton's laws, and to doubt these laws was heresy never attempted."

I think that this statement of Born's is a bit misleading, for even if students were being taught that Newton was unchallengeable, their teachers were not so dissuaded from questioning the Newtonian laws of dynamics. Lorentz (1895) had shrugged off the failure of Newton's third law in his theory, though this difficulty was later resolved in the early twentieth century.[44] Poincaré had speculated in the early 1900's about a "new mechanics" which might have the speed of light as a limiting velocity, and the proponents of an electromagnetic view of nature were questioning the validity of the Newtonian scheme. In his (1909) work, Poincaré considered in a fairly systematic manner some of the implications of the Lorentz theory

[41] The relation of Maxwell's theory to mechanics is discussed extensively in my (1970), chapters 4 and 5.
[42] See Jammer (1961), McCormmach (unpublished), and Goldberg (1970b).
[43] Important discussions of the theoretical context at this time are given by Holton (1967–68), by Klein (1967) and, of course, in the most important autobiographical notes of Einstein (1949).
[44] See note 28, and especially see Lorentz (1909), pp. 30–32.

as amended by his own extensions (Poincaré, 1906). Though Poincaré did think that the Lorentz theory would entail revisions in the traditional mechanics, much as Einstein's theory would, such revisions are less revolutionary than Einstein's revisions, because they are based on extensions of accepted electromagnetic theory, e.g., forces are taken to be (and to transform as) electrical forces, and mass is conceived of as velocity dependent because of the self-induction of the electron. Such modifications are accordingly alterations in dynamics.

Even though traditional *dynamics* was being questioned, however, traditional *kinematics*, the science of space and time, was not under fire. Furthermore, there was little criticism of the notion of the electrodynamic ether, though physicists had begun to find it more convenient to work with Maxwell's and Lorentz's equations, per se, and not worry about further mechanical ethers underlying these field theories.

It was into this theoretical context that Einstein's special theory of relativity intruded, eliminating the ether, revolutionizing the basic concepts of space and time, and implying that Newton was, at best, only approximately right. Einstein's radical analysis of simultaneity and the concept of time was the reason why his theory was so novel and exciting, but also the reason why it was so suspect and difficult to accept. Born (1956) wrote concerning this point:

In 1907 . . . I returned to my home city Breslau, and there at last I heard the name of Einstein and read his papers. I was working at that time on a relativistic problem and [a friend] . . . directed my attention to Einstein's articles. . . . Although I was quite familiar with the relativistic idea [apparently in the Lorentz sense] and the Lorentz transformations, Einstein's reasoning was a revelation to me. . . . For me—and many others—the exciting feature of the paper [Einstein, 1905b] was not so much its simplicity and completeness, but the audacity of challenging Isaac Newton's established philosophy, the traditional concepts of space and time, that distinguishes Einstein's work from his predecessors.[45]

Lorentz also found Einstein's theory quite different from his own, and in 1909 characterized it as "very interesting" and remarked at its "fascinating boldness" and at its "simplicity." Nevertheless, he could not yet accept it. In 1915, however, he wrote in the second edition of *The Theory of Electrons*:

If I had to write the last chapter ["On Optical Phenomena in Moving Bodies"] now, I should certainly have given a more prominent place to Einstein's theory of relativity . . . by which the theory of electromag-

[45] Born (1956), pp. 193, 195.

341

netic phenomena in moving systems gains a simplicity that I had not been able to attain. The chief cause of my failure was my clinging to the idea that the variable t only can be considered as the true time and that my local time t' must be regarded as no more than an auxiliary mathematical quantity. In Einstein's theory, on the contrary, t' plays the same part as t.[46]

Similar resistance to the revolutionary analysis of time is to be found in the willingness of some scientists of that period to give up Maxwell's equations if traditional kinematics could be maintained. W. Ritz's ballistic theory was such an attempt. W. Pauli (1958) commented on the motives behind the Ritz theory, writing:

The constancy of the velocity of light, in combination with the relativity principle, leads to a new concept of time. For this reason W. Ritz, and independently, Tolman, Kunz, and Comstock have raised the following question: Could one not avoid such radical deductions and yet retain agreement with experiment, by rejecting the constancy of the velocity of light and retaining only the first postulate [i.e., relativity]? It is clear that one would have to abandon not only the idea of the existence of an aether but also Maxwell's equations for the vacuum, so that the whole of electrodynamics would have to be constructed anew. Only Ritz has succeeded in doing this in a systematic theory.[47]

The conclusion which can be drawn from these examples is that, with respect to a decision made in 1905 on the *theoretical context sufficiency* of the Lorentz and the Einstein theories, Lorentz's was clearly favored. It not only retained the ether, but also retained Newtonian kinematics, and did not, therefore, contradict in the same fundamental manner the Newtonian dynamics which still appeared, to most, to be a fundamentally sound and still viable theory.

I shall defer consideration of the comparative "experimental adequacy" of the Einstein and Lorentz theories for a moment and shall turn to an examination of the third of the categories of the logic of comparative theory evaluation.

B. Simplicity

On the grounds of fitness and of ontological and system simplicity, Einstein's theory was clearly better than Lorentz's.[48] Even before Einstein's theory was available as a contrasting approach, Poincaré had criticized Lorentz's pre-1904 attempts at a theory of the electrodynamics of moving

[46] Lorentz (1915), p. 321.
[47] Pauli (1958), pp. 5–6.
[48] This is a point which is heavily stressed by Holton (1960), though he uses a different analysis of simplicity than the modified Hertzian analysis presented here.

342

bodies, noting that Lorentz seemed to require a new hypothesis for his theory each time a new experiment was done.[49] The Lorentz theory had to postulate in an unconnected manner, and in a way that increased the ad hoc character of the postulates, many special transformations of electromagnetic parameters. Special postulates for electron contraction, velocity addition, charge density modification, and the like, as were discussed above, were added in piecemeal fashion in attempts to outflank the results of prima facie falsifying experiments. In contrast, Einstein's theory began with two postulates and after deductively developing the various kinematical transformations, was applied to the fundamental equations of the Maxwell and Lorentz theories, resulting in the derivation of the electrodynamical transformations which Lorentz had to postulate. With respect to system simplicity, then, Einstein's theory was better than Lorentz's.

This was acknowledged by the scientists of the time. M. Laue (1911), in the first textbook on the special theory of relativity, wrote: "Though a true experimental decision between the theory of Lorentz and the theory of relativity is indeed not to be gained, and that the former, in spite of this, has receded into the background, is chiefly due to the fact that, close as it comes to the theory of relativity, it still lacks the great simple universal principle, the possession of which lends the theory of relativity . . . an imposing appearance."[50] Lorentz himself, as quoted above on page 342, also acknowledged that Einstein's theory had a "simplicity that I had not been able to attain."

Einstein's theory is also *ontologically* simpler than Lorentz's inasmuch as it dispenses with that strange entity, the ether. In Lorentz's theory the ether still has the role of the special framework for Maxwell's and Lorentz's equations, and it apparently also exerts causal influences, such as the contraction phenomena. In Einstein's theory, however, as in Lorentz's recasting of his own theory in the last few pages of his (1909) monograph, such contractions are reciprocal for two reference frames moving with a constant velocity, and accordingly cannot be ascribed to the asymmetrical ether rest frame.

With respect to fitness, Einstein's theory is also in a better position than is Lorentz's. Though the potential scope of Einstein's theory was considerably greater than Lorentz's theory, Einstein's theory did not introduce distinctions at the theoretical level which clearly had no further empirical

[49] Cf. Poincaré (1900).
[50] Cf. Laue (1911), as quoted and translated in Cassirer (1923), p. 376.

consequences associated with them. Both Maxwell's theory and Lorentz's analysis make a distinction between the case of a magnet stationary in the ether and a circuit moving near it, and the case of a circuit stationary in the ether with the magnet moving through the ether. These theoretically distinct states of affairs have the same observable consequences, and Einstein, at the very beginning of his fundamental (1905b) paper, called attention to these "asymmetries which do not appear to be inherent in the phenomena" to make his own symmetrical theory more plausible. There are other asymmetrical effects which Lorentz's theory permits, but which have no experimental support. Lorentz apparently thought, until about 1910, that observations on Jupiter's moons might reveal an ether wind effect. Pauli (1958) commented on this unrealized hope, writing:

In Lorentz's theory it might be possible to obtain first-order effects of the "aether wind" by considering gravitation. Thus, as mentioned by Maxwell, the motion of the solar system relative to the aether would produce first order differences in the times of the eclipses of the Jupiter satellites; but it was found by C. V. Burton (*Phil. Mag.* 19 (1910), 417; cf. also H. A. Lorentz . . . (1914), p. 21) that the inherent observational errors would be as large as the expected magnitude of the effect. Observations on the satellites would therefore not help in deciding for or against the old aether theory.[51]

Einstein's theory denied the possibility of any asymmetrical "ether wind" effects, and as part of the systematic consequences of the theory, obtained a set of consistent transformations which allowed full covariance. Lorentz only obtained approximate covariance using his own transformations, and the *departures* from full covariance did not have any experimental support, though they were not at the time ruled out by experiments. Clearly, however, a theory that claims the existence of such asymmetries which are not empirically accessible is deficient in fitness, as compared with a theory which does not make any such claims.

In regard to what was said earlier about the interaction of the categories, however, we can also see why Lorentz did not feel that the complexities of his theory were necessarily a poor feature. In Lorentz's judgment, the ether was still a viable concept and not one to be easily relinquished. In terms of the theoretical context of this judgment, it is a legitimate one. Moreover, Lorentz saw good reasons for wishing to retain the eminently successful Newtonian kinematics. In the light of such judgments, the ontological simplicity of a theory without the ether was not necessarily desirable.

[51] Pauli (1958), p. 1, n. 3.

Lorentz also seemed willing to tolerate a certain amount of complexity and ad hoc hypotheses in his theory if he could maintain the ether and traditional kinematics. From his point of view, certainly until about 1910, one could rationally anticipate experimental effects of the ether wind. After this time, Lorentz still seemed to hope for some unexpected but possible ether wind effect; at least he wished to keep the possibility open. To the extent that he did this, he did not accept relativity.

Lorentz could appreciate the simplicity of the Einstein theory over his own theory, and in fact, as we saw, admitted such simplicity. Nevertheless he felt that conflicts with the theoretical context were too great to permit him to accept Einstein's theory.

In spite of the existence of an interfering influence of judgments based on other categories, we can conclude that Einstein's theory enjoyed a considerable advantage with respect to simplicity, as displayed in its higher degree of fitness, its less complex ontology, and its higher degree of systematic interrelation of transformation equations.

C. Experimental Adequacy

A determination of the relative merits of the Einstein and Lorentz theories in regard to experimental adequacy is rather straightforward, though an account of how it can be so involves the complex issues of the interaction of theory and observation discussed earlier. First it should be noted that the consensus view of physicists in the years 1905–15 was that no experimental evidence was available which would permit a clear choice between the Lorentz and the Einstein theories. (See the quotation from Laue (1911) above.) This was in interesting contrast to Hertz's theory of the electrodynamics of moving bodies which had been eliminated by its failure to account for the Fresnel partial dragging coefficient and thus for Fizeau's (1851) moving water experiment. Ritz's theory also was eliminated on experimental grounds, principally by de Sitter's (1913) experiment on the light from double stars, but partly also because no natural explanation of the Fresnel coefficient was forthcoming from Ritz's theory.[52] Experiments with clearly associated asymmetrical ether wind effects would have decided for Lorentz's theory against Einstein's, but as pointed out above, the lack of such experiments did not falsify the Lorentz theory; it only diminished its fitness.

Mutual experimental adequacy, however, does not imply that both

[52] See Pauli (1958), pp. 5–9, and also Panofsky and Phillips (1955), chapter 14, for a discussion of the experimental inadequacy of the Ritz theory.

theories account for the results of experiments in the same way. On the contrary, the Lorentz analysis and the Einstein analysis of the Michelson-Morley experiment, and most other experiments associated with the theories, involve rather different and incompatible analyses of what the light rays and the instruments are purportedly doing when readings are being taken. This is not to deny the thesis developed earlier that there is overlap between the theories: both the Lorentz and the Einstein theories make contact with experience by using observation terms which have identical primary senses. Moreover, both theories even overlap to some degree in more theoretical sectors. Both utilize Maxwell's and Lorentz's equations for bodies at rest, though the two theories diverge in that Lorentz ultimately associates his equations with an ether and an absolute reference frame, whereas Einstein does neither. Both theories also find that the behavior of reflected intersecting orthogonal light rays, such as exemplified in the Michelson-Morley experiment, is relevant as a test of their contentions about electromagnetic phenomena associated with moving bodies. It is the overlap of the observation terms, and the ability of both absolutist and relativist to agree about the primary sense of the observation terms and O-sentences, that permits *both* theorists to use the Michelson-Morley experimental results.

Let us also refer to a rather different experiment, the one mentioned above which was responsible for the elimination of the Hertz theory, and also for suspicion concerning the Ritz theory. Lorentz, as was mentioned in section VI, formulated an explanation of the Fizeau experiment by finding a way to derive the Fresnel partial dragging coefficient from the equations of his electron theory (Lorentz, 1892a). The Lorentz explanation involved physical hypotheses concerning the interaction of light in the moving water (in the Fizeau experiment) with the electrons of the water, and the reradiation of light waves by these electrons. The net result is a light wave which moves with a velocity of $c/n + v(1 - 1/n^2)$, where c is the velocity of light in a vacuum, n is the absolute index of refraction of the water, and v is the absolute velocity of the water. In Lorentz's derivation, he used standard Newtonian kinematics in adding velocities.

Einstein himself did not offer an explanation for the partial dragging coefficient, but M. Laue (1907) was able to provide one by showing that the required observed effects followed simply from the relativistic velocity addition theorem. In this explanation, *partial* drag was not assumed; the moving water completely dragged the light with it. The velocity of the

346

light in the water and the velocity of the water, however, added relativistically so as to give the appearance of a partial dragging effect. Again, in carrying out the Fizeau (1851) experiment or the later Michelson-Morley (1886) more precise version of the experiment, both a relativist and an absolutist would agree on the primary sense of the observation sentences reporting the results of the experiment. In this experiment there is a positive fringe shift of measurable amount. The two theories would, however, as this account of O-sentences permits, take the O-sentences as evidence for, or indexes of, quite different aspects of nature's regularity.[53]

Accordingly, we see in connection with our example how it is possible for two inconsistent and perhaps incommensurable theories to have common points of contact with experiment.

With regard to the question of the *relevance* of experiments vis-à-vis the issue of experimental control over theory acceptance, one can, I believe, conclude that if Einstein's theory had been unable to account for the Fizeau experiment and the "partial" dragging coefficient, Lorentz's theory would have ranked over it as regards experimental adequacy. For the Einstein theory asserts itself to be a theory of the electrodynamics and optics of moving bodies, and it gives no reasons why it should not account for first-order experiments in this domain. Thus the Einstein theory (taken together with the theoretical context of optics) informs a physicist that such an experiment is relevant. Lorentz's theory does the same, and accordingly, even though the two theories disagree on what is occurring in the experimental device, the experiment does constitute a common standard for the judgment of the competing theories' experimental adequacy.

In summary, then, the three categories of comparative theory evaluation which we have been considering yield a split decision in, say, 1905–7: theoretical context sufficiency supports Lorentz's theory, relative simplicity supports Einstein's, and experimental adequacy, because of the flexibility of the interpretation of "observation" statements, selects neither theory as the better one.

But science was not a static enterprise in these years, and judgments made in 1905 are not quite what they would be were they made in 1906 or 1907. In 1906 M. Planck (1906) reanalyzed Einstein's suggestions concerning slowly accelerated electrons, a topic to which Einstein had turned his attention at the very end of his (1905b) paper. Planck was able to show that a simpler form of the mass dependency on velocity would follow if a

---

[53] Cf. Kuhn (1962), p. 129, from whom this sentence is closely paraphrased.

347

Kenneth F. Schaffner

different, and entirely plausible, definition of force were employed. In 1907, Minkowski (1908) worked out a revision of mechanics in an Appendix to a systematic representation of relativistic electrodynamics. In this Appendix Minkowski referred to the contradictions in the relation of the special theory of relativity to the theoretical context (especially with Newtonian mechanics) when he noted that many authors were arguing that "classical mechanics contradicts the postulate of relativity which is here taken as the basis of electrodynamics." Minkowski went on to assert that "it would be extremely unsatisfactory if the new conception of time should be valid only in a particular branch of physics."[54]

Minkowski's reformulation of mechanics, which now made mechanics agree with the special theory of relativity, was subsequently followed by Born's (1909) refinement as well as by Lewis and Tolman's (1909) famous paper on elastically colliding masses. The latter paper was the first non-electrodynamical derivation of the relativistic mass-velocity effect.

Clearly, then, we have evidence of a change in the theoretical context of Einstein's special theory of relativity in these four years from 1905 to 1909. The reluctance of physicists to accept Einstein's theory because of its conflicts with Newtonian kinematics and Newtonian dynamics lessened as mechanics itself was gradually altered so as to agree with the new theory. Other derivations from the relativistic theory followed, one of the most important being the mass-energy relation first proposed by Einstein (1905c) shortly after the publication of his fundamental relativistic paper. Experimental tests of this implication were not available for many years afterwards, however, so that this perhaps most famous implication of relativity theory was not instrumental in affecting its acceptance.

What seems to be responsible for the acceptance of the Einstein theory over the Lorentz theory by, say, 1910 in Germany, and a bit later in other countries, is the complete experimental adequacy of the theory, its high degree of simplicity, and the revision in the theoretical context caused largely by itself. Such a revision took time, as did the derivation of further consequences from the Einstein theory. Nevertheless it does appear that many scientists did relinquish the Lorentz theory in favor of Einstein's, and the reasons which are often given appear to agree with the analysis of the logic of comparative theory evaluation sketched in these pages. There were, to be sure, some physicists who did not relinquish the older ether theories. Apparently they did not do so because they saw no clear experi-

[54] Minkowski (1908), as translated and quoted in Hirosige (1968), p. 47.

mental reasons for doing so, and they could comprehend neither a physics of fields without an ether nor a mechanics without absolute simultaneity. But such physicists represent another instance of the "subjective fluctuations" from the consensus of physics, fluctuations which here indicate not that theory change is irrational, but rather that some physicists are irrational.

Accordingly, we see that judgments made using these categories of comparative theory evaluation are themselves dependent on what experiments have been done and what revisions in theory scientists have attempted. But such a relativization does not imply the subjectivity of different scientific schools which some of the historical philosophers of science have suggested. Rather the time dependency of the assessment of strengths of different and competing scientific theories seems to reflect the dynamism and actual progress of science through history.

IX. Implications of the Above Tricategorical Analysis for the Relation of the History and Philosophy of Science

Let us now consider whether the philosophical analysis sketched here might admit of application by the historians of science, and in particular, how the analysis might clarify the controversy in the history of recent physics over who had priority in the development of the special theory of relativity. The controversy arose when Whittaker (1953) argued that the theory of relativity really belonged to Lorentz and Poincaré, and that Einstein had simply made some minor contributions to the subject.[55]

It should be clear that if the antecedent theoretical meaning of the terms of the two theories is taken cognizance of there are certainly two different types of "relativity": Lorentz's theory maintains *absolute* velocities, Newtonian universal time, an ether, and fundamental asymmetries; Einstein's theory argues for *relative* velocities, a new time concept which is incompatible with Newtonian kinematics, repudiation of the ether, and a fundamental symmetry of all transformations. Thus attention to the antecedent theoretical meaning of the two theories already discloses a number of basic differences between the two theories.

If we now apply the tricategorical logic of comparative theory evaluation to the two theories, and of course I shall now only summarize what has been argued in more detail earlier, further differences between the two theories are brought forth. The conflicts between the theoretical context

---

[55] Cf. Born (1956), Holton (1960), and Grünbaum (1961) for earlier critiques of Whittaker's position.

Kenneth F. Schaffner

and the Lorentz theory were not nonexistent, but they were less fundamental than were the conflicts between the accepted theories of the time and Einstein's theory. Lorentz's theory, if extended to all mechanics as Poincaré (1909) suggested, would result in a *new* mechanics owing to science taking account of new forces and force transformation laws. But the potential conflict with previous mechanics entailed by such an extension was not as immediate and fundamental as that between Einstein's new kinematics and the then accepted theories of physics. Einstein's theory also eliminated the ether, whereas Lorentz's theory did not, and this was at the time a significant difference for many physicists who considered the ether a precondition of the intelligibility of any field theory.

Though both theories were comparable as regards experimental adequacy, they did differ in the ways in which they accounted for the various experiments which supported them. The account of the interrelations of theory and experiment given above in sections II and IV B, and applied in section VIII C, suggests further differences between the Einstein and Lorentz theories. Finally the application of the category of relative simplicity in its dimensions of fitness, ontological simplicity, and system simplicity also indicates some interesting differences between the two theories. The relevance and logical status of Einstein's objections against Maxwell's and Lorentz's theories, on the grounds that they involved asymmetries in the phenomena of induction, are disclosed. Further, the difference between Einstein's full covariance and Lorentz's *approximate* covariance is best measured according to this logic in terms of the subcategory of relative fitness. (This particular difference does not apply to the relation between Einstein's and Poincaré's modification of Lorentz's theory.) The systematization, or system simplicity, which Einstein achieved in comparison with Lorentz, and the different roles of the principle of relativity—for Einstein a fundamental postulate, for Lorentz an approximately true *theorem*—also indicate differences between the two theories.

With the aid of such philosophical distinctions, as informed with a detailed knowledge of the historical case which is being studied, a historian could perhaps understand better why Whittaker might have made a mistake in attributing "relativity" theory to Lorentz and Poincaré, though such a combination of historical and philosophical knowledge will not account for Whittaker's refusal to acknowledge Einstein's fundamental and revolutionary contributions to the subject.

Though I argue the case for the relevance of philosophical distinctions

350

and philosophical analyses for the historian of science, I should not want to overreact and claim that the philosopher of science has little *more* to learn from the historian of science than he now knows. Certainly the historical school's philosophical thesis of the irrationality and incommensurability of the relations between historically successive scientific theories is not borne out by the careful study of the relations between the Lorentz and Einstein theories, and in spite of what I find to be the fruitfulness of some of the claims which the historical school makes, I think the claims would have been better founded on a more accurate view of the history of science.

The relation of the Lorentz and Einstein theories as set out here is based on and supported by a number of historical studies. It is hoped that the outlines of the logic of comparative theory evaluation sketched out in these pages might in turn admit of some further application by historians attempting to reconstruct the chronological record of science.

BIBLIOGRAPHY

Born, M. (1909). "Die träge Masse und das Relativitätsprinzip," *Annalen der Physik,* 28:571–584.
———— (1956). "Physics and Relativity," in his *Physics in My Generation.* London: Pergamon. Pp. 189–206.
Carnap, R. (1956). "The Methodological Character of Theoretical Concepts," in *Minnesota Studies in the Philosophy of Science,* vol. I, ed. H. Feigl and M. Scriven. Minneapolis: University of Minnesota Press. Pp. 38–76.
Cassirer, E. (1923). *Einstein's Theory of Relativity,* trans. W. C. Swabey and M. C. Swabey. Chicago: Open Court.
Clark, J. T. (1962). "The Philosophy of Science and the History of Science," in *Critical Problems in the History of Science,* ed. M. Clagett. Madison: University of Wisconsin Press. Pp. 103–140.
d'Abro, A. (1927). *The Evolution of Scientific Thought.* New York: Dover.
de Sitter, W. (1913). "Ein Astronomischer Beweis für Die Konstanz der Licht geschwindigkeit," *Physikalische Zeitschrift,* 14:429.
Drude, P. (1902). *The Theory of Optics,* trans. C. R. Mann and R. A. Millikan. New York: Longmans, Green.
Duhem, P. (1906). *The Aim and Structure of Physical Theory,* trans. P. P. Wiener. Princeton, N.J.: Princeton University Press. (The English edition bears the date 1954.)
Einstein, A. (1905a). "Über einen die Erzeugung und Verwandlung des Lichtes betreffenden heuristischen Geischtspunkt," *Annalen der Physik,* 17:132–148.
———— (1905b). "Zur Elektrodynamik bewegter Körper," *Annalen der Physik,* 17: 891–921. A translation by W. Perrett and G. B. Jeffery appears in A. Einstein et al. (1923). *The Principle of Relativity.* New York: Dover.
———— (1905c). "Ist die Trägheit eines Körpers von seinem Energiegehalt abhängig," *Annalen der Physik,* 18:639–641. Also in the Dover volume; see Einstein (1905b).
———— (1949). "Autobiographical Notes," in *Albert Einstein: Philosopher-Scientist,* ed. P. A. Schilpp. New York: Tudor. Pp. 2–95.

Kenneth F. Schaffner

Feyerabend, P. (1962). "Explanation, Reduction, and Empiricism," in *Minnesota Studies in the Philosophy of Science*, vol. III, ed. H. Feigl and G. Maxwell. Minneapolis: University of Minnesota Press. Pp. 28–97.

————— (1965). "Problems of Empiricism," in *Beyond the Edge of Certainty*, ed. R. Colodny. Englewood Cliffs, N.J.: Prentice-Hall. Pp. 145–260.

Fizeau, H. (1851). "Sur les hypotheses relatives à l'éther luminieux, et sur une experiènce qui parâit demontrer que le mouvement des corps change la vitesse avec laquelle la lumière se propage dans leur interieur," *Comptes Rendus*, 33:349–355.

Foster, M. H., and M. L. Martin, eds. (1966). *Probability, Confirmation, and Simplicity*. New York: Odyssey Press.

Goldberg, S. (1969a). "Early Response to Einstein's Special Theory of Relativity, 1905–1911: A Case Study in National Differences." Unpublished Ph.D. thesis, Harvard University.

————— (1969b). "The Lorentz Theory of Electrons and Einstein's Relativity," *American Journal of Physics*, 37:982–994.

————— (1970a). "In Defense of Aether: The British Response to Einstein's Relativity, 1905–1911," *Historical Studies in Science*, in press.

————— (1970b). "The Abraham Theory of the Electron: The Symbiosis of Theory and Experiment," *Archives for the History of the Exact Sciences*, in press.

Goodman, N. (1966). *The Structure of Appearance*. 2nd ed., Indianapolis: Bobbs-Merrill.

Grünbaum, A. (1961). "The Genesis of the Special Theory of Relativity," in *Current Issues in the Philosophy of Science*, ed. H. Feigl and G. Maxwell. New York: Holt, Rinehart and Winston. Pp. 43–55.

————— (1963). *Philosophical Problems of Space and Time*. New York: Knopf.

————— (1964). "The Bearing of Philosophy on the History of Science," *Science*, 143:1406–12.

Hanson, N. R. (1958). *Patterns of Discovery*. Cambridge: Cambridge University Press.

Hempel, C. G. (1965). *Aspects of Scientific Explanation*. New York: Free Press.

Hertz, H. (1894). *The Principles of Mechanics*, trans. D. E. Jones and J. T. Walley. New York: Dover.

Hesse, M. (1961). *Forces and Fields*. London: Nelson.

Hirosige, T. (1962). "Lorentz's Theory of Electrons and the Development of the Concept of Electromagnetic Field," *Japanese Studies in the History of Science*, 1:101–110.

————— (1966). "Electrodynamics before the Theory of Relativity, 1890–1905," *Japanese Studies in the History of Science*, 5:1–49.

————— (1968). "Theory of Relativity and the Ether," *Japanese Studies in the History of Science*, 7:37–53.

Holton, G. (1960). "On the Origins of the Special Theory of Relativity," *American Journal of Physics*, 28:627–636.

————— (1967–68). "Influences on Einstein's Early Work in Relativity Theory," *American Scholar*, 37:60–79.

Jammer, M. (1961). *Concepts of Mass*. Cambridge, Mass.: Harvard University Press.

Klein, M. J. (1967). "Thermodynamics in Einstein's Thought," *Science*, 157:509–516.

Kuhn, T. S. (1962). *The Structure of Scientific Revolutions*. Chicago: University of Chicago Press.

Lakatos, I. (1968). "Criticism and the Methodology of Scientific Research Programmes," *Proceedings of the Aristotelian Society*, 69:149–186.

————— (1970). "Falsificationism and the Methodology of Scientific Research Programmes," in *Criticism and the Growth of Knowledge*, ed. I. Lakatos and A. Musgrave. Cambridge: Cambridge University Press, in press.

Laue, M. (1907). "Die Mitführung des Lichtes durch bewegte Körper nach dem Relativitätsprinzip," *Annalen der Physik*, 23:989–990.

————— (1911). *Das Relativitätsprinzip*. Braunschweig: Vieweg und Sohn.

A LOGIC OF COMPARATIVE THEORY EVALUATION

Lewis, G. N., and R. C. Tolman (1909). "The Principle of Relativity and Non-Newtonian Mechanics," *Philosophical Magazine*, 18:510–523.

Lorentz, H. A. (1892a). "La Théorie Électromagnétique de Maxwell et Son Application aux Corps Mouvants," *Archives Néerlandaises*, 25:363–552.

——— (1892b). "De Relatieve beweging van de arde en den aether," *Verhandelingen Akademie van Wetenschappen, Amsterdam*, 1:74–79.

——— (1895). *Versuch Einer Theorie der elektrischen und optischen Erscheinungen in bewegten Körpern*. Leyden: Brill.

——— (1904a). "Elektronentheorie," in *Encyklopädie der mathematischen Wissenschaften*, vol. II. Leipzig.

——— (1904b). "Electromagnetic Phenomena in a System Moving with Any Velocity Less Than That of Light," *Proceedings of the Academy of Sciences of Amsterdam*, English edition, 6:809.

——— (1909) (1915). *The Theory of Electrons*. 1st ed. (1909), New York: Columbia University Press. 2nd ed. (with same text pagination but divergencies in notes), New York: Dover.

——— (1914). *Das Relativitätsprinzip. Drei Vorlesungen, gehalten in Teylers Stiftung zu Haarlem*. Leipzig.

McCormmach, R. (1970). "Einstein, Lorentz, and the Electron Theory," *Historical Studies in the Physical Sciences*, 2, in press.

——— (unpublished). "H. A. Lorentz and the Electromagnetic View of Nature," privately circulated.

Maxwell, J. C. (1867). "On the Dynamical Theory of Gases," *Philosophical Transactions of the Royal Society of London*, 157:49–88.

Michelson, A. A., and E. W. Morley (1886). "Influence of Motion of Medium on the Velocity of Light," *American Journal of Science*, 31:377–386.

——— (1887). "The Relative Motion of the Earth and the Luminiferous Ether," *American Journal of Science*, 34:333–345.

Minkowski, H. (1908). "Die Grundgleichungen für die elektromagnetischen Vorgänge in bewegten Körpern," *Göttingen Nachrichten*, pp. 53–111.

Nagel, E. (1961). *The Structure of Science*. New York: Harcourt, Brace and World.

Panofsky, W., and M. Phillips (1955). *Classical Electricity and Magnetism*. Reading: Addison-Wesley.

Pauli, W. (1958). *The Theory of Relativity*. Oxford: Pergamon. (Originally published in German in 1921.)

Planck, M. (1906). "Das Prinzip der Relativität und die Grundgleichungen der Mechanik," *Verhandlungen der Deutschen Physikalischen Gesellschaft*, 7:136–141.

Poincaré, H. (1900). "Relations entre la physique expérimentale et la physique mathématique," *Rapports du Congrès de Physique de 1900*. Paris. Pp. 1–29.

——— (1901). *Électricité et Optique*. Paris: Gauthier-Villars.

——— (1906). "Sur la dynamique de l'électron," *Rendiconti del Circolo Matematico di Palermo*, 21:1206–76.

——— (1909). "La Méchanique Nouvelle," in *Mathematischen Vorlesungen an der Universität Göttingen*. Leipzig: Teubner. Pp. 49–58.

Popper, K. R. (1959). *The Logic of Scientific Discovery*. New York: Basic Books. (A translation and enlargement of *Logik der Forschung*, 1934.)

——— (1966). "A Note on the Difference between the Lorentz-Fitzgerald Contraction and the Einstein Contraction," *British Journal for the Philosophy of Science*, 16:332–333.

Putnam, H. (1962). "The Analytic and the Synthetic," in *Minnesota Studies in the Philosophy of Science*, vol. III, ed. H. Feigl and G. Maxwell. Minneapolis: University of Minnesota Press. Pp. 358–397.

Reichenbach, H. (1957). *The Philosophy of Space and Time*, trans. M. Reichenbach and J. Freund. New York: Dover. (Originally published in German in 1928.)

Kenneth F. Schaffner

Rudner, R. (1961). "An Introduction to Simplicity," *Philosophy of Science*, 28:109–115.

Schaffner, K. F. (1969a). "The Lorentz Electron Theory and Relativity," *American Journal of Physics*, 37:498–513.

———— (1969b). "Correspondence Rules," *Philosophy of Science*, 36:280–290.

———— (1969c). "The Watson-Crick Model and Reductionism," *British Journal for the Philosophy of Science*, 20:325–348.

———— (1970). *Nineteenth Century Aether Theories*. Oxford: Pergamon, in press.

Scheffler, I. (1967). *Science and Subjectivity*. Indianapolis: Bobbs-Merrill.

Shapere, D. (1964). "The Structure of Scientific Revolutions," *Philosophical Review*, 73:383–394.

———— (1965). *Philosophical Problems of Natural Science*. New York: Macmillan.

———— (1966). "Meaning and Scientific Change," in *Mind and Cosmos*, ed. R. Colodny. Pittsburgh: University of Pittsburgh Press. Pp. 41–85.

———— (1970). "Explanation and Scientific Progress," forthcoming in a volume of the *Boston Studies in the Philosophy of Science*, ed. R. S. Cohen and M. W. Wartofsky.

Sommerfeld, A. (1952). *Mechanics*, trans. M. O. Stern. New York: Academic Press.

Toulmin, S. (1961). *Foresight and Understanding*. Bloomington: Indiana University Press.

Williams L. P. (1965). *Michael Faraday*. New York: Basic Books.

———— (1966). *Origins of Field Theory*. New York: Random House.

Whittaker, E. T. (1953). *A History of the Theories of Aether and Electricity*, vol. II, *The Modern Theories*. London: Thomas Nelson.

Wittgenstein, L. (1953). *Philosophical Investigations*, ed. G. E. M. Anscombe and B. Rhees. Oxford: Blackwell.

COMMENT BY HOWARD STEIN

1. A point of historical-theoretical explication: The issue of action-reaction between electrons and the ether, or of electromagnetic momentum, was in one way (cf. note 18 of my paper in this volume) a stage in the elimination of the ether. For according to Lorentz, as Professor Schaffner has said, the ether is—i.e., all its parts are—immobile. If the ether were a material medium interacting with charged bodies, its parts would receive momentum through such interaction; this is what Maxwell conceived. Lorentz's dilemma was this: if the ether is rigorously immobile, either the third law of motion (i.e., conservation of momentum) fails, or ethereal (i.e., electromagnetic) momentum is unrelated to ethereal velocity. Either constitutes a serious falling-off from the classical notion of a dynamical theory: in Professor Schaffner's terms, a failing in theoretical context sufficiency.

2. In discussing the role of a positivist attitude in Einstein's work, I think it important to distinguish between what might be called critical positivism on the one hand and a kind of blanket epistemological positivism on the other. In the history of science, as it seems to me, critique of a (roughly) positivist or empiricist type has on several occasions been ex-

tremely valuable—and I think the Einstein-Lorentz case is an example; but what has been of exceedingly dubious value is the attempt to describe a certain procedure of concept construction as a requirement on all notions to be introduced into theoretical science. That is to say—and this is a general normative moralistic remark—a concept ought not to be required to exhibit its credentials in order to be entertained in a theory; but when one has a theory, and one examines the mathematical structure of the theory (for example) and shows that a certain notion appearing in it can be eliminated, then one has shown that that notion is devoid of content. This I think is what Einstein did vis-à-vis Lorentz—in that version of the Lorentz theory which is mathematically equivalent to his own. Einstein's critique, eliminating the "velocity vector attached to empty space," of course involved still more fundamentally the time interval between two events as an absolute notion, and in particular the notion of absolute time-interval zero (i.e., absolute simultaneity). Einstein did not simply impugn this notion on epistemological grounds: he showed that the notion could be sacrificed in a certain definite way, without violence to any known facts, and with notably improved justice to some facts. (I say *improved* justice because the version of Lorentz that was equivalent to Einstein had after all not yet been formulated; and although, as Professor Schaffner has said, the issue of experimental adequacy was uncertain, the *elusiveness* of "motion relative to the ether" did already stand as a very notable circumstance.)

Again, as to the elimination of that velocity vector attached to empty space, I think one should not overemphasize the epistemological or ontological importance for Einstein of this elimination per se: he was, after all, perfectly willing to attach the electromagnetic field tensor to empty space. In one version of the controversy about the ether, we do really face a verbal quibble: whether to call the electromagnetic field itself "the ether"—as if it were a serious issue whether one uses that word or not. Setting this verbal question aside, the possibility of vectors or fields or what have you attached to empty space is certainly one that Einstein was happy with: it always formed an essential part of his notion of the structure of physics, and hence of reality. His critique consisted in asking, of a theoretical term or principle, whether it does in fact have an effect—and an effect which contributes positively to the theory's experimental adequacy.

355

COMMENT BY ARNOLD KOSLOW
I. The Three Criteria

Professor Schaffner has generalized Hertz's criteria of *Zulässigkeit* (permissibility), *Richtigkeit* (correctness), and *Zweckmässigkeit* (appropriateness) in order to show how certain theories of the physical sciences may be compared and ranked. His stated aim is to provide criteria which "will resolve some of the paradoxes raised by the historical school of the philosophy of science."

He has very ably described those "paradoxes" and there is no need to rehearse them once more. My first point is that contrary to Professor Schaffner's claim, the criteria introduced by him do not resolve the critical issues raised by the "historical school." These criteria, it seems to me, cannot be applied to the theories in question unless some resolution is made or answer is given to the criticisms of the historical school. The application of Schaffner's criteria do not provide a resolution, they presuppose that one has been given. Perhaps some examples will make the point more clear. Schaffner's parallel for Hertz's *permissibility* is *theoretical context sufficiency*, by which he refers to "the amount of concordance between a theory to be assessed and the corpus of accepted scientific knowledge." Hertz's criterion of permissibility required consistency, and Schaffner's generalized version refers to the concordance of theories. Obviously it will not do to speak of the amount of consistency, but "concordance" and "the amount of concordance" seem to me to require explication if we are to regard theoretical context sufficiency as a serviceable notion. If, however, Schaffner does mean consistency, or something which at least requires consistency, then it seems to me that he has begged those issues raised about the commensurability of theories. When Einstein compared his theory with Lorentz's, he stated that "the introduction of a 'luminous aether' will prove to be superfluous inasmuch as the view here to be developed will not require an 'absolutely stationary space' provided with special properties, nor assign a *velocity* vector to a point of the empty space in which *electromagnetic* processes take place."[1] In the context of the quotation, Einstein uses the terms "material point," "body," "velocity-vector," and "electromagnetic process." However, if certain members of the "historical school" are correct, the terms mentioned above have different meanings, even though they are shared by both theories. Indeed, in one

[1] *The Principle of Relativity*, trans. W. Perrett and G. G. Jeffery (New York: Dover, 1923), p. 38.

356

passage, Schaffner indicates that there are important differences of meaning for the term "velocity" in both theories. I do not wish to defend these theses of the historical school because I think that most of them are false or ill-grounded. But I think that a simple application of the criterion of theoretical context sufficiency does not resolve the basic problems or paradoxes of that school.

The same kind of point can be raised against Schaffner's use of the criteria of experimental adequacy and appropriateness or simplicity. If we take the puzzles of the historical school seriously, then according to some of its members, all the terms of different theories (relative of course to some criteria for individuating theories which have so far remained unspecified) will differ. In particular, "body" will have different meanings in the theories of Lorentz and Einstein, and the purportedly contrasting assertions about the way bodies behave may not contrast at all. (I say "may," since two sentences sharing a term may be contradictory even though the term occurs in each sentence with a different meaning; indeed they may be contradictory even though they have no term with the same meaning in common.) All the problems central to the historical school—the issues of meaning variance and the consequent incommensurability of theories— are untouched by Schaffner's criteria. These issues would first have to be settled if the criteria are to be applied at all.

I find it surprising that one could resolve the paradoxes of the historical school by granting, "absorbing," and not contradicting any of that school's major theses. It does sound as if Schaffner is saying that with suitable adjustments—presumably acceptable to that school, but maybe not—he can settle some of their outstanding paradoxes. It is hard to understand how Schaffner can accept, absorb, and not contradict all the major theses of that school, and still argue, counter to their view, for the comparative evaluation of theories. He does not contradict their logic, their epistemology, or their pronouncements on meaning. He does present some arguments concerning what he calls O-sentences and C-sentences. But from what I can detect, nothing that he says constitutes a challenge to the historical school. I have some difficulty understanding his characterization of the primary and secondary senses of O-sentences partly because there is a shift in his exposition between O-sentences, O-terms, and O-predicates. I take it that Schaffner's main concern in making these distinctions is to show how, given two theories, there could be an intertheoretic common base. But his distinction is far from being a clear one. For example, he con-

357

Kenneth F. Schaffner

siders two competing theories of optics and introduces O_a, an O-sentence which describes a fringe pattern on a surface illuminated by a monochromatic beam through a double-slit screen. According to Schaffner, the primary sense of the O-predicate is "the referent [sic], the bandlike appearance one experiences in learning the meaning [sic] of 'optical fringe phenomena.'" He also states that the "second sense is that which is associated with the O term by virtue of the term's being embedded or hooked into a theory through correspondence rules or C-sentences."

It would appear that not only do certain terms have a meaning, they also have primary and secondary senses as well. Schaffner's distinction of two senses therefore does not obviate all reference to meaning; it depends upon some notion of meaning. Further, I am not convinced that his notion of the primary sense of an O-term does the work he requires of it. He would like to say that different theories such as Lorentz's and Einstein's overlap in their observation terms, and that the proponents of those theories agree in the primary sense of those terms. But why should it be the case that two physicists have had the same experience when they learned the meaning of a shared O-term? Trivially, one of them could have learned the meaning through pictures, another through slides, another through diagrams, and another by means of glances through the telescope of an interferometer properly set up. There is no special reason why the proponents of the same theory would agree on the primary sense of an O-term, as Schaffner has described it. Further, Schaffner claims that it is only with respect to the second sense of an O-sentence that we speak of that sentence as evidence for one theory or another. I assume that Schaffner would also accept the related view that only in the secondary sense of an O-sentence do we speak of wave optics, for example, explaining why the sentence is true. Without further qualification, Schaffner's description of secondary senses permits as many secondary senses of a specific O-sentence as there are "different" theories which that sentence is "embedded or hooked into through correspondence rules or C-sentences." Thus if it is true that M's being evidence for a theory R rests on the relation of a secondary sense of M to T, then we may find that M cannot be evidence for both of two theories T and T'. For in the former case, it is some secondary sense of M that stands in a certain relation to T; in the latter, it is again some secondary sense of M which stands in that relation to T'. If "different" theories T and T' imply different secondary senses for O-sentences related to the theories by C-sentences, then T and T' have no common evidential base. According to a similar argument, different theories can have no explanations in common. Schaffner maintains that both theorists can use the Michelson-Morley experimental results. Thus he writes, "It is the overlap of the observation terms, and the ability of both absolutist and relativist to agree about the primary sense of the observation terms and O-sentences, that permits *both* theorists to use the Michelson-Morley experimental results." And later, in the same vein, "In this [Michelson-Morley] experiment there is a positive fringe shift of measurable amount. The two theorists would, however, as this account of O-sentences permits, take the O-sentences as evidence for, or indexes of, quite different aspects of nature's regularity."

Professor Schaffner states that both theorists can use the Michelson-Morley experimental results. But these results, as he refers to them, are not part of any intertheoretic common base between Lorentz's and Einstein's theories. In fact, the results are irrelevant to the support of either theory, for they cannot even be used for evidence. No common description of the experimental results would figure in a record of the support, the evidence, the explanatory or predictive successes of either theory. If we cannot use experimental results such as the outcome of the Michelson-Morley experiments to convince ourselves and others of the merit of a theory, then what use of experiment can Schaffner have in mind?

II. Ontology: The Ether and Simplicity

Professor Schaffner's use of relative simplicity raises an interesting question. He includes differences in the ontological commitment of theories under the criterion of simplicity. He seems to think of the repudiation of the ether as an ontological loss for Einstein's theory, and a gain in relative simplicity. I have some questions about how these two observations are related. Is it supposed to be true that an ontological loss is always a gain in simplicity? I think not. There are the obvious objections that the statements of the theory may become terribly complex with a change in ontology. In Quine's terminology, ontological shifts may be offset by ideological ones.[2] Moreover there is in the Lorentz-Einstein situation a very special complicating factor. It may count against the simplicity of a theory if it reduces the totality of causal problems under investigation. Galileo

---

[2] Having said this, I should also add that I think that the general principle put forward by Schaffner is false. It is not true that "a broader and more complex theory will be acceptable over a simpler and narrower theory." Suppose that T is broader than T', and more complex. We may *not* prefer T to T', if T yields very "shallow," not very detailed answers to a broad spectrum of problems, while T', though narrower in scope, gives good "in-depth" answers.

reduced the number, according to a well-known story, by claiming that only deviations from inertial motions required causal explanation. Lorentz, and many with him, devoted their energies to a host of problems involving the explanation of electromagnetic processes in terms of the interaction of the ether and charged bodies. Einstein's ontological move, if it was one, implied that a certain class of scientists, Lorentz included, had a good deal less explanatory work on their hands than they supposed. It is therefore not clear to me that an ontological reduction results automatically in a simpler theory. Second, is it so plain that Einstein's maneuver with respect to the ether is best understood as an ontological shift? Has there been a reduction in, say, Quine's sense?[3] As far as I know, there has been no careful investigation which establishes the existence of a suitable proxy function. And I am not convinced that the omission of a singular referring term like "ether" results in an automatic ontological shift downward. One reason for skepticism in this matter has to do with the belief that ontology concerns kinds of things, not specific items. It might be objected that the term "is the ether" does represent a "kind of thing" in the minimal sense that it is a general term. But I do not think so, since one consideration relevant here is whether there are principles of individuation. Are there criteria for determining where one ether begins and another leaves off? I think not, since the ether, in the nature of the case, neither begins nor leaves off. A similar argument could be given for the notion of electron in various theories of electrons, and the notion of body in certain nineteenth-century hydromechanical theories of matter. But this is a large problem which cannot be pursued here. All I wish to register here is my skepticism that the removal of a singular referring term from a theory represents an ontological simplification.

The problem is complicated somewhat by Einstein's later "reintroduction" of the ether in 1920.[4] The reintroduction makes it hard to understand the claim, often put forward, that the theory of relativity implies that there is no ether. If the difference is, as Pauli described it in 1921,[5] that Lorentz believed that the ether had mechanical properties, that it

---

[3] W. V. Quine, "Ontological Reduction and the World of Numbers," in The Ways of Paradox and Other Essays (New York: Random House, 1966); also "Ontological Relativity," in Ontological Relativity and Other Essays (New York: Columbia University Press, 1969).

[4] A. Einstein, Äther und Relativitäts-Theorie: Rede gehalten an der Reichs-Universität zu Leiden (Berlin: Springer, 1920).

[5] W. Pauli, The Theory of Relativity, trans. G. Field (Oxford: Pergamon, 1958; original German publication 1921), p. 4.

was a *substance*, but Einstein did not, then it is not apparent that the difference is a simple one over the existence of the ether. It looks as if the difference concerns the properties of the ether, not whether the ether exists. Einstein's later view of the matter is surely not an irrelevant consideration. In *Äther und Relativitäts-Theorie* (1920) he plays down the idea that the special theory of relativity differs in ontology from Lorentz's theory of 1909. He offers instead, a picture of the continuous development of physical theory which, in its latest phase, involved programs intended to reduce electromagnetic theory to mechanics, and the reverse program of the believers in an electromagnetic world view. Hertz's theory, according to Einstein, was experimentally inadequate and the Hertzian ether as the bearer of both mechanical properties and electromagnetic fields was basically no different from the ponderable matter in it. Einstein argued that Lorentz's great contribution was his bringing of theory into accord with experiment by a remarkable simplification of the theoretical foundations. He says that Lorentz "erzielte diesen wichtigsten Fortschritt der Elektrizitätstheorie seit Maxwell, indem er dem Äther seine mechanischen, der Materie ihre elektromagnetischen Qualitäten wegnahm." Einstein thought that Lorentz left the ether with only one mechanical property, namely, immobility, and that the entire change which the special theory of relativity brought to the conception of the ether was that it denied even that quality to the ether. The concept of the ether, however, was not banished from theory in this view, but all hope for the program of reducing electromagnetics to mechanics vanished once the ether was denied any mechanical properties. And the proponents of electromagnetic world views no longer had to worry that electromagnetic theory would always be parisitic upon mechanics. For it now looked as if it was not true that states of matter were states of a something which has *some* mechanical properties. No wonder, then, that for these and other reasons, backers of electromagnetic world views could regard Einstein (after a while) as one of their own. Einstein's picture of the sequence of theories formulated by Maxwell, Hertz, Lorentz, and himself is one which, in one dimension, is marked by a continuous development. This is not of course to deny the revolutionary character of his theory: the special theory of relativity as Einstein described it in 1920 was both revolutionary and traditional. His reference to the concept of the ether was one way in which he described the tradition to which the special theory belonged. It should be noted that in a lighter vein, Philipp Frank once said that at a Solvay Congress, Einstein was particularly disturbed by the arguments which many younger physicists were using: if the effect cannot be detected, then it does not exist. When Frank reminded Einstein that these arguments were only variants of Einstein's, he replied: "Yes. But a good joke shouldn't be repeated too often." Despite the vagueness of some of the preceding remarks, it seems clear that in the absence of careful representations of the theories under consideration, and without any fairly definite criterion for the ontological commitment of a physical theory, the claim that there is an ontological difference between the theories of Lorentz and Einstein is not yet a result which ought to be taken for granted.

III. Schaffner's Criteria and the Ranking of Theories

Aside from problems related to the application of Schaffner's criteria, there are some questions about the role he thinks they have in the evaluation of theories. I think that he does not wish to say that *theoretical context sufficiency, experimental adequacy,* and *simplicity* are relevant only to the Lorentz-Einstein discussion. He intends that they be relevant to all cases of theory evaluation and choice, and the Lorentz-Einstein discussion is used to illustrate how his criteria work. But how do they function? What is Schaffner's view about the way in which these criteria are related to the choices scientists make, or even the preferences that they have between theories? If a scientist prefers theory T to T', must T be ranked over T' by all of the three criteria, one, or none? Are the three criteria jointly necessary and sufficient for T to be preferred over T', or only sufficient? Further, there is a view upon which the criteria may be neither necessary nor sufficient, taken singly or jointly. It is the view that a scientist's preferences are set or determined by some combination or "weighting" of the three criteria. I shall call this view *combinatorial.* It seems to me that Professor Schaffner has not clarified how the preference between theories is related to the three criteria he has isolated. There is an additional problem which is not settled by a mere listing of criteria, no matter how carefully described. When a list of relevant criteria is given, is it intended that these criteria will be the only ones which enter into or are relevant to a preference, or that these are the only central or "dominant" criteria? Is it excluded that, over time, the list may be altered, new criteria introduced and old ones dropped? If Schaffner's view is combinatorial, then does he allow that even though the criteria might be the same over time, the "weighting" might vary? If the weighting can shift over time, can it also shift over people—can different members of a scientific community employ significantly different combinations of shared criteria? These, it seems to me, are some of the important questions for a logic of comparative theory evaluation; they cannot be settled by a careful characterization of the criteria which enter into such a logic.

Consider Professor Schaffner's summation toward the end of his paper, "In summary, then, the three categories of comparative theory evaluation which we have been considering yield a split decision, in, say, 1905-7: theoretical context sufficiency supports Lorentz's theory, relative simplicity supports Einstein's, and experimental adequacy, because of the flexibility of the interpretation of 'observation' statements, selects neither theory as the better one."

Professor Schaffner calls this situation a "split decision." But who is split on this? Certainly not Lorentz, and certainly not Einstein. For each, even in the light of these rankings, preferred his own theory over the other, at the time in question. On a combinatorial view, these remarks seem to show that if Lorentz and Einstein thought that the only relevant considerations are the three introduced by Schaffner, then they weighted the criteria differently; sufficiently so that their preferences were the reverse of each other. Schaffner's researches, coupled with some combinatorial view of theory evaluation, yield interesting conclusions about the way in which scientists aggregate or weigh the agreed-upon relevant criteria.

If it is not Lorentz's split decision, and not Einstein's, then it seems that it is ours or Schaffner's, or possibly some contemporary of Einstein's. Why is the decision split? This is not evident just from the tally which Schaffner offers. On one score (experimental adequacy), neither theory at the time outranks the other. The other two criteria give high marks, one to each theory. Schematically, we might construct a table:

	Lorentz	Einstein
Theoretical context sufficiency	$+$	$-$
Experimental adequacy	$=$	$=$
Relative simplicity	$-$	$+$

where it is supposed that Lorentz, Einstein, and others who had to make a decision would agree that this table represents the results of the three comparisons at the time in question. Why does Schaffner describe his result as a split decision? Is it because each theory has held its own, or done better, on two of the three rankings, or is it split because neither theory

Kenneth F. Schaffner

has the advantage with respect to experimental adequacy, and an advantage in theoretical context sufficiency cancels a disadvantage in relative simplicity? Schaffner has described some of the criteria which practicing scientists take into account in making their preferences between theories. He has not described the preference pattern of scientists. I should like to say that he has described the components upon which preferences depend, but he has not described how the total preference of working scientists depends upon those components. To that extent, his logic of comparative theory evaluation falls short of the stated task.

COMMENT BY PETER A. BOWMAN

Professor Schaffner evidently assigns equal weight to the three categories of comparative theory evaluation, but one can question whether that is justified. Surely, practicing scientists give more weight to a theory's being experimentally adequate than to its being theoretical context sufficient. We see a clear example of this in the case at hand through 1905, when Lorentz and Poincaré are going to great lengths to modify the theory of electrons so as to be able to incorporate the results of the Michelson-Morley and other second-order experiments; i.e., they adjust the theoretical context to fit experimental results. Further, it would seem that consideration of relative simplicity comes into play only if considerations of the other two categories prove indecisive. For example, in 1915 Lorentz can only consider the simplicity of Einstein's approach if it is given that (a) there is little chance of an experiment's revealing an asymmetry between the ether frame and the moving frame; thus, it is unlikely that any experiment will show Lorentz's theory to be more adequate than Einstein's; (b) the concomitant development of atomic theory renders risky at best the sort of basic assumptions about the constitution of matter made in the theory of electrons; thus, it is no longer desirable to make the detailed theoretical assumptions about the electronic context of electrodynamics which the latter theory, but not Einstein's, does.

Therefore, we are forced, I think, to assign priority to Professor Schaffner's categories in the following order: (1) experimental adequacy, (2) theoretical context sufficiency, (3) relative simplicity. Then, assuming that his assessment of how the two rival theories compare in each category is correct, we have to conclude that, considerations as to (1) being ambivalent in 1905, a scientist would be justified in accepting Lorentz's theory on the basis of (2) vis-à-vis Einstein's theory on the basis of (3).

REPLY BY KENNETH F. SCHAFFNER

In replying to the comments on my paper, I think that it would be best to begin with Professor Stein's contribution, which I consider amplificatory of my own position, and then turn to Professors Koslow's and Bowman's comments, which are more critical.

Professor Stein makes two points, both of which I agree with. His comments about the failure of Lorentz's theory to agree in toto with Newtonian mechanics concern something which I drew attention to in my paper, though I did not put it in quite the way Stein does. Stein's second point is indirectly concerned with my views on "antecedent theoretical meaning," but it is more explicitly relevant to a discussion by Professors McMullin, Cohen, Buchdahl, and Feigl, which ensued after the paper was presented, and which revolved around the extent to which the positivistic and operationalistic attitude of Einstein was responsible for the special theory of relativity, with its denial of the ether.

Apropros of the notion of antecedent theoretical meaning which was advanced in my paper in section II, and earlier elsewhere (Schaffner, 1969b), one can say, in part, that the ether was prima facie scientifically meaningful by virtue of (1) the antecedently understood properties definitionally associated with it, and (2) the hypothetical characteristics imputed to the ether in terms of these and other antecedently understood properties. For Lorentz, the stationary ether possessed experimental consequences, sentences expressing these consequences being derivable from antecedently meaningful statements concerning the ether, the electromagnetic field and electrons, and parts of associated theories, such as mechanics. Lorentz believed that Einstein, by *accepting* a principle of relativity, eliminated the *possibility* of the ether possessing experimentally accessible consequences; it was not the case that the ether had *necessarily*, by 1905, been stripped of its consequences by experiments such as that of Michelson and Morley. I went to some lengths to indicate what the situation in 1905 was as regards the ether and experiments in my paper, and do not think it necessary to rehearse that material again.

In the unreproduced discussion which followed my paper, Professor Buchdahl suggested that "Einstein . . . believed the suggestion of a velocity vector with respect to the ether *had become meaningless and must be given up* [my italics] and wasn't *this* what was viewed by the positivists as being the antithesis of an ontological [or physical realist's] position?" Professor Feigl supported this interpretation, historically, mentioning that the Vienna Circle, including Schlick, Carnap, and Reichenbach, and in this country Bridgman and C. I. Lewis, believed that the last-ditch defense of the ether (in 1904 and beyond) by Lorentz and Larmor, for example, had made the ether *in principle* immune to the verification or falsification of its existence, and that, accordingly, its elimination by Einstein "fell under the critique of our meaning criterion" (from Professor Feigl's comment).

From the point of view of the history of philosophy this is apparently the case, but it seems to me that the positivists' position is not defensible as a reconstruction of the history of early twentieth-century physics. My position is based on the comments which I made in the previous paragraph concerning Lorentz's conceptions of the ether and the principle of relativity. Similar comments were more extensively presented in my paper. Lorentz, himself, did not think that the ether's effects had been forever masked; had they *in principle* been so masked, he would have been quite willing to relinquish the ether.

Accordingly, though I cannot subscribe to what Professor Stein refers to as a "blanket epistemological positivism" because of my views concerning antecedent theoretical meaning and associated metascientific notions, I would agree with his defense of a "critical positivism," and in fact, think that my subcategory of "fitness" incorporates this "critical positivism" within the analysis presented in my paper.

I must now turn to consider the interesting and incisive comments of Professor Koslow. I shall also reply to Mr. Bowman's comments in the context of my rejoinder to Koslow.

Koslow criticizes several of my theses. First, he contends that my generalized Hertzian criteria for the ranking of competing scientific theories cannot be applied without first resolving the paradoxes concerning the relation between theories and between theories and experiments which have been articulated by the historical school. Koslow writes that "the application of Schaffner's criteria do not provide a resolution; they presuppose that one has been given." He seems persuaded that it is impossible both to accept *many* of the insights of the historical school concerning the epistemology of science and at the same time to discover theory-transcendent criteria. Koslow elaborates on his critique, arguing that my acceptance of some meaning variance of important scientific terms in competing scientific theories raises the possibility that "purportedly contrasting assertions about the way . . . [entities] behave may not contrast at all," if the sense

366

of the concepts concerning the entities changes. Koslow also does not seem persuaded that my rather different analysis of theory and observation—different from the historical school's account, that is—will do the task it is intended to do, since the distinction which I make between the "primary" and "secondary" senses of "observation" terms is, according to him, both (1) too weak, as it does not even require that "proponents of the same theory would agree on the primary sense of an O-term," and (2) too strong, as it is only by virtue of the secondary, more theoretical and non-common sense, that the observation term is characterizable as evidence for some theory.

Let me reply to these criticisms before going on to Koslow's other points.

First, it does not seem to me that the acceptance of aspects of the historical school's epistemology necessarily entails that theory-transcendent criteria cannot be provided. It is true that both Kuhn and Feyerabend do think that this is so, but this is no reason for accepting what they consider is the consequence of their epistemologies. Let me show how Koslow's more specific difficulties in this regard are met and outflanked by the meta-scientific analysis presented in my paper.

Koslow is hesitant on his point that the acceptance of meaning variance may render the theories under question incapable of clashing with the theoretical context and, though this is not quite as important, with each other. He notes that "two sentences sharing a term may be contradictory even though the term occurs in each sentence with a different meaning."

I think that Koslow is correct in his hesitation on this point, because it does appear that this conflict occurs. As I understand the historical school's thesis of meaning variance, the meaning of a term t, such as "velocity" or "mass," changes when it is understood as a part of theory T_1 rather than T_2, if and only if those characteristics which are definitionally true for the term are incompatibly different in T_1 and T_2. There are apparently, in all cases considered by authors such as Kuhn and Feyerabend, sufficiently common concepts or categories which can be taken as a common base to demonstrate this incompatibility.[1]

That conflict between the theoretical context and a new theory does occur in Feyerabend's analysis is indicated by his remark that "it is . . . impossible to define the exact classical concepts in relativistic terms or to

[1] References to the literature, unless otherwise indicated, are to the bibliography appended to my paper in this volume. Discussions which are relevant to this issue are presented by Scheffler (1967), pp. 36–44, and by Shapere (1966), pp. 77–78.

Kenneth F. Schaffner

relate them with the help of an *empirical generalization* [my italics]. Any such procedure would imply the false assertion that the velocity of light is infinitely large."[2] T. S. Kuhn similarly states: "Newtonian mass is conserved; Einsteinian is convertible with energy. Only at low relative velocities may the two be measured in the same way, and even then they are not conceived to be the same."[3]

Koslow is perhaps misled by Feyerabend's comment about the relation between Newtonian mechanics and relativity which is based on the argument cited above. Feyerabend wrote: "The order introduced into our experiences by Newton's theory is retained and improved upon by relativity. This means that the concepts of relativity are sufficiently rich for the formulation of all the facts which were stated before with the help of Newtonian physics. Yet the two sets of concepts are completely different and bear no logical relation to each other."[4]

Assuming that the "logical relation" spoken of here refers to those two relations characterized in Feyerabend's earlier citation, though the historical school can be interpreted as claiming that no *positive* logical relation may be constructible, conflict between the Newtonian theory and the theory of relativity is admitted. In sum, then, it appears that the difference in meaning can and does permit the type of conflict which I refer to in my paper as "the amount of concordance." Since Professor Koslow also finds this locution somewhat unclear, let me attempt to explicate it again.

It is impossible in these pages to attempt to construct a *calculus* which would provide a metrical analysis of the notion of "amount of concordance," though it is not certain that such a calculus is a priori unconstructible. The idea can be made rather clear in a comparative sense, however, though I thought that I had sufficiently done so in the context of the Lorentz-Einstein example discussed in my paper. There, the relative *amount of concordance* between the theoretical context and a new theory, either Lorentz's 1904 theory or Einstein's 1905 theory in this case, is dependent on the fundamentality and extent of the contradiction between the theoretical context and each of the new theories. Einstein's theory, as was stated, conflicted with the very fundamental and extensively employed concepts of Newtonian space and time, whereas Lorentz's theory only contradicted—primarily by enlarging on—some of the dynamical postulates

---

[2] Feyerabend (1962), pp. 80–81.
[3] Kuhn (1962), p. 101.
[4] Feyerabend (1962), p. 89.

concerning moving ions or electrons and molecules. I do not believe that there is any doubt among historians of this period either about the difference in the depth and extent of the conflicts between Einstein's theory and the accepted classical theories, or about the thesis that Einstein's theory was in greater conflict with traditional physics than was Lorentz's theory.

Professor Koslow has also criticized my critique of the historical school's account of the relation between theoretical and observational terms. Here I believe that Koslow does make an important point concerning the "primary" sense of an observational term, when he suggests that it may not be necessary for two physicists to have the same experience when they learn the primary sense of an O-term. This does appear to be a consequence of my analysis as given in the paper. On further reflection, I think it important to be able to generalize the referent to a referent class, and to permit "pictures," "slides," "diagrams," and the like, any and all of which can convey the primary sense of an O-term such as "optical fringe phenomena," or an O-sentence in which such a term appears. Though there are interesting epistemological problems associated with determining the boundaries of this class of referents, I do not think that they are peculiar or crucial to this inquiry.

But this point being granted, I do not see that the distinction into primary and secondary senses does not do the job for which it was proposed. Specifically, Koslow suggests that the distinction cannot fulfill its task because it is only by virtue of the secondary non-common sense that an O-sentence stands in relation to a theory. Consequently, Koslow notes, "if different theories T and T' imply different secondary senses for O-sentences related to the theories by C-sentences, then T and T' have no common evidential base."

Koslow considers this difficulty more specifically in the context of the Michelson-Morley experiment, and the support which it provides for both the Lorentz theory and Einstein's theory. He contends, contrary to my own position, that the interferometer experiment's results are not part of any intertheoretic common base between Lorentz's and Einstein's theories.

I believe that Professor Koslow has not completely understood my account of the relation of the O-sentences to theoretical sentences. Let me present, first, a most elementary example, and then re-present the interferometer example.

369

Kenneth F. Schaffner

Consider the case of an apple which is found each morning on a teacher's desk. The apple can be described as "red," "round," and, if tasted, as having an "apple-like taste." Perhaps simply describing it as an "apple" will be sufficient. The apple can *also* be taken as evidence for the existence of (1) a thoughtful, concerned student, or (2) a diabolical student who is attempting to worry the teacher about the reason for the appearance of the apple each morning. We can assume that (1) and (2) are inconsistent with each other. We can also, depending on our assessment of the situation, ascribe either the secondary sense, "evidence for a thoughtful concerned student," to the apple, or the sense associated with the alternative explanation of the apple's appearance. The adjectives "red" and "round" and the noun "apple" are part of the primary vocabulary in this case and capture the primary sense of the referent, the apple. (This is not to say, of course, that "red" and "round" and "apple" are necessarily atheoretical terms.) A description of the apple in these terms is accepted by all parties concerned, and such a description can be accepted on hearsay by proponents of either explanation (1) or (2), even if a "neutral" observer records the appearance of the apple each morning.

The same type of analysis can be made regarding the Lorentz and Einstein accounts of the Michelson-Morley experiment. The Lorentz and Einstein theories are both theories which are proposed to account for the electrodynamics (and optics) of moving bodies. The earth is a moving body, in both theories, and the interferometer experiment, in both theories, tests for a "second-order" effect of this motion in a very precise fashion. The O-sentence reporting the result of a rotation of the interferometer through 90° states that "no experimentally significant optical fringe shift was discernible." This sentence expresses the primary sense associated with the observation just as a sentence characterizing the apple on the teacher's desk expressed the primary sense associated with that observation. The interferometer result just given, can, with the (possibly tacit) aid of auxiliary theories and C-sentences connecting the observation with the theories, be taken to support either the Lorentz theory, cum L-F contraction, or the Einstein theory, by being shown to be a consequence of these theories, auxiliary theories, and appropriate initial conditions. The O-sentence can then have associated with it, or grafted onto it, the secondary sense which interprets the observation *as support for* this or that theory.

It should be clear that this is not the epistemology of the historical school, which is at this point being defended by Koslow. The historical school would not accept the distinction between the primary and secondary senses ascribable to an O-sentence. What I am defending here is the thesis that both the Lorentzian and the Einsteinian can and do withdraw to a level, characterized by the primary vocabulary, where both agree on the observational evidence. That the Michelson-Morley data were used both by Lorentzians and relativists is a hard fact of scientific history; the rationale for that fact has been re-presented here.

I shall now turn briefly to Koslow's claim that Einstein's theory may not be simpler than Lorentz's because it eliminates a "singular referring term," the ether, which has only one property, immobility. First let me note that the ether, though it is principally characterized by Lorentz as possessing immobility, and as having velocity vectors ascribable to it, in the sense of the ether "wind," also exerts specific causal influences, such as the contraction effect, and may exert other causal influences as yet unrevealed by experiment. Accordingly, it is not simply a substantiated property of immobility, but a more complex entity, at least as regards its effects.

Further, Lorentz did not dispute the hypothesis that the ether had some structure, for as I noted on page 332 of my paper, Lorentz ascribed the existence of electrical and magnetic forces to its state. It is true that Lorentz did not think there was much value in the mechanical analyses of the ether which had been pursued throughout the nineteenth century, but his characterization of the ether did not preclude further analysis of this entity, which he himself, in 1909, regarded "as endowed with a certain degree of substantiality, however different it may be from all ordinary matter" (my italics).

I would not deny Koslow's suggestion, following Quine, that a theory with a simpler ontology but a more complex ideology might be, on balance, more complex than a theory with a more complex ontology. I would hope, however, that the various senses of relative simplicity, including fitness, discussed in my paper, might prove sufficiently sensitive in such a situation and accord with both Quine and our intuitions. I do not see that Koslow has provided any arguments to show that this is not the case.

Finally, let me consider Koslow's difficulty with the joint application of the assessments made in terms of the three theory-evaluative categories. It should be clear from my paper that the total judgment functions in a combinatorial fashion. Contrary to Mr. Bowman's suggestions, it seems to me that theoretical context sufficiency and experimental adequacy have rough-

Kenneth F. Schaffner

ly equal weight, with relative simplicity becoming an important factor when the assessment made by the application of the previous categories is contradictory or indeterminate. (It should be remembered, however, that the theories constituting the theoretical context are themselves tested by experiments, so that *indirectly* experimental adequacy enjoys a paramount position.) I perhaps also ought to mention that though I believe the *categories* to be weighted as I suggested in my paper and again in this rejoinder, in different specific cases of theory conflict there will be differences in the importance of judgments made in the three categories. This will depend on the extent and the fundamentality of the contradictions between a new theory and the theoretical context, and between the new theory, experimental results, and the expectations one has of the theory's need to be able to account for specific experimental results. As a case in point, the acceptance of the old quantum theory probably depends more heavily on black-body radiation *experiments*, and not significantly on relative simplicity.

I shall reply in two different ways to the question which Professor Koslow raises concerning "who" is making a split decision in 1905–7 as regards the worth of the competing Lorentz and Einstein theories. First, insofar as the tricategorical logic proposed in my paper constitutes a *logic*, the question does not have to be answered by naming a particular person or group of persons. It is similar to the question "who believes that the syllogism in Barbara is valid?" Though I have admitted that the proposed tricategorical logic does not admit of any easy and logically "effective" *application*, once the application has been made it is up to Koslow to indicate where an incorrect premise has been admitted, or a fallacious inference made. Since he has not done this, I feel fairly secure that my analysis still stands, and that the "who" from *this* point of view is superfluous.

But if the tricategorical logic is to be relevant to the history of science, it should also be possible to find a group of scientists who did indeed make a split decision in 1905–7, and who came to be convinced of the superiority of Einstein's theory by 1909–10. Also, since both Lorentz and Einstein, as Koslow points out, did not make such a decision themselves, it is necessary to account for their polar positions on the issue.

In my paper I introduced the notion of a "subjective fluctuation" from the scientific consensus. For reasons which I presented in some detail in the paper, both Einstein and Lorentz constituted subjective fluctuations. (See especially pages 337 and 344. Parenthetically related to this idea is the

372

conjecture that great scientific creativity necessarily involves such "subjective fluctuations." Though there are interesting implications which might be based on this conjecture, to consider them further would take us beyond the scope of this rejoinder.)

Were the sociology of science a well-developed discipline instead of a still relatively new branch of science, it would be tempting to compare the results of my "logical" analysis with the sociologically determined consensus of science in the years 1905–10. Since this is not a viable possibility, I will close by going out on a limb, and suggesting that the verdict of a "split decision" as regards the comparative worth of Lorentz's and Einstein's theories in 1907, say, is to be predicated of the "group mind" of the scientific elite primarily in Germany—France, England, etc., not having yet been in sufficient journalistic contact with Einstein's theory at this time. The sense of the term "group mind" is not very different from E. Durkheim's term *l'âme collective*, which is discussed in his classic *The Rules of Sociological Method.*[5] According to Durkheim, such a "group mind" ought to be reflected in an appropriate statistical sampling composed of individual members of this elite, and should be revealed in scientific journal articles and in correspondence. To the best of my knowledge, which is based on reading such material written by Planck, Sommerfeld, von Laue, and Minkowski, among others, this is in fact the case.

[5] Cf. Emil Durkheim, *The Rules of Sociological Method,* trans. S. A. Solovay and J. H. Mueller, ed. G. E. G. Catlin (New York: Free Press, 1938; originally published in French, 1895), pp. 8ff.

INDEXES

Name Index

Name Index

Name Index

Name Index

Subject Index

381

Subject Index

Subject Index

Milton Keynes UK
Ingram Content Group UK Ltd.
UKHW011316190923
428957UK00011B/417